Nucleic Acids and Molecular Biology

18

Series Editor
H. J. Gross

Hans-Jürgen Bandelt Vincent Macaulay
Martin Richards (Eds.)

Human Mitochondrial DNA and the Evolution of *Homo sapiens*

With 31 Figures, and 10 Tables

Springer

Professor Hans-Jürgen Bandelt
University of Hamburg
Department of Mathematics
Bundestr. 55
20146 Hamburg
Germany

Dr. Martin Richards
University of Leeds
Institute of Integrative
& Comparative Biology
Faculty of Biological Sciences
Leeds, LS2 9JT
UK

Dr. Vincent Macaulay
University of Glasgow
Department of Statistics
University Avenue
Glasgow, G12 8QQ
UK

ISSN 0933-1891
ISBN-10 3-540-31788-0 Springer Berlin Heidelberg New York
ISBN-13 978-3-540-31788-3 Springer Berlin Heidelberg New York

Library of Congress Control Number: 2006922802

Springer is a part of Springer Science+Business Media

springeronline.com

Production and typesetting: LE-TEX Jelonek, Schmidt & Vöckler GbR, 04229 Leipzig
Cover design: *design & production* GmbH, 69126 Heidelberg

Printed on acid-free paper 31/3180/YL - 5 4 3 2 1 Spin 12080397

Preface

Mitochondrial DNA (mtDNA) is, compared to our total genome, only a tiny molecule. Yet deciphering its evolution has profoundly changed our perception about how modern humans spread across our planet. Analysis of mtDNA variation has matured in the course of the past 20 years and has become a versatile tool in the study of our species for the time horizon of the last 100,000 years or so, as well as our relationship to other species. In this time, it has effectively come of age, although the process has perhaps been rather more fraught than any growing marker system might have the right to expect. By comparison, the Y chromosome has matured in a rather more genteel fashion. We might say that whereas mtDNA's youth was more one of wild living and riotous conflict, the Y chromosome was born middle-aged.

These two markers together, though, deserve their special role in unravelling the evolutionary history of humanity. In contrast to traditional genetic markers residing in the nuclear DNA of the autosomal chromosomes, mtDNA is not reshuffled from generation to generation, but is inherited purely along the maternal line of descent (except in most unusual circumstances). Just as with the non-recombining part of the Y chromosome inherited down the paternal line, we have a uniparental marker at our disposal that evolves along a genealogical tree, which can then become the major target of investigation. Although it is impossible to reconstruct the complete mtDNA genealogy of humankind in time and space, parts of it— a kind of shadow—are frozen into the mutational pattern that is in turn reflected in the variation we see in samples of modern mtDNA sequences. That variation allows the reconstruction of an mtDNA tree (or "phylogeny" as we would say, by a slight abuse of the traditional language), the maternal genealogy as reflected in a glass darkly.

When people began to work on mtDNA in the 1980s after the publication of the first complete mtDNA genome sequence—the famous Cambridge reference sequence of Anderson and colleagues—they first employed just a few restriction enzymes, using them to estimate very simple trees. The genealogical resolution of these trees—that is, the amount of genealogy or coalescent tree detail they estimated—was so low that they gave quite misleading results, and led to a variety of (not particularly well-publicized) proposals that supported either an 'out-of-Asia' origin for modern humans, or even the multiregional model. It wasn't until Allan Wilson's group, in the late 1980s, began to use a higher-resolution restriction analysis system, that mtDNA became both famous and notorious when used as evidence supporting the 'out-of-Africa' model. Then, in the early 1990s, PCR and control-region sequencing, again pioneered by the Wilson lab, suddenly made mtDNA studies technically straightforward and accessible to a wider community of researchers.

In the control region (especially the first hypervariable segment, HVS-I) a high proportion of mtDNA's variation is concentrated in a very small and easily sequenced stretch, and for this reason it has exerted an enormous hold on the research community ever since. Whilst we would not wish to deny the value of the early work that was done, there are serious limitations in focusing on the control region, even (as is really the bare minimum that is necessary to make anything of the mtDNA data) when supplemented with some important coding-region sites as well. The problems are well known: the high level of variation in the control region is accompanied by high levels of recurrent mutation, blurring the structure of the tree; and at the same time there is nevertheless insufficient variation to distinguish many important ancient branches within the tree.

Why does it matter? Every branching event, and indeed every mutation that occurs in mtDNA during the branching genealogical process, is a historical event that took place at a certain time and at a certain place. It is the mutational variation that allows us to study the branching events, and it is the geographical variation and time depth of the branches that help us to recover aspects of our evolutionary past, especially processes of dispersal and colonization. From the perspective of this book, the most useful aspect of mtDNA is this possibility of being able to reconstruct where and when particular branching events took place, as the first step in the reconstruction of prehistoric dispersals—that is to say, the possibility of being able to draw *phylogeographic* inferences from mtDNA sequences.

Since 2000, though, there has been a new phase of mtDNA studies: the blossoming of complete genome analyses. Of course, sequencing complete mtDNA genomes is much more expensive and requires much more labour. However, the efforts are paying off as the resolution of the mtDNA tree has been improved many-fold. There are now more than 2000 complete mtDNA genomes published and the basal branching structure of mtDNA variation in many—perhaps most—parts of the world is now rather well understood.

It is still not all plain sailing, however. Recurrent mutations do occur in the coding region of mtDNA, and the precise branching order will in some cases never be fully resolved. There are also a number of multifurcations in the tree, arising, we believe, from a very rapid series of dispersals and expansions as populations spread out from East Africa in the late Pleistocene. This means that there is still insufficient variation, even in the whole mtDNA genome, to resolve some important branching events, such as some of the branching that would reflect both the partial re-population of Africa and the peopling of the rest of the world. Still, we are convinced that a lot can and will be achieved with complete mtDNA genome sequences. The shortcomings of the system itself—e.g., that it can only help us to reconstruct the female side of the story and that it is a molecule of finite length—are no longer the main issue. The major ongoing problems are more insidious.

Firstly, there is the problem of poor quality data. It has become apparent that published mtDNA sequences, both control regions and complete genomes, are often not of very high quality. In the early days of manual sequencing this was probably more often than not due to technical shortcomings in the laboratory process, but in more recent times it seems to be largely due to database-handling steps further down the production line. It is not clear that mtDNA workers are any worse than other DNA researchers in this respect, but the tendency in population studies to sequence only one strand of the molecule may have exacerbated the problem. Thus, mistakes have emerged, and are indeed often rather common, even when both strands have apparently been sequenced. Unfortunately (and perhaps of even greater concern), even the forensics community has not been immune from this. Moreover, the situation is even worse in regard to the now increasingly popular research based on ancient mtDNA, which regularly announces exciting new findings, but all too often cannot get past a mere faith in the authenticity of the sequencing results.

Then there is the question of methods of analysis. Analytical problems have beset mtDNA studies from the early days of the Wilson lab's first mtDNA trees to the present. Nowadays, though, the debate centres much less on phylogenetic reconstruction, in part because it is so much more straightforward with complete sequence data—although neighbour-joining trees of control-region data continue, inexplicably, to see the light of day—but on more fundamental questions concerning what approach to take. A number of researchers coming from a traditional human population genetics background still argue that mtDNA is best analysed using the battery of statistical tools developed for classical markers, in effect ignoring the genealogical information in the data. They are then, if they are cautious, troubled that little can be learned from these approaches. If they are less cautious, they may put forward interpretations of the data that are hard to sustain, for example, seeing homogeneity of mtDNA variation across Europe or East Asia.

How can we move from mere branching nodes in the tree to dispersal and colonization times? After all, as Guido Barbujani and colleagues have commented, "suppose that some Europeans colonize Mars next year: If they successfully establish a population, the common mitochondrial ancestor of their descendants will be Paleolithic. But it would not be wise for a population geneticist of the future to infer from that a Paleolithic colonization of Mars." We can discuss this question in the context of the settlement of Eurasia, as described in the chapters by Metspalu et al. and Richards et al. Suppose population geneticists from Mars were to evaluate the pros and cons of a southern route dispersal of modern humans from Africa ~ 65,000 years ago *versus* later migration(s), say, in the Earthling Neolithic period. They would correctly infer that the coalescence times of non-African samples (which coalesce in the African root of the pan-Afroeurasian haplogroup L3, more than 80,000 years ago), or of the major two or three macro-haplogroup constituents *alone*, would be ambiguous about this, and most likely compatible with *any* hypothesis. Indeed they might even erect a First Law of Phylogeography: "The time of colonization of a geographical region cannot be inferred from the coalescence times of genetic lineages in that region", with which their Earthling colleagues would doubtless all agree.

However, what Martian population genetics would be failing to exploit is the geographical specificity of hierarchical levels of the human mtDNA distribution. Each ancestral node in the mtDNA phylogeny had a unique time and place of origin. The former can be rather well (relying upon a realistic calibration of the mutation rate) estimated from the node's descendants if sufficiently many (nearly) independent lineages diverged from it. The latter can be inferred from the geographic distribution of the descendants (provided that subsequent events have not obscured the pattern too much). When, for example, the thus reconstructed and dated ancestral mtDNA types all appear to have given rise to essentially autochthonous branches of the mtDNA phylogeny with approximately equal coalescence times in several sub-continents, then one could speak of common founder types involved in one colonization event, as seems to be the case with the three founder mtDNAs of Eurasia (see the chapter by Richards et al.).

A weakness running through much of the work taking the traditional population genetics approach—which was never a weakness in the classical studies of the father of human population genetics, Luca Cavalli-Sforza—is a lack of embeddedness within the anthropological context. The traditional population genetics approach is very much both to propose and to test hypotheses in an atheoretical vacuum, without any regard for what is known from other disciplines—archaeology, anthropology, palaeontology or whatever—in the name of scientific objectivity. One advantage of phylogeographic approaches to the data is that they have tended to make good use of non-genetic as well as genetic evidence, whilst attempting to maintain some level of independence, so as to avoid circularity. So an understanding of both the strengths and limitations of mtDNA as a marker system—reflecting, we must always remember, just a single line of descent—is an underlying theme of this book. In addition, we have found it important to place any mtDNA inferences about the human past in the context of a solid foundation of knowledge about the molecule and the segregation and evolutionary processes that are responsible for the variation that can be observed. Perhaps, having agreed with our Martian colleagues on the First Law of Phylogeography, we might also suggest to them an even more fundamental principle: "Make use of as much of the available evidence as you can."

In this volume we have focused on the evolution and spread of modern humans in the decisive period from about 100,000 to 40,000 years ago, which thrust modern humans on to a wider stage outside Africa. In a subsequent volume we will discuss the developments that set in afterwards, which eventually led to the peopling of the entire landmass of the earth (except for the Antarctic and extreme environments on other continents). Another volume will then deal with the medical aspects of mtDNA and its role in pathogenesis and ageing.

After first planning the present volume in March 2002, it became clear that more information was needed from whole mtDNA genome sequencing before we could set out to fill the book chapters. The picture is now coming much more into focus, and so we are slightly more confident that our words will not become stale before the ink is dry.

We are grateful to Professor H.J. Gross from Würzburg who first suggested a book in a Springer series about the human mitochondrial genome in pathogenesis and evolution. Antonio Torroni, who was in fact approached by him, then further delegated this task to us, and we are most grateful to him for his encouragement and critical guidance of the work. We thank all contributors and co-authors for their enthusiasm and patience in completing this collaborative effort. We are very grateful to Antonio Torroni for commenting on a late version of the manuscript; the mistakes that remain are our own.

Hamburg,
Glasgow,
Leeds,
March 2006

Hans-Jürgen Bandelt
Vincent Macaulay
Martin Richards

Contents

Part I Prerequisites and Caveats

Mitochondrial DNA in *Homo Sapiens*
P. F. Chinnery 3
1 Introduction 3
2 Mitochondria—Structure and Function 3
3 Mitochondrial Biogenesis 5
4 Human mtDNA 6
5 mtDNA Transcription, Translation, and Replication 8
6 Pathogenic Mutations of mtDNA, Heteroplasmy, and the Threshold Effect 9
7 The Role of Homoplasmic mtDNA Mutations in Human Disease 10
8 Conclusions 11
References 12

The Transmission and Segregation of Mitochondrial DNA in *Homo Sapiens*
P. F. Chinnery 17
1 Introduction 17
2 mtDNA Mutations and Human Disease 17
3 Mechanisms That Can Change the Level of Heteroplasmy . . 18
4 The Inheritance of mtDNA 19
4.1 Maternal Inheritance 19
4.2 The Mitochondrial Genetic Bottleneck 20
5 Segregation During Early Development 24
6 Conclusions and Future Perspectives 25
References 26

Numts Revisited
C. M. Bravi, W. Parson, H.-J. Bandelt 31
1 Introduction 31
2 Numts in Humans 33

3 A Numt from an Egyptian Mummy 34
4 A Numt from the Sperm's Head? 36
5 A Bouquet of Numts? 36
6 Adverse Laboratory Conditions 40
7 Conclusion 41
References 42

Estimation of Mutation Rates and Coalescence Times: Some Caveats
H.-J. Bandelt, Q.-P. Kong, M. Richards, V. Macaulay 47
1 Introduction 47
2 Prerequisite: a Global mtDNA Tree 49
3 Transition/Transversion Rate Ratio 55
4 Spectrum of Relative Mutational Rates Along the Molecule . 59
5 Pitfalls with Estimation of Positional Rate Variability 63
6 Calibration of the Mutational Clock 73
7 Pitfalls with Age Estimations 80
8 Conclusion 84
References 85

Postmortem Damage of Mitochondrial DNA
M. T. P. Gilbert 91
1 Introduction 91
2 Ancient DNA 91
3 DNA Damage in Ancient Samples 92
3.1 DNA Degradation Immediately Following Cell Death 92
3.2 Long-term DNA Degradation 93
3.3 Damage-Driven DNA Miscoding Lesions 97
4 Insights from Miscoding Lesions
into in Vivo mtDNA Mutation 102
4.1 Mutational Hotspots and mtDNA Recombination 102
4.2 Effects of mtDNA Secondary and Tertiary Structure 103
4.3 Hotspots for Postmortem mtDNA Damage 103
4.4 Sequence Motifs with Limited DNA Damage 108
5 Implications of Postmortem Damage Hotspots
on Sequence Authenticity 109
6 Conclusion 110
References 111

Lab-Specific Mutation Processes
H.-J. Bandelt, T. Kivisild, J. Parik, R. Villems, C. Bravi,
Y.-G. Yao, A. Brandstätter, W. Parson 117
1 Introduction 117
2 The Sequencing Process and Data Handling 118
3 Sources of Error 119

4 Pitfalls of mtDNA Sequencing 123
5 'Ancient' DNA 131
6 Conclusion 134
References 140

Part II Evolution of Human mtDNA

The World mtDNA Phylogeny
T. Kivisild, M. Metspalu, H.-J. Bandelt, M. Richards, R. Villems . 149
1 Haplotypes and Trees 147
2 Haplogroup Structure and Definition 153
3 Phylogeographic Inferences 157
4 African mtDNA Variation and Haplogroup Structure 163
5 mtDNA Variation Outside Africa 167
6 The Role of Selection on mtDNA Variability 170
References 172

The Pioneer Settlement of Modern Humans in Asia
M. Metspalu, T. Kivisild, H.-J. Bandelt, M. Richards, R. Villems . 181
1 Introduction 181
2 Palaeoclimatological Context 182
3 Archaeological and Palaeontological Evidence of the Peopling of Asia by AMH 183
4 How to Infer 'Pioneer Settlement' from Extant mtDNA Variation? 185
5 The Peopling of Asia as Seen Through the Lens of mtDNA Diversity 186
6 A Route Through Northern Asia? 190
7 Conclusion 194
References 194

Ancient DNA and the Neanderthals
W. Goodwin, I. Ovchinnikov 201
1 Introduction 201
2 Origins of the Neanderthals 202
3 DNA Analysis of Neanderthal Specimens 203
4 The Limitations of Ancient DNA Analysis 205
5 DNA Degradation 207
6 The Extraction and Analysis of Neanderthal DNA: Assessing the Preservation 207
7 Environmental Factors 208

8 Molecular Preservation 209
9 Isolation of Neanderthal DNA 210
10 DNA Extraction 210
11 Amplification and Sequence Analysis of Neanderthal DNA . 211
12 Ancient DNA Artefacts 213
13 Authentication 214
14 Evaluation of Neanderthal DNA 215
15 Phylogenetic Analysis 216
16 Admixture between Modern Humans and Neanderthals . . . 218
17 Neanderthal Diversity 220
18 The Age of Divergence 220
19 Conclusions 221
References 222

A Model for the Dispersal of Modern Humans out of Africa
M. Richards, H.-J. Bandelt, T. Kivisild, S. Oppenheimer 225
1 Introduction: Setting the Agenda 225
2 Genetics and the Traditional Debate over Human Origins . . 226
2.1 The Debate Continues 228
2.2 Moving on 232
3 How Many Dispersals of Modern Humans from Africa? . . . 233
3.1 The mtDNA Evidence and Multiple Dispersals 234
3.2 Y-Chromosome Founders 238
4 Which Way out of Africa? 241
4.1 mtDNA and the Indian Staging Post 243
4.2 A Far Eastern Ancestry? 245
4.3 Opening up the West 247
4.4 Y-Chromosome Passage to India and Beyond 248
4.5 In Context 252
References 255

Subject Index 267

Part I
Prerequisites and Caveats

Nucleic Acids and Molecular Biology, Vol. 18
Hans-Jürgen Bandelt, Vincent Macaulay, Martin Richards (Eds.)
Human Mitochondrial DNA and the Evolution of *Homo sapiens*

Mitochondrial DNA in *Homo Sapiens*

Patrick F. Chinnery

Mitochondrial Research Group, The University of Newcastle upon Tyne,
Framlington Place, Newcastle upon Tyne NE2 4HH, UK
p.f.chinnery@newcastle.ac.uk

1
Introduction

The first complete sequence of human mitochondrial DNA (mtDNA) was published in 1981 (Anderson et al. 1981). This acted as a catalyst for an explosion of interest in the mitochondrial genome and its role in human evolution and disease. Early evolutionary studies focussed on polymorphic restriction sites, but with the rapid development of semiautomated complete mitochondrial genome sequencing, recent work has incorporated full-genome analysis. This has posed new bioinformatics challenges, but is has also raised intriguing new questions about mtDNA evolution that we are only just beginning to understand. This chapter presents a brief overview of the structure and function of the human mitochondrial genome in health and disease, providing basic knowledge required for the rest of the book. The reader should refer to recent review articles for a more detailed discussion of the vertebrate mitochondrial genome (Pereira 2000), mitochondrial biology (Scheffler 2000; Bayona-Bafaluy et al. 2002), and a discussion of human mitochondrial diseases (Smeitink et al. 2001; Chinnery and Schon 2003; DiMauro and Schon 2003).

2
Mitochondria—Structure and Function

Mitochondria are double-membraned intracellular organelles present within all nucleated mammalian cells. On a structural level, the traditional view that mitochondria are small cigar-shaped or bean-shaped structures is probably naïve, and it is more accurate to think of mitochondria as a budding and fusing network similar to the endoplasmic reticulum (Iborra et al. 2004). Mitochondria are intimately involved in cellular homeostasis. They play a part in intracellular signalling and apoptosis (programmed cell death), intermediary metabolism (such as the Krebs or tricarboxylic acid cycle), and in the metabolism of amino acids, lipids, cholesterol, steroids, and nucleotides, among other functions. Mitochondria also have a fundamental role in cellular energy metabolism. This includes fatty acid β oxidation, the urea cycle and the final common pathway for adenosine triphosphate (ATP) production—the respiratory chain.

The mitochondrial respiratory chain is a group of five enzyme complexes situated on the inner mitochondrial membrane (Fig. 1). Each complex is composed of multiple subunits, the largest being complex I with over 40 polypeptide components. Reduced cofactors (NADH and $FADH_2$) generated from the intermediary metabolism of carbohydrates, proteins, and fats donate electrons to complex I and complex II. These electrons flow

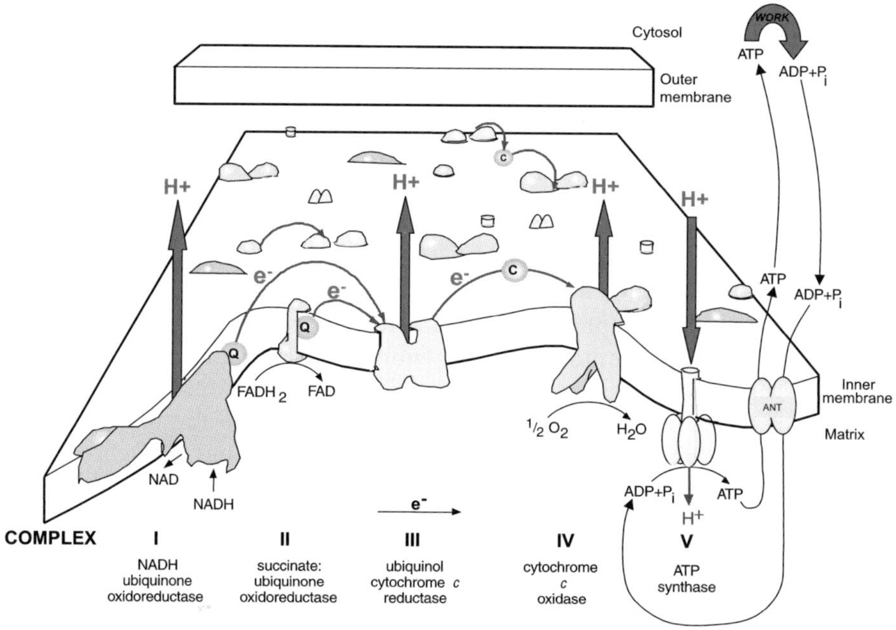

Fig. 1 The respiratory chain. Reduced cofactors (*NADH* and *$FADH_2$*) are produced from the intermediary metabolism of carbohydrates, proteins, and fats. These cofactors donate electrons to complex I (NADH–ubiquinone oxidoreductase) and complex II (succinate–ubiquinone oxidoreductase). These electrons flow between the complexes down an electrochemical gradient (*black arrows*), shuttled by ubiquinone (*Q*) and cytochrome *c* (*C*), involving complex III (ubiquinol–cytochrome *c* oxidase reductase) and complex IV (cytochrome *c* oxidase, or COX). Complex IV donates an electron to oxygen, which results in the formation of water. Protons are pumped from the mitochondrial matrix into the intermembrane space (*red arrows*). This proton gradient generates the mitochondrial membrane potential which is harnessed by complex V to synthesise adenosine triphosphate (*ATP*) from adenosine diphosphate (*ADP*) and inorganic phosphate (P_i). *ANT* adenine nucleotide translocator which exchanges ADP for ATP across the mitochondrial membrane. (With thanks to Z. Chrzanowske-Lightowlers)

between the complexes down an electrochemical gradient, shuttled by complexes III and IV and by two mobile electron carriers, ubiquinone (ubiquinol, coenzyme Q10) and cytochrome *c*. The electron-transfer function of complexes I–IV is accomplished via subunits harbouring prosthetic groups (e.g. iron–sulfur groups in complexes I, II, and III, and heme iron in cytochrome *c* and in complex IV). The liberated energy is used by complexes I, III, and IV to pump protons out of the mitochondrial matrix into the intermembrane space. This proton gradient, which generates the bulk of the mitochondrial membrane potential, is harnessed by complex V to synthesise ATP from adenosine diphosphate (ADP) and inorganic phosphate. (The asymmetric distribution of ions, such as Na^+, K^+, and Ca^{2+}, across the inner membrane makes up the 'chemical' portion of the electrochemical gradient). The overall process is called oxidative phosphorylation (OXPHOS). ATP is the high-energy source used for essentially all active metabolic processes within the cell, and it must be released from the mitochondrion in exchange for cytosolic ADP. This is carried out by the adenine nucleotide translocator (ANT), which has a number of tissue specific isoforms.

3 Mitochondrial Biogenesis

Two distinct genetic systems encode mitochondrial proteins: mtDNA and nuclear DNA (nDNA). Nuclear genes code for the majority of mitochondrial respiratory chain polypeptides (Shoubridge 2001). These polypeptides are synthesised in the cytoplasm with a mitochondrial targeting sequence that directs them through the translocation machinery spanning the outer and inner membranes. The targeting sequence is then cleaved before the subunit is assembled with its counterparts on the inner mitochondrial membrane. The components of the import machinery (trans inner membrane, TIM, and trans outer membrane, TOM, proteins), the importation processing enzymes, and the respiratory chain assembly proteins are all the products of nuclear genes (Shadel 2004).

Nuclear genes are also important for maintaining the mitochondrial genome, including those encoding the mtDNA polymerase γ (*POLG1*) (Van Goethem et al. 2001) and products that maintain an appropriate balance of free nucleotides within the mitochondrion (*TP, TK, DGK* and *ANT1*) (Nishino et al. 1999; Kaukonen et al. 2000; Mandel et al. 2001; Saada et al. 2001). A recently described gene, *C10orf2*, codes for a helicase-like protein called Twinkle that also appears to be important for mtDNA maintenance (Spelbrink et al. 2001). nDNA also codes for essential factors needed for intramitochondrial transcription and translation, including *TFAM, TFBM1, and TFBM2* (Larsson et al. 1998; Falkenberg et al. 2002). A disruption of both nuclear and mitochondrial genes can therefore cause mitochondrial dysfunction.

4 Human mtDNA

mtDNA is a small 16.6-kilobase (kb) circle of double-stranded DNA which codes for 13 essential components of the respiratory chain (Fig. 2). *ND1–ND6*, and *ND4L* encode seven subunits of complex I (NADH–ubiquinone oxidoreductase). *Cyt b* is the only mtDNA encoded complex III subunit (ubiquinol-cytochrome *c* oxidase reductase). COI– COIII encode for three of the complex IV (cytochrome *c* oxidase, or COX) subunits, and the *ATP 6* and *ATP 8* genes encode for two subunits of complex V (ATP synthase). Two ribosomal RNA (rRNA) genes (12S and 16S rRNA), and 22 transfer RNA (tRNA) genes are interspaced between the protein-encoding genes. These provide the necessary RNA components for intramitochondrial protein synthesis.

The mtDNA displacement loop (D-loop, or control region) is a 1.1-kb non-coding region which is involved in the regulation of transcription and replication of the molecule. The D-loop extends from position 16024 to position 576 of the mtDNA and is the largest region not directly involved in the synthesis of respiratory chain polypeptides. There are a number of short segments and single base pairs of the mitochondrial genome that are not directly involved in coding for RNAs or proteins (such as the adenine nucleotide at position 7517, and the 5-bp sequence from 5580–5586 between the $tRNA^{Trp}$ and $tRNA^{Ala}$ genes). The D-loop contains three short regions which, in comparison to the rest of the genome, have a highly variable sequence at the population level: hypervariable sequence (HVS) HVS-I, HVS-II, and HVS-III, corresponding to HVR1, HVR2, and HVR3 in some sources (Brandstätter et al. 2004a). The precise definition of the different hypervariable sequences does vary from context to context. The forensic community traditionally took 16024–16365 to be HVS-I, 73–340 to be HVS-II, and 438–576 to be HVS-III (Brandstätter et al. 2004a). By contrast, more recent population genetics studies (Brandstätter et al. 2004b) take wider ranges, particularly in HVS-I in order to capture the phylogenetically important positions 16390, 16391, and 16399 (HVS-I 16024–16400, HVS-II 44–340, and HVS-III 438–576; Fig. 3).

The complete function of these hypervariable regions is not known, but they appear to be important for genome replication and transcription, and either contain or are near to the origins of heavy-strand and light-strand mtDNA replication: O_H and O_L (Sect. 5).

The mtDNA genetic code differs from the universal genetic code in a number of ways (Anderson et al. 1981). In human mtDNA UGA codes for tryptophan and not for termination, AUG codes for methionine and not for isoleucine, and AGA and AGG are termination codons, rather than arginine codons. Finally, AUA and AUG are both initiation codons. Early studies identified the strand asymmetry of the molecule, with a guanosine-rich 'heavy' strand or H-strand, and a cysosine-rich 'light' strand or L-strand. Traditionally the human mtDNA is numbered with reference to the light strand,

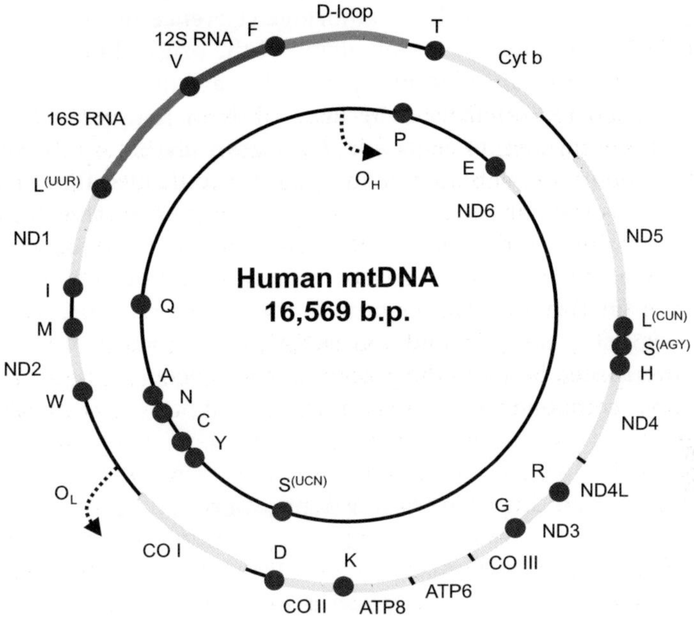

Fig. 2 The human mitochondrial genome. *ND*1–*ND*6 and *ND4L* are complex I genes (NADH–ubiquinone oxidoreductase). Cyt *b* is a complex III subunit gene (ubiquinol–cytochrome *c* oxidase reductase). *CO I–CO III* are complex IV (cytochrome *c* oxidase, or COX) subunit genes, and *ATP* 6 and *ATP* 8 are complex V subunit genes (ATP synthase). Two ribosomal RNA genes (12*SRNA* and 16*S RRNA*), and 22 transfer RNA genes are interspersed between the protein-encoding genes, designated by *single letters* (using standard amino acid nomenclature). *D-loop* is the 1.1-kb non-coding region, which is involved in the regulation of transcription and replication of the molecule (Fig. 3). O_H and O_L are the origins of heavy-strand and light-strand mitochondrial DNA (*mtDNA*) replication

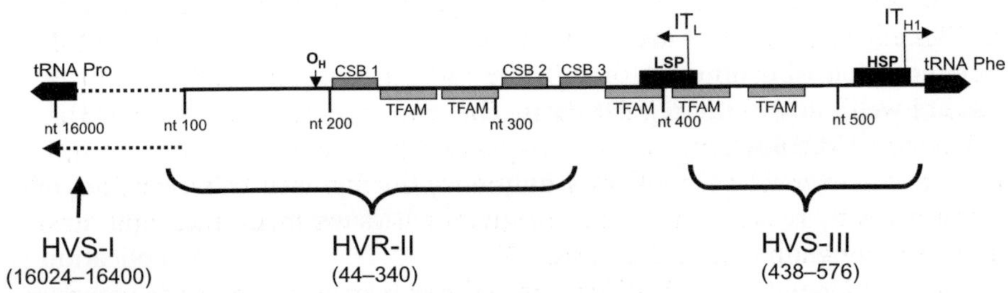

Fig. 3 The mtDNA control region. Note that the orientation of this diagram is opposite to that of Fig. 2. The mtDNA control region is contained within the non-coding D-loop and contains hypervariable sequences *HVS-I* (positions 16024–16400), *HVS-II* (positions 44–340), *HVS-III* (positions 438–576) (Brandstätter et al. 2004b), and conserved sequence blocks *CSB1*, *CSB2*, and *CSB3*. The light-strand promoter (*LSP*), heavy-strand promoter (*HSP*), and mtDNA transcription factor A (*TFAM*) binding sites are also shown. IT_L transcription initiation site for the light strand, IT_{H1} major transcription initiation site for the heavy strand

according to the original so-called Cambridge reference sequence (Anderson et al. 1981) of mtDNA. Recent resequencing of the original placental mtDNA sample used by the Cambridge group revealed a number of errors, either because of sequencing artefacts or because a bovine sample was used when it was technically difficult to sequence the human placental mtDNA sample (Andrews et al. 1999). The resequencing exercise confirmed that the original placental DNA sample belonged to mtDNA haplogroup H, revealed ten substitution errors, and showed that the original reference sequence had two cytosine residues at positions 3106 and 3107, where there should only have been one cytosine residue. The authors of the 'revised Cambridge reference sequence' suggested that to avoid confusion, the original light-strand numbering system should be retained (Andrews et al. 1999).

5
mtDNA Transcription, Translation, and Replication

The basic mechanism of human mtDNA transcription was elucidated in the 1980s (Clayton 1992). Transcription is initiated at two major sites: IT_{H1} and IT_L, which are both within 150 bp of one another in the non-coding region (Fig. 3). Heavy-strand transcription starts at position 561 located within the heavy-strand promoter, and light-strand transcription starts at position 407 within the light-strand promoter, with enhancer elements just upstream that bind mitochondrial transcription factor A (TFAM, or mtTFA). TFAM promotes bidirectional transcription, unwinding the DNA template. Transcription factors BM1 and BM2 bind to the core polymerase (POLRMT) in a 1 : 1 stoichiometry, and play a crucial role in initiating transcription (Falkenberg et al. 2002). The light strand is transcribed as a single polycistronic precursor, which is subsequently processed and modified before the mtDNA encoded proteins are synthesised within the mitochondrial matrix on mitochondrial ribosomes. Although the basic mechanisms behind this process are well known, the regulatory mechanisms are only just being clarified (Gagliardi et al. 2004).

Unlike nDNA, which replicates only once during each cell cycle, mtDNA is continuously recycled, even in non-dividing tissues such as skeletal muscle and brain (Bogenhagen and Clayton 1977; Birky 2001). mtDNA replication is therefore independent of the cell cycle (i.e. it is relaxed). The precise mechanism of mtDNA replication is currently a topic of great debate. The traditional view is that replication is strand-asymmetric. According to this model (Clayton 1982), the heavy strand leads the replication cycle beginning at O_H with the cleavage of a primary transcript synthesised from the light-strand promoter. Replication of the heavy strand continues in a clockwise direction until the origin of light-strand replication (O_L) is exposed, and the light (or lagging) strand is then synthesised in the counterclockwise direction. However, recent experimental evidence supports an alternative strand-symmetric, or 'rolling circle', model, whereby the replication of mtDNA begins at numerous points in a 5.5-kb critical region between the D-loop and the *ND4* gene (referred to as ORI) (Bowmaker et al. 2003). These replication 'bubbles' then proceed in both directions, stopping at O_H, and stalling briefly in the region of O_L before completing the replication cycle, with the lagging strand catching up with the ligation of Okazaki fragments.

6
Pathogenic Mutations of mtDNA, Heteroplasmy, and the Threshold Effect

The first pathogenic mtDNA mutations were identified in the late 1980s. Large-scale deletions of mtDNA, removing both tRNA and protein coding genes, were found in the skeletal muscle of patients with chronic progressive external ophthalmoplegia (CPEO) and the Kearns Sayre syndrome (Holt et al. 1988; Zeviani et al. 1988), and point mutations of mtDNA were found in a patient with mitochondrial encephalomyopathy with strokelike episodes (MELAS) (Goto et al. 1990) and Leber hereditary optic neuropathy (LHON) (Wallace et al. 1988). In a recent survey, over 200 different point mutations and deletions had been associated with disease (Servidei 2003). tRNA gene mutations and deletions cause a respiratory chain deficiency through a generalised effect on protein synthesis, whereas missense, non-sense and mutations which generate premature stop codons affect specific respiratory chain complexes. Duplications of mtDNA have also been described, but these are not thought to be primarily pathogenic, and only cause disease indirectly through the subsequent generation of deleted mtDNA molecules (Manfredi et al. 1997).

The majority of primary pathogenic mtDNA mutations are heteroplasmic in affected individuals—with varying proportions of mutated mtDNA within the same individual. Whilst most human cells contain two copies of nDNA, they contain many more copies of mtDNA (from 1000 to 100000, depending on the cell type). These are often all identical in a healthy individual at birth (*homoplasmy*), but some mixture (*heteroplasmy*) may occur, especially involving hypervariable sites and length polymorphisms of polycytosine stretches. In particular, patients harbouring pathogenic mtDNA defects often have a mixture of mutated and wild-type mtDNA (Holt et al. 1990; Wallace 1997). The percentage of mutated mtDNA can vary widely among different patients, and also from organ to organ, and even between cells within the same individual. In vitro studies using "transmitochondrial cytoplasmic hybrid (cybrid)" cells (King and Attardi 1988), containing different amounts of mutated mtDNA, have shown that most mtDNA mutations are highly recessive. In other words, the cells were able to tolerate high percentage levels of mutated mtDNA (typically 70–90%) before they developed a biochemical respiratory chain defect. The precise threshold for biochemical expression varies from mutation to mutation, and from tissue to tissue. Large retrospective studies have shown that the percentage level of mutated mtDNA in clinically relevant tissues does correlate with the severity of disease (Chinnery et al. 1997; White et al. 1999).

From the population genetics perspective, the presence of heteroplasmy implies a recent mutation event, and the independent occurrence of pathogenic mtDNA mutations at the same site has been confirmed by D-loop sequencing in many cases (Man et al. 2003). A full discussion of mtDNA mutations and human disease is given in several recent reviews (Smeitink et al. 2001; Chinnery and Schon 2003; DiMauro and Schon 2003).

7
The Role of Homoplasmic mtDNA Mutations in Human Disease

The most common homoplasmic mtDNA disease is LHON (Chinnery et al. 2000), which is primarily due to mutations in mtDNA complex I (*ND*) genes and is characterised by subacute bilateral visual failure presenting in early adult life (Howell 1997). LHON is intriguing because it is essentially an organ-specific disease that principally affects the retinal ganglion cells and the optic nerve (Saadati et al. 1998). LHON also has a markedly reduced penetrance with a clear sex bias, with only approximately 50% of men and approximately 10% of women developing visual failure (Newman et al. 1991; Nikoskelainen 1994; Riordan-Eva et al. 1995). Most patients with LHON are homoplasmic mutated for one of three mtDNA *ND* gene mutations (Man et al. 2003), so heteroplasmy cannot explain the varied disease penetrance, and a number of unknown additional factors appear to be important.

Two of the three principal LHON mtDNA mutations (14484T>C in the *ND6* gene and 11778G>A in the *ND4* gene) are preferentially associated with haplogroup J, which is found in approximately 10–15% of northern Europeans (Torroni et al. 1997). The reason for this association is not known, but it seems likely that haplogroup J increases the penetrance of the 14484T>C and 11778G>A mutations (Howell et al. 1995). It therefore appears that the mitochondrial genetic background can influence disease expression—but this cannot explain the gender bias in LHON.

The segregation pattern of disease in some LHON families suggests that there may be a nuclear genetic modifier locus modulating the clinical expression of the LHON mtDNA mutations. A recessive visual loss susceptibility locus on the X chromosome would explain the gender bias in LHON (Bu and Rotter 1991), but attempts to identify the locus have not been successful (Chalmers et al. 1996). Environmental factors may also play a part in LHON. There are many anecdotal reports of visual failure following alcohol intoxication, starvation, heavy smoking, and head trauma (Riordan-Eva et al. 1995), but large studies have yielded conflicting results (Tsao et al. 1999; Kerrison et al. 2000).

In many ways LHON is best considered as a complex trait, where the disease phenotype arises through multiple genetic factors (both mitochondrial and nuclear) interacting with the environment. A similar mechanism might explain the variable penetrance of other homoplasmic mtDNA mutations that cause organ-specific disease—such as the 1555A>G mtDNA mutation in the 12S rRNA gene that causes maternally inherited susceptibility to aminoglycoside-induced deafness, and possibly the 4300A>G mtDNA mutation in $tRNA^{Ile}$ that causes maternally inherited cardiomyopathy (Carelli et al. 2003). Similar nuclear–mitochondrial interactions are also likely to contribute to the varied phenotype seen in other mitochondrial disorders—be they due to primary nDNA or primary mtDNA defects.

The role of homoplasmic mtDNA mutations in human disease is a developing area. There is accumulating evidence that polymorphic variants may contribute to the risk of late-onset neurodegenerative disorders, including idiopathic Parkinson's disease (van der Walt et al. 2003), Alzheimer's disease (van der Walt et al. 2004), male infertility (Ruiz-Pesini et al. 2000), late-onset (type II) diabetes (Poulton et al. 2002), and idiopathic cardiomyopathy (Khogali et al. 2001). With the exception of male infertility (Ruiz-Pesini et al. 1998), there have not been any convincing studies to show a direct functional link between the genetic variants and mitochondrial respiratory chain function. This does not, of course, mean that the genetic variations are not relevant—but providing convincing evidence will be difficult to achieve.

8
Conclusions

Following the publication of the first complete sequence of human mtDNA, and the widespread use of semiautomated mtDNA sequencing, there has been a major growth of interest in mtDNA and its role in human evolution and disease. Although superficially these two topics may appear to be unrelated, there is increasing evidence that human mtDNA evolution has an important role in disease expression, both for monogenic disorders (such as LHON) and possibly for complex traits (such as Parkinson's disease). Conversely, these and other diseases, coupled with the environment, may have shaped the evolution of mtDNA in concert with the nuclear genome.

Acknowledgements The author is a Wellcome Trust Senior Fellow in Clinical Science. He also receives funding from Ataxia (UK), the Alzheimer's Research Trust, the Association Francaise contre les Myopathies, and the European Union under the FP6 framework.

References

Anderson S, Bankier AT, Barrell BG, de Bruijn MH, Coulson AR, Drouin J, Eperon IC, Nierlich DP, Roe BA, Sanger F, Schreier PH, Smith AJ, Staden R, Young IG (1981) Sequence and organization of the human mitochondrial genome. Nature 290:457–465

Andrews RM, Kubacka I, Chinnery PF, Lightowlers RN, Turnbull DM, Howell N (1999) Reanalysis and revision of the Cambridge reference sequence for human mitochondrial DNA [letter]. Nat Genet 23:147

Bayona-Bafaluy MP, Fernandez-Silva P, Enriquez J-A (2002) The thankless task of playing genetics with mammalian mitochondrial DNA: and 30-year review. Mitochondrion 2:3–26

Birky CW Jr (2001) The inheritance of genes in mitochondria and chloroplasts: laws, mechanisms, and models. Annu Rev Genet 35:125–148

Bogenhagen D, Clayton DA (1977) Mouse L cell mitochondrial DNA molecules are selected randomly for replication throughout the cell cycle. Cell 11:719–727

Bowmaker M, Yang MY, Yasukawa T, Reyes A, Jacobs HT, Huberman JA, Holt IJ (2003) Mammalian mitochondrial DNA replicates bidirectionally from an initiation zone. J Biol Chem 278:50961–50969

Brandstätter A, Niederstätter H, Parson W (2004a) Monitoring the inheritance of heteroplasmy by computer-assisted detection of mixed basecalls in the entire human mitochondrial DNA control region. Int J Legal Med 118:47–54

Brandstätter A, Peterson CT, Irwin JA, Mpoke S, Koech DK, Parson W, Parsons TJ (2004b) Mitochondrial DNA control region sequences from Nairobi (Kenya): inferring phylogenetic parameters for the establishment of a forensic database. Int J Legal Med 118:294–306

Bu X, Rotter JI (1991) X chromosomal-linked and mitochondrial gene control of Leber hereditary optic neuropathy: evidence from segregation analysis for dependence on X-chromosome inactivation. Proc Natl Acad Sci USA 88:8198–8202

Carelli V, Giordano C, d'Amati G (2003) Pathogenic expression of homoplasmic mtDNA mutations needs a complex nuclear-mitochondrial interaction. Trends Genet 19:257–262

Chalmers RM, Davis MB, Sweeney MG, Wood NW, Harding AE (1996) Evidence against an X-linked visual loss susceptibility locus in Leber hereditary optic neuropathy. Am J Hum Genet 59:103–108

Chinnery PF, Schon EA (2003) Mitochondria. J Neurol Neurosurg Psychiatry 74:1188–1199

Chinnery PF, Howell N, Lightowlers RN, Turnbull DM (1997) Molecular pathology of MELAS and MERRF. The relationship between mutation load and clinical phenotypes. Brain 120:1713–1721

Chinnery PF, Johnson MA, Wardell TM, Singh-Kler R, Hayes C, Brown DT, Taylor RW, Bindoff LA, Turnbull DM (2000) Epidemiology of pathogenic mitochondrial DNA mutations. Ann Neurol 48:188–193

Clayton DA (1982) Replication of animal mitochondrial DNA. Cell 28:693–705

Clayton DA (1992) Replication and transcription of vertebrate mitochondrial DNA. Annu Rev Cell Biol 7:453–478

DiMauro S, Schon EA (2003) Mitochondrial respiratory-chain diseases. N Engl J Med 348:2656–2668

Falkenberg M, Gaspari M, Rantanen A, Trifunovic A, Larsson NG, Gustafsson CM (2002) Mitochondrial transcription factors B1 and B2 activate transcription of human mtDNA. Nat Genet 31:289–294

Gagliardi D, Stepien PP, Temperley RJ, Lightowlers RN, Chrzanowska-Lightowlers ZM (2004) Messenger RNA stability in mitochondria: different means to an end. Trends Genet 20:260–267

Goto Y, Nonaka I, Horai S (1990) A mutation in the tRNA(Leu)(UUR) gene associated with the MELAS subgroup of mitochondrial encephalomyopathies. Nature 348:651–653

Holt I, Harding AE, Morgan-Hughes JA (1988) Deletion of muscle mitochondrial DNA in patients with mitochondrial myopathies. Nature 331:717–719

Holt IJ, Harding AE, Petty RK, Morgan-Hughes JA (1990) A new mitochondrial disease associated with mitochondrial DNA heteroplasmy. Am J Hum Genet 46:428–433

Howell N (1997) Leber hereditary optic neuropathy: mitochondrial mutations and degeneration of the optic nerve. Vision Res 37:3495–3507

Howell N, Kubacka I, Halvorson S, Howell B, McCullough DA, Mackey D (1995) Phylogenetic analysis of the mitochondrial genomes from Leber hereditary optic neuropathy pedigrees. Genetics 140:285–302

Iborra FJ, Kimura H, Cook PR (2004) The functional organization of mitochondrial genomes in human cells. BMC Biol 2:9

Kaukonen J, Juselius JK, Tiranti V, Kyttala A, Zeviani M, Comi GP, Keranen S, Peltonen L, Suomalainen A (2000) Role of adenine nucleotide translocator 1 in mtDNA maintenance. Science 289:782–785

Kerrison JB, Miller NR, Hsu F, Beaty TH, Maumenee IH, Smith KH, Savino PJ, Stone EM, Newman NJ (2000) A case-control study of tobacco and alcohol consumption in Leber hereditary optic neuropathy. Am J Ophthalmol 130:803–812

Khogali SS, Mayosi BM, Beattie JM, McKenna WJ, Watkins H, Poulton J (2001) A common mitochondrial DNA variant associated with susceptibility to dilated cardiomyopathy in two different populations. Lancet 357:1265–1267

King MP, Attardi G (1988) Injection of mitochondria into human cells leads to a rapid replacement of the endogenous mitochondrial DNA. Cell 52:811–819

Larsson NG, Wang J, Wilhelmsson H, Oldfors A, Rustin P, Lewandoski M, Barsh GS, Clayton DA (1998) Mitochondrial transcription factor A is necessary for mtDNA maintenance and embryogenesis in mice. Nat Genet 18:231–236

Man PY, Griffiths PG, Brown DT, Howell N, Turnbull DM, Chinnery PF (2003) The epidemiology of leber hereditary optic neuropathy in the north East of England. Am J Hum Genet 72:333–339

Mandel H, Szargel R, Labay V, Elpeleg O, Saada A, Shalata A, Anbinder Y, Berkowitz D, Hartman C, Barak M, Eriksson S, Cohen N (2001) The deoyguanosine kinase gene is mutated in individuals with depleted hepatocerebral mitochondrial DNA. Nat Genet 29:337–341

Manfredi G, Vu T, Bonilla E, Schon EA, DiMauro S, Arnaudo E, Zhang L, Rowland LP, Hirano M (1997) Association of myopathy with large-scale mitochondrial DNA duplications and deletions: which is pathogenic? Ann Neurol 42:180–188

Newman NJ, Lott MT, Wallace DC (1991) The clinical characteristics of pedigrees of Leber's hereditary optic neuropathy with the 11778 mutation. Am J Ophthalmol 111:750–762

Nikoskelainen EK (1994) Clinical picture of LHON. Clin Neurosci 2:115–120

Nishino I, Spinazzola A, Hirano M (1999) Thymidine phosphorylase gene mutations in MNGIE, a human mitochondrial disorder. Science 283:689–692

Pereira SL (2000) Mitochondrial genome and vertebrate phylogenetics. Genetics and Molecular Biology 23:745–752

Poulton J, Luan J, Macaulay V, Hennings S, Mitchell J, Wareham NJ (2002) Type 2 diabetes is associated with a common mitochondrial variant: evidence from a population-based case-control study. Hum Mol Genet 11:1581–1583

Riordan-Eva P, Sanders MD, Govan GG, Sweeney MG, Da Costa J, Harding AE (1995) The clinical features of Leber's hereditary optic neuropathy defined by the presence of a pathogenic mitochondrial DNA mutation. Brain 118:319–337

Ruiz-Pesini E, Diez C, Lapena AC, Perez-Martos A, Montoya J, Alvarez E, Arenas J, Lopez-Perez MJ (1998) Correlation of sperm motility with mitochondrial enzymatic activities. Clin Chem 44:1616–1620

Ruiz-Pesini E, Lapena A-C, Diez-Sanchez C, Perez-Martos A, Montoya J, Alvarez E, Diaz M, Urries A, Montoro L, Lopez-Perez MJ, Enriquez J-A (2000) Human mtDNA haplogroups associated with a high or reduced spermatozoa motility. Am J Hum Genet 67:682–696

Saada A, Shaag A, Mandel H, Nevo Y, Eriksson S, Elpeleg O (2001) Mutant mitochondrial thymidine kinase in mitochondrial DNA depletion myopathy. Nat Genet 29:342–344

Saadati HG, Hsu HY, Heller KB, Sadun AA (1998) A histopathologic and morphometric differentiation of nerves in optic nerve hypoplasia and Leber hereditary optic neuropathy. Arch Ophthalmol 116:911–916

Scheffler IE (2000) A century of mitochondrial research: achievements and perspectives. Mitochondrion 1:3–31

Servidei S (2003) Mitochondrial encephalomyopathies: gene mutation. Neuromuscul Disord 13:848–853

Shadel GS (2004) Coupling the mitochondrial transcription machinery to human disease. Trends Genet 20:513–519

Shoubridge EA (2001) Nuclear genetic defects of oxidative phosphorylation. Hum Mol Genet 10:2277–2284

Smeitink J, van den Heuvel L, DiMauro S (2001) The genetics and pathology of oxidative phosphorylation. Nat Rev Genet 2:342–352

Spelbrink JN, Li FY, Tiranti V, Nikali K, Yuan QP, Wanrooij S, Garrido N, Comi GP, Morandi L, Santoro L, Toscano A, Fabrizi GM, Somer H, Croxen R, Beeson D, Poulton J, Suomalainen A, Jacobs HT, Zeviani M, Larsson C (2001) Human mitochondrial DNA deletions associated with mutations in the gene encoding Twinkle, a phage T7 gene 4-like protein localised in mitochondria. Nat Genet 28:223–231

Torroni A, Petrozzi M, D'Urbano L, Sellitto D, Zeviani M, Carrara F, Carducci C, Leuzzi V, Carelli V, Barboni P, De Negri A, Scozzari R (1997) Haplotype and phylogenetic analyses suggest that one European-specific mtDNA background plays a role in the expression of Leber hereditary optic neuropathy by increasing the penetrance of the primary mutations 11 778 and 14 484. Am J Hum Genet 60:1107–1121

Tsao K, Aitken PA, Johns DR (1999) Smoking as an aetiological factor in a pedigree with Leber's hereditary optic neuropathy. Br J Ophthalmol 83:577–581

van der Walt JM, Nicodemus KK, Martin ER, Scott WK, Nance MA, Watts RL, Hubble JP, et al. (2003) Mitochondrial polymorphisms significantly reduce the risk of Parkinson disease. Am J Hum Genet 72:804–811

van der Walt JM, Dementieva YA, Martin ER, Scott WK, Nicodemus KK, Kroner CC, Welsh-Bohmer KA, Saunders AM, Roses AD, Small GW, Schmechel DE, Murali Doraiswamy P, Gilbert JR, Haines JL, Vance JM, Pericak-Vance MA (2004) Analysis of European mitochondrial haplogroups with Alzheimer disease risk. Neurosci Lett 365:28–32

Van Goethem G, Dermaut B, Lofgren A, Martin J-J, Van Broeckhoven C (2001) Mutation of POLG is associated with progressive external ophthalmoplegia characterized by mtDNA deletions. Nat Genet 28:211–212

Wallace DC (1997) Mitochondrial DNA in aging and disease. Sci Am 277:40–47

Wallace DC, Singh G, Lott MT, Hodge JA, Schurr TG, Lezza AM, Elsas LJD, Nikoskelainen EK (1988) Mitochondrial DNA mutation associated with Leber's hereditary optic neuropathy. Science 242:1427–1430

White SL, Collins VA, Woolfe R, Cleary MA, Shanske S, DiMauro S, Dahl HM, Thorburn DR (1999) Genetic counseling and prenatal diagnosis for the mitochondrial DNA mutations at nucleotide 8993. Am J Hum Genet 65:474–482

Zeviani M, Moraes CT, DiMauro S, Nakase H, Bonilla E, Schon EA, Rowland LP (1988) Deletions of mitochondrial DNA in Kearns-Sayre syndrome. Neurology 38:1339–1346

Nucleic Acids and Molecular Biology, Vol. 18
Hans-Jürgen Bandelt, Vincent Macaulay, Martin Richards (Eds.)
Human Mitochondrial DNA and the Evolution of *Homo sapiens*

The Transmission and Segregation of Mitochondrial DNA in *Homo Sapiens*

Patrick F. Chinnery

University of Newcastle upon Tyne, The Medical School, Framlington Place, Newcastle upon Tyne NE2 4HH, UK
p.f.chinnery@ncl.ac.uk

1 Introduction

The phylogenetic analysis of human mitochondrial DNA (mtDNA) is based upon the comparison of mtDNA sequences within and between different human populations. These sequence differences are a consequence of mtDNA mutations that originally arose within individual cells within individual organisms. The mechanisms that led to the fixation of the new mutations within a maternal founder are therefore fundamental to the process of mtDNA evolution within populations. This chapter will focus on the molecular and cellular mechanisms behind the segregation of new mtDNA mutations during human life, and the processes that underpin mtDNA transmission and lead to the fixation of new mutations within individuals of the species *Homo sapiens*.

Much of our understanding of the segregation and transmission of mtDNA in organisms comes from work done on humans and human pedigrees transmitting pathogenic or disease-causing mtDNA mutations. A basic knowledge of mtDNA and human disease therefore provides the background for our understanding of mtDNA transmission and segregation. Although there is emerging evidence that some aspects of segregation and transmission may be the same for pathogenic and neutral mtDNA mutations, this may not always be the case. Selective forces acting at the level of the organism, cell, mitochondrion, or possibly the individual genome may be important, particularly during mtDNA segregation. Great care should be taken when extrapolating from pathogenic to neutral non-pathogenic mutations when reading this chapter and the literature.

2 mtDNA Mutations and Human Disease

Recent epidemiological studies have established that human disease due to pathogenic mtDNA mutations is much more common than was previously thought (Majamaa et al. 1998; Chinnery et al. 2000a). Pathogenic mtDNA mutations are found in at least 1 in 8000 Europeans, and mtDNA disease affects at least 1 in 15 000 adults. The incidence of mtDNA disease in children is probably much lower, and most children with mitochondrial dysfunction have a mutation in a nuclear gene that is important for respiratory chain function (Uusimaa et al. 2000). Pathogenic mtDNA mutations fall into two groups: rearrangements (deletions of the molecule) and point mutations (single base changes, see Chap. 1 for more details) (Schon et al. 1997; Wallace 1999; DiMauro and Schon 2001). For reasons that are not understood, large-scale mtDNA deletions are generally not transmitted (Poulton and Turnbull 2000). They therefore do not become fixed in maternal pedigrees and do not contribute to mtDNA evolution on a population level. By contrast, many point mutations are transmitted from mother to offspring (Poulton and Turnbull 2000).

Each diploid human cell contains thousands of copies of mtDNA. At birth, these are usually identical (*intracellular homoplasmy*), but many patients with mtDNA disease harbour a mixture of mutant and wild-type mtDNA (*intracellular heteroplasmy*) (Larsson and Clayton 1995). The proportion of mutant mtDNA (mutation load) varies between individual cells. It is possible to generate cultured cells containing different mutation loads (cybrid fusions; King and Attardi 1989), and also to measure mitochondrial function and the proportion of mutant mtDNA in single cells from human biopsy material (Moraes et al. 1992). These studies have shown that the proportion of mutant mtDNA must exceed a critical threshold level before a cell expresses a biochemical defect of the respiratory chain (typically between 50 and 85% mutant) (Schon et al. 1997). This threshold is mutation-specific and it varies from tissue to tissue. These factors are thought to explain the varied clinical features seen in pedigrees transmitting the same pathogenic mtDNA mutation (Wallace et al. 1998).

3 Mechanisms That Can Change the Level of Heteroplasmy

Two mechanisms can lead to changes in mutation load in human cells in vivo.

1. *Relaxed replication.* Unlike nuclear DNA, which replicates once during each cell cycle, mtDNA is destroyed and replicated continuously, even in non-dividing tissues (Bogenhagen and Clayton 1977). Replication is considered to be 'relaxed' because it occurs independently of the cell cycle whilst the total amount of mtDNA remains constant (Chap. 1). Since individual molecules appear to be randomly selected for destruction and replication, in a heteroplasmic cell this process can lead to changes in the proportion of mutant and wild-type mtDNA molecules over a period of

time through random *intracellular* genetic drift (Birky 1994; Chinnery and Samuels 1999). This mechanism may be responsible for the changes in heteroplasmy that occur in postmitotic tissues such as skeletal muscle and the brain (Chinnery and Samuels 1999). These tissues bear the brunt of the pathology in patients with mtDNA disease (DiMauro and Schon 2001).

2. *Vegetative segregation*. The unequal partitioning of mutant and wild-type mtDNA that occurs during cell division can also lead to changes in the level of heteroplasmy in a proliferative tissue (Birky 1994), such as blood leucocytes or cells in culture (Lehtinen et al. 2000).

In non-dividing tissues (muscle and neurons), relaxed replication may prevent the accumulation of deleterious somatic mutations that will be lost through random intracellular drift (Elson et al. 2001b). However, the same mechanism may also lead to very high levels of somatic mutations within individual cells (so-called clonal expansion) (Elson et al. 2001b). These cells accumulate throughout life (Brierley et al. 1998; Elson et al. 2001b), and the acquired mitochondrial deficiency may contribute to the human aging process (Wallace et al. 1998). Vegetative segregation is also an effective process for removing deleterious somatic mtDNA mutations (Lehtinen et al. 2000), but it will also occasionally lead to high levels of mutant mtDNA within individual cells. Cells containing levels of mutant mtDNA that are above the critical threshold will develop a bioenergetic defect. It is likely that there will be selection against these cells, resulting in the loss of mutant mtDNA from the cell population over time. This has been observed in serial blood samples taken from patients with mtDNA disease (Larsson et al. 1990; 't Hart et al. 1996).

In many human tissues, and particularly during human development, both relaxed replication and vegetative segregation will lead to changes in the level of heteroplasmy. If a mutation occurs within the germ line, or in a developing organism, then the number of mutant molecules may increase to high levels in that individual, or enter the germ line and be transmitted to subsequent generations. This may result in the loss of the mutation, usually within a few generations, or fixation of the new mutation in the maternal line.

4
The Inheritance of mtDNA

4.1
Maternal Inheritance

The traditional view is that mtDNA is exclusively inherited down the maternal line (Giles et al. 1980), but this has recently been challenged. An excess of homoplasies (parallel mutations) in geographically distinct populations

(Hagelberg et al. 1999), and an apparent inverse relationship between the intermolecular mtDNA recombination rate and genetic distance (Awadalla et al. 1999), raised the issue of paternal transmission and questioned the role of paternal mtDNA in studies of human evolution (Awadalla et al. 1999). However, a number of subsequent studies cast doubt on the interpretation of these data. Different measures of recombination on different data sets show no convincing evidence of recombination (Macaulay et al. 1999; Elson et al. 2001a), and sequencing artefacts provide an explanation for the excess homoplasies described in earlier reports (Hagelberg et al. 2000). The recent report of a small pathogenic mtDNA deletion of paternal origin has established that paternal 'leakage' of mtDNA may occur (Schwartz and Vissing 2002). However, many large families with mtDNA disease have been studied throughout the world over the last decade, and there are no other published examples of paternal transmission. Moreover, there is little evidence to support any intermolecular recombination between mtDNAs in vivo (Howell 1997). Recent work by Kraytsberg and colleagues provides tantalising evidence which the authors believe demonstrates recombination in vivo (Kraytsberg et al. 2004), but the techniques used are prone to generate artefacts that could give misleading results (Bandelt et al. 2005). Manfredi and colleagues recently reported human mtDNA recombination human cell lines (Euromit VI, Nijmegen, July 2004), but in vivo human mtDNA recombination has yet to be demonstrated. Moreover, the available evidence therefore indicates that paternal leakage of mtDNA is exceptionally rare, and even if it does occur, it is extremely unlikely that there will be significant recombination between paternal and maternal mtDNAs. There therefore seems no reason at present to question the traditional dogma of maternal transmission, at least from the population genetics point of view.

The precise molecular mechanisms behind strict maternal transmission have yet to be clarified. Although it was originally thought that paternal mitochondria did not enter the oocyte, this is not the case. Paternal mtDNA molecules have been detected in early human preimplantation embryos (St. John et al. 2000), but the paternal mitochondria appear to be destroyed by an active mechanism that involves ubiquitination (Sutovsky et al. 1999).

4.2
The Mitochondrial Genetic Bottleneck

Rapid intergenerational changes in mitochondrial genotype were first observed in Holstein cows transmitting mtDNA polymorphisms (Upholt and Dawid 1977; Hauswirth and Laipis 1982; Olivo et al. 1983). Similar results were subsequently described in many mammalian species, including humans transmitting pathogenic mtDNA mutations (Holt et al. 1989; Vilkki et al. 1990; Larsson et al. 1992). These observations led to the suggestion that only a small number of mtDNA molecules were passed on from mother to offspring—the mitochondrial 'genetic bottleneck' (Hauswirth and Laipis 1982). Our understanding of this process was greatly advanced by studies of heteroplasmic mice transmitting neutral mtDNA polymorphisms (Jenuth et al. 1996; Meirelles and Smith 1997, 1998). By measuring the percentage level of heteroplasmy in the offspring, and comparing this to the level in developing oocytes, Jenuth et al.(1996) showed that the variation in heteroplasmy amongst the offspring of a single female was determined by random genetic drift at an early stage during oogenesis in the developing mother, before the

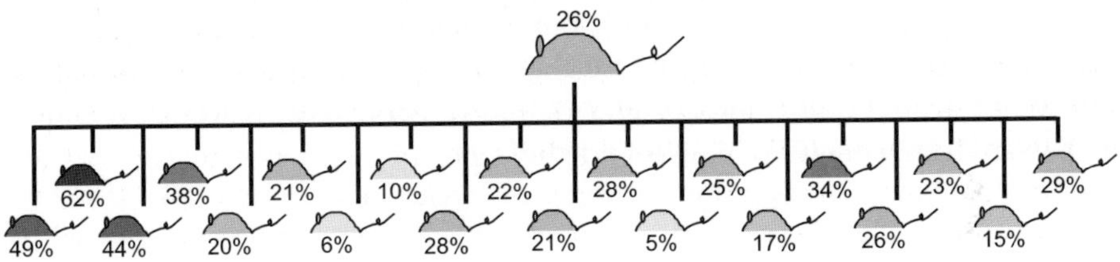

Fig. 1 The transmission of mitochondrial DNA (*mtDNA*) heteroplasmy in NZB/BALB heteroplasmic mice. Random genetic drift during the transmission of NZB/C57Bl6 heteroplasmic mice. The percentage level of C57Bl6 genotype is indicated adjacent to each mouse. The mean percentage level in all of the offspring is equal to the level in the mother, and there are approximately as many offspring with greater than the mean level as there are offspring with less than the mean percentage level of C57Bl6. (Courtesy of S. White, H.H. Dahl, and D.R. Thorburn. Modified from Chinnery et al. 2000b)

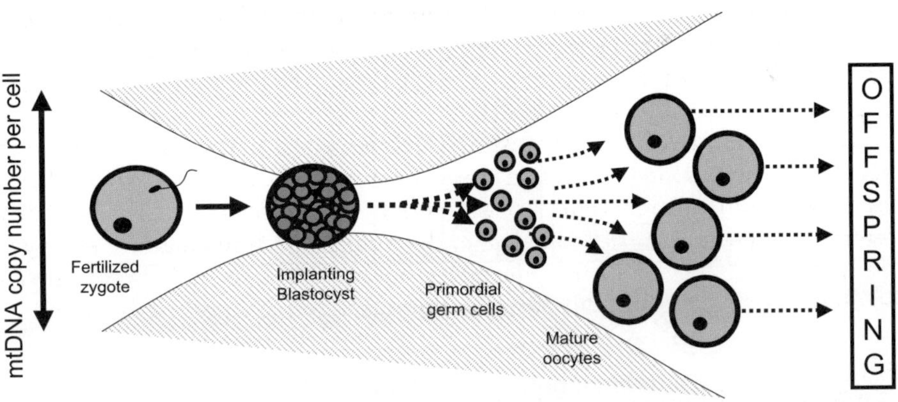

Fig. 2 The mitochondrial genetic bottleneck hypothesis. A reduction in the effective number of mitochondrial genomes occurs during early embryogenesis in the developing female germ line. This generates the variability in mutation load seen amongst the offspring of a single female. This variability is present within the primary oocytes of the female mouse that will become the mother of the next generation. It is not known whether the bottleneck is due to a physical reduction in the number of mitochondrial genomes within individual cells, a reduction in the effective population size due to the compartmentalisation of genomes into homoplasmic segregating units, or the preferential amplification of specific genotypes

formation of the primary oocytes (Figs. 1, 2). Large studies of many human pedigrees suggested that the same mechanism also operates during the transmission of pathogenic mtDNA mutations (Fig. 3) (Chinnery et al. 2000b), but problems with pedigree ascertainment bias through an affected individual, and changing levels of mutant mtDNA in accessible human tissues, influence the interpretation of the data. In an attempt to circumvent this problem, the level of mutant mtDNA was measured in a large number of primary oocytes from a woman harbouring a known pathogenic mtDNA point mutation (the A3243G tRNA$^{Leu(UUR)}$ mutation; Fig. 4) (Brown et al. 2001). The frequency distribution of oocytes corresponded to a binomial distribution, as would be expected for a random sampling or 'bottleneck' process. Further studies are needed to confirm that the same mechanism operates during the transmission of other mutations, particularly at higher percentage levels where there may be selection against the mutant genotype.

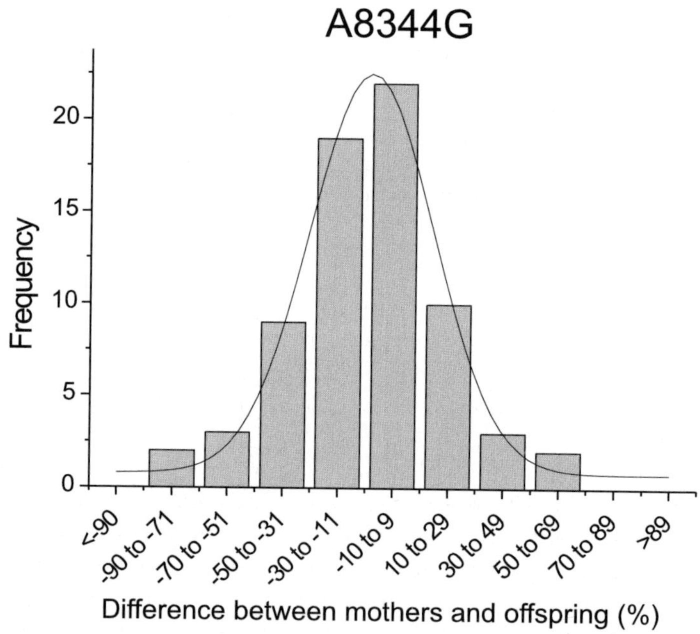

Fig. 3 The transmission of pathogenic mtDNA mutations in human pedigrees. Frequency distribution of the difference in the percentage level of the pathogenic A8344G mtDNA tRNA^{Lys} mutation. The offspring–mother difference was calculated by subtracting the percentage level of mutated mtDNA in the offspring's blood from the percentage level of mutated mtDNA in the corresponding mother. The frequency distribution for 70 such offspring pairs is plotted. For this mutation the bell-shaped distribution is symmetrical about a mean value that was not statistically significant from zero (mean – 7.44% mutant), as expected for a process that is primarily governed by random genetic drift; see Chinnery et al. (2000b) for details

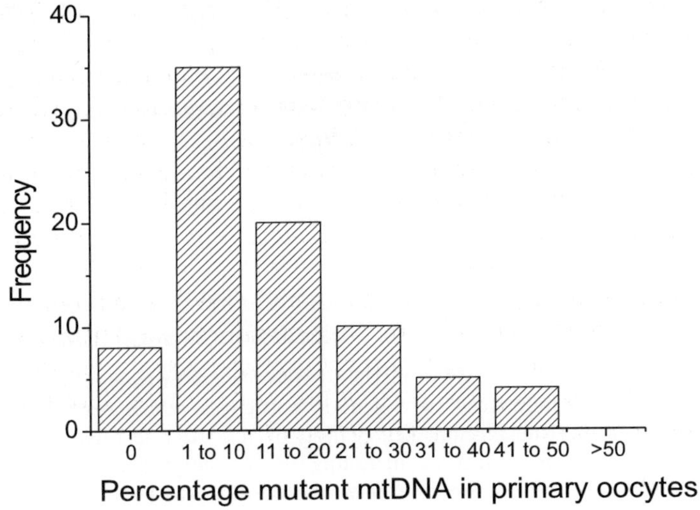

Fig. 4 Frequency distribution of A3243G mtDNA tRNA$^{Leui(UUR)}$ mutant mtDNA in primary oocytes. The 82 primary oocytes were dissected from a single ovary from a woman known to harbour the pathogenic A3243G mtDNA tRNA$^{Leu(UUR)}$ mutation (Brown et al. 2001). The data correspond to a binomial distribution as would be expected for a bottleneck process, with approximately as many oocytes having a mutation load greater than the mean as there are oocytes with a mutation load less than the mean. This indicates that the transmission process in this individual was primarily governed by random genetic drift

Attempts to determine the size and mechanism behind the bottleneck have been less forthcoming. Various authors have modelled the bottleneck process using an adapted population genetics bottleneck model (Wright 1969; Ashley et al. 1989; Howell et al. 1992; Marchington et al. 1997, 1998; Poulton et al. 1998), or a model based upon probabilistic deductions from the observation of intergenerational variance (Bendall et al. 1996). Whilst these models enable predictions to be made about the behaviour of the bottleneck process in specific pedigrees, they may not accurately reflect the underlying biological process. In particular, it has not been possible to distinguish between a single 'tight' bottleneck and multiple less stringent bottlenecks which can both produce the same variation amongst offspring (Howell et al. 1992; Marchington et al. 1998). In addition, it is not known whether the bottleneck is due to a physical restriction in the number of mtDNA molecules within the developing germ cells, the physical selection of a subgroup of molecules, the compartmentalisation of mtDNA into homoplasmic 'segregating units' (resulting in a reduction in the effective population size), or the preferential replication of a subgroup of molecules as has been shown in cultured cells (Davis and Clayton 1996). These are important issues, particularly in the light of recent advances in reproductive cloning, where the cloned offspring

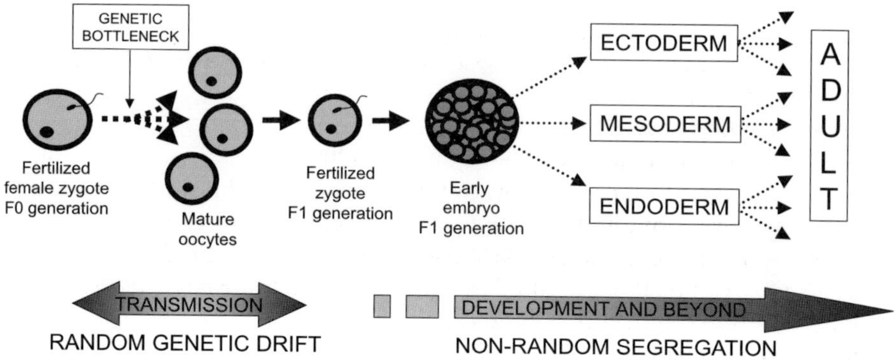

Fig. 5 Factors influencing the level of mtDNA heteroplasmy in *Homo sapiens*. The transmission of mtDNA is influenced by a genetic bottleneck which results in accelerated random genetic drift. Non-random segregation occurs after fertilisation and may be influenced by a number of factors, including the phenotypic effects of a mutation, differences in mtDNA replication rate between the different genotypes, differences in the mtDNA degradation rate between different genotypes, nuclear genes that influence mtDNA maintenance, or other currently unknown nuclear genetic factors (Chinnery 2002). Changes in the level of heteroplasmy may occur throughout human life

may harbour a mixture of nuclear-donor and enucleated-recipient mtDNA (although the first cloned sheep, Dolly, only harboured enucleated-recipient mtDNA) (Steinborn et al. 2000).

In conclusion, there is accumulating evidence that the actual process of transmission of mtDNA heteroplasmy is largely determined by random genetic drift through a hypothetical bottleneck (Fig. 5). It has been suggested that the process evolved in order to counteract the relentless accumulation of mildly deleterious mutations in an asexual organelle (Muller's ratchet), and thus prevent 'mutational meltdown' and the extinction of a maternal line (Muller 1964). The bottleneck will lead to rapid random genetic drift, the loss of most novel mutations, and the rapid fixation of some. If severely deleterious, these mutations will be lost through natural selection acting at the level of the organism. If these mutations are only weakly pathogenic, advantageous, or truly neutral, they will contribute to the molecular evolution of the mitochondrial genome in subsequent maternal descendents.

5 Segregation During Early Development

Although the maternal transmission of mtDNA appears to be governed by random genetic drift, the tissue segregation of mutant mtDNA is anything but random. Tissue segregation studies of established pathogenic mtDNA mutations often show high levels of mutant mtDNA in non-dividing tissues (brain, muscle) and low levels in proliferating tissues (Macmillan et al. 1993; Chinnery et al. 1999). For some time it was thought that this was due to selection against dividing cells containing high levels of mutant mtDNA (see before), but many proliferative tissues also contain high percentage levels of mutant, for example urinary epithelium (Dubeau et al. 2000) and sperm (Spiropoulos et al. 2002), so additional factors must be important. Again, studies of heteroplasmic mice are providing us with further insight (Jenuth et al. 1997). These mice were generated by karyoplast transfer from one inbred strain to another (NZB and BALB, also called cytoplasmic transfer, when the cytoplasm containing mitochondria from one cell is fused with an early embryonic cell from another organism). The resultant heteroplasmic mice contained varying proportions of two naturally occurring mouse mitochondrial genomes which differed at 15 sites. Despite the fact that the 15 sequence variants were considered to be phenotypically neutral, the two genotypes segregated in a tissue-specific manner, irrespective of the nuclear genetic background. Detailed investigations showed that the tissue-specific segregation pattern was not due to a respiratory chain defect, nor was it due to different rates of cell division or mtDNA replication. It was concluded that the level of heteroplasmy was somehow being regulated at the level of the individual mitochondrial genome, possibly involving factors important for the long-term maintenance of mtDNA (Battersby and Shoubridge 2001; Chinnery 2002).

It seems likely that nuclear genetic factors modulate the level of heteroplasmy and thereby play an important role in determining the clinical phenotype of mtDNA disease. This is important because they may influence the transmission of heteroplasmic mtDNA defects, and may also influence the evolution of mtDNA at the population level.

6 Conclusions and Future Perspectives

Recent advances in our understanding of the segregation and transmission of mutant mtDNA have important implications for the study human mtDNA evolution. We now have a better understanding of the mechanism behind the transmission of mtDNA heteroplasmy and the fixation of new mutations within pedigrees, and by studying the factors that modulate heteroplasmy we may gain more insight into additional nuclear genetic factors that interact with the mitochondrial genome and thereby influence the way that mtDNA evolves within populations.

There is emerging evidence from a number of species that nuclear and mitochondrial genes coevolve, possibly through mechanisms that do not directly involve the respiratory chain (Blier et al. 2001). These could act at the population level, on the organism, cell, organelle, or even on the genome itself (Battersby and Shoubridge 2001; Chinnery 2002). There may be many confounding factors, including the mitochondrial and nuclear genetic background (Torroni et al. 1996; Battersby and Shoubridge 2001), and the environment (Blier et al. 2001); and the effects of a particular mutation may be so subtle that it will not be detectable in single organisms within their natural environment or in the laboratory. As in other species (Blier et al. 2001), it may only be possible to detect the consequences of the mutations at the population level. In this way, our understanding of the genetic evolution of *Homo sapiens* will advance our understanding of the segregation of mtDNA within individual organisms, and have particular relevance for the investigation of mtDNA and human disease.

Studies of mtDNA segregation and transmission within individuals, and mtDNA evolution within populations, are therefore symbiotically interrelated. In the future it is likely that advances in one area will complement novel findings in the other, providing valuable insights for intergenomic coadaptation in health and disease.

Acknowledgements The author is a Wellcome Trust Senior Fellow in Clinical Science. He also receives funding from Ataxia (UK), The Alzheimer's Research Trust, the Association Francaise contre les Myopathies, and the European Union under the FP6 framework.

References

Ashley MV, Laipis PJ, Hausworth WW (1989) Rapid segregation of heteroplasmic bovine mitochondria. Nucleic Acids Res 17:7325–7331

Awadalla P, Eyre-Walker A, Maynard Smith J (1999) Linkage disequilibrium and recombination in hominid mitochondrial DNA. Science 286:2524–2525

Bandelt H-J, Kong Q-P, Parson W, Salas A (2005) More evidence for non-maternal inheritance of mitochondrial DNA? J Med Genet 42:957–960. DOI 10.1136/jmg.2005.033589

Battersby BJ, Shoubridge E (2001) Selection of a mtDNA sequence variant in hepatocytes of heteroplasmic mice is not due to difference in respiratory chain function or efficiency of replication. Hum Mol Genet 10:2469–2479

Bendall KE, Macaulay VA, Baker JR, Sykes BC (1996) Heteroplasmic point mutations in the human mtDNA control region. Am J Hum Genet 59:1276–1287

Birky CW (1994) Relaxed and stringent genomes: why cytoplasmic genes don't obey Mendel's laws. J Hered 85:355–365

Blier PU, Dufresne F, Burton RS (2001) Natural selection and the evolution of mtDNA-encoded peptides: evidence for intergenomic co-adaptation. Trends Genet 17:400–406

Bogenhagen D, Clayton DA (1977) Mouse L cell mitochondrial DNA molecules are selected randomly for replication throughout the cell cycle. Cell 11:719–727

Brierley EJ, Johnson MA, Lightowlers RN, James OF, Turnbull DM (1998) Role of mitochondrial DNA mutations in human aging: implications for the central nervous system and muscle. Ann Neurol 43:217–223

Brown DT, Samuels DC, Michael EM, Turnbull DM, Chinnery PF (2001) Random genetic drift determines the level of mutant mitochondrial DNA in human primary oocytes. Am J Hum Genet 68:535–536

Chinnery PF (2002) Modulating heteroplasmy. Trends Genet 18:173–176

Chinnery PF, Samuels DC (1999) Relaxed replication of mtDNA: a model with implications for the expression of disease. Am J Hum Genet 64:1158–1165

Chinnery PF, Zwijnenburg PJG, Howell N, Lightowlers RN, Bindoff L, Taylor RW, Walker M, Turnbull DM (1999) Non-random tissue distribution of mutant mitochondrial DNA. Am J Med Genet 85:498–501

Chinnery PF, Johnson MA, Wardell TM, Singh-Kler R, Hayes C, Brown DT, Taylor RW, Bindoff LA, Turnbull DM (2000a) Epidemiology of pathogenic mitochondrial DNA mutations. Ann Neurol 48:188–193

Chinnery PF, Thorburn DR, Samuels DC, White SL, Dahl HM, Turnbull DM, Lightowlers RN, Howell N (2000b) The inheritance of mitochondrial DNA heteroplasmy: random drift, selection or both? Trends Genet 16:500–505

Davis AF, Clayton DA (1996) In situ localisation of mitochondrial DNA replication in intact mammalian cells. J Cell Biol 135:883–893

DiMauro S, Schon EA (2001) Mitochondrial DNA mutations in human disease. Am J Med Genet 106:18–26

Dubeau F, De Stefano N, Zifkin BG, Arnold DL, Shoubridge EA (2000) Oxidative phosphorylation defect in the brains of carriers of the $tRNA^{leu(UUR)}$ A3243G mutation in a MELAS pedigree. Ann Neurol 47:179–185

Elson JL, Andrews RM, Chinnery PF, Lightowlers RN, Turnbull DM, Howell N (2001a) Analysis of European mtDNAs for recombination. Am J Hum Genet 68:145–153

Elson JL, Samuels DC, Turnbull DM, Chinnery PF (2001b) Random intracellular drift explains the clonal expansion of mitochondrial DNA mutations with age. Am J Hum Genet 68:802–806

Giles RE, Blanc H, Cann HM, Wallace DC (1980) Maternal inheritance of human mitochondrial DNA. Proc Natl Acad Sci USA 77:6715–6719

Hagelberg E, Goldman N, Lió P, Wheelan S, Schiefenhovel W, Clegg JB, Bowden DK (1999) Evidence for mitochondrial DNA recombination in the human population of island Melanesia. Proc R Soc Lond Ser B 266:485–492

Hagelberg E, Goldman N, Lió P, Whelan S, Schiefenhovel W, Clegg JB, Bowden DK (2000) Evidence for mitochondrial DNA recombination in a human population of island Melanesia: correction. Proc R Soc Lond Ser B 267:1595–1596

Hauswirth WM, Laipis PJ (1982) Mitochondrial DNA polymorphism in a maternal lineage of Holstien cows. Proc Natl Acad Sci USA 79:4686–4690

Holt IJ, Miller DH, Harding AE (1989) Genetic heterogeneity and mitochondrial DNA heteroplasmy in Leber's hereditary optic neuropathy. J Med Genet 26:739–743

Howell N (1997) mtDNA recombination: what do the in vitro data mean? Am J Hum Genet 61:18–22

Howell N, Halvorson S, Kubacka I, McCullough DA, Bindoff LA, Turnbull DM (1992) Mitochondrial gene segregation in mammals: is the bottleneck always narrow? Hum Genet 90:117–120

Jenuth J, Peterson AC, Fu K, Shoubridge EA (1996) Random genetic drift in the female germ line explains the rapid segregation of mammalian mitochondrial DNA. Nat Genet 14:146–151

Jenuth JP, Peterson AC, Shoubridge EA (1997) Tissue-specific selection for different mtDNA genotypes in heteroplasmic mice. Nat Genet 16:93–95

King MP, Attardi G (1989) Human cells lacking mtDNA: repopulation with exogenous mitochondria by complementation. Science 246:500–503

Kraytsberg Y, Schwartz M, Brown TA, Ebralidse K, Kunz WS, Clayton DA, Vissing J, Khrapko K (2004) Recombination of human mitochondrial DNA. Science 304:981

Larsson NG, Clayton DA (1995) Molecular genetic aspects of human mitochondrial disorders. Ann Rev Genet 29:151–178

Larsson NG, Holme E, Kristiansson B, Oldfors A, Tulinius M (1990) Progressive increase of the mutated mitochondrial DNA fraction in Kearns-Sayre syndrome. Pediatr Res 28:131–136

Larsson NG, Tulinius MH, Holme E, Oldfors A, Andersen O, Wahlstrom J, Aasly J (1992) Segregation and manifestations of the mtDNA tRNA(Lys) A→G(8344) mutation of myoclonus epilepsy and ragged-red fibers (MERRF) syndrome. Am J Hum Genet 51:1201–1212

Lehtinen SK, Hance N, El Meziane A, Juhola MK, Juhola KM, Karhu R, Spelbrink JN, Holt IJ, Jacobs HT (2000) Genotypic stability, segregation and selection in heteroplasmic human cell lines containing np 3243 mutant mtDNA. Genetics 154:363–380

Macaulay V, Richards M, Sykes B (1999) Mitochondrial DNA recombination—no need to panic. Proc R Soc Lond B Biol Sci 266:2037–2039; discussion 2041–2042

Macmillan C, Lach B, Shoubridge EA (1993) Variable distribution of mutant mitochondrial DNAs (tRNA(Leu[3243])) in tissues of symptomatic relatives with MELAS: the role of mitotic segregation. Neurology 43:1586–1590

Majamaa K, Moilanen JS, Uimonen S, Remes AM, Salmela PI, Karppa M, Majamaa-Volti KAM, Rusanen H, Sorri M, Peuhkurinen KJ, Hassinen IE (1998) Epidemiology of A3243G, the mutation for mitochondrial encephalomyopathy, lactic acidosis, and strokelike episodes: prevalence of the mutation in an adult population. Am J Hum Genet 63:447–454

Marchington DR, Hartshorne GM, Barlow D, Poulton J (1997) Homopolymeric tract heteroplasmy in mtDNA from tissues and single oocytes: support for a genetic bottleneck. Am J Hum Genet 60:408–416

Marchington DR, Macaulay V, Hartshorne GM, Barlow D, Poulton J (1998) Evidence from human oocytes for a genetic bottleneck in an mtDNA disease. Am J Hum Genet 63:769–775

Meirelles F, Smith LC (1997) Mitochondrial genotype segregation in a mouse heteroplasmic lineage produced by embyonic karyoplast transplantation. Genetics 145:445–451

Meirelles F, Smith LC (1998) Mitochondrial genotype segregation during preimplantation development in mouse heteroplasmic embryos. Genetics 148:877–883

Moraes CT, Ricci E, Bonilla E, DiMauro S, Schon EA (1992) The mitochondrial tRNA(Leu(UUR)) mutation in mitochondrial encephalomyopathy, lactic acidosis, and strokelike episodes (MELAS): genetic, biochemical, and morphological correlations in skeletal muscle. Am J Hum Genet 50:934–949

Muller HJ (1964) The relation of recombination to mutational advance. Mutat Res 1:2–9

Olivo PD, Van de Walle MJ, Laipis PJ, Hauswirth WW (1983) Nucleotide sequence evidence for rapid genotypic shifts in the bovine mitochondrial DNA D-loop. Nature 306:400–402

Poulton J, Turnbull DM (2000) 74th ENMC international workshop: mitochondrial diseases. Neuromuscul Disord 10:460–462

Poulton J, Macaulay V, Marchington DR (1998) Mitochondrial genetics '98: Is the bottleneck cracked? Am J Hum Genet 62:752–757

Schon EA, Bonilla E, DiMauro S (1997) Mitochondrial DNA mutations and pathogenesis. J Bioenerg Biomembr 29:131–149

Schwartz M, Vissing J (2002) Paternal inheritance of mitochondrial DNA. N Engl J Med 347:576–580

Spiropoulos J, Turnbull DM, Chinnery PF (2002) Can mitochondrial DNA mutations cause sperm dysfunction? Mol Hum Reprod 8:719–721

Steinborn R, Schinogl P, Zakhartchenko V, Achmann R, Schernthaner W, Stojkovic M, Wolf E, Muller M, Brem G (2000) Mitochondrial DNA heteroplasmy in cloned cattle produced by fetal and adult cell cloning. Nat Genet 25:255–257

St John J, Sakkas D, Dimitriadi K, Barnes A, Maclin V, Ramey J, Barratt C, De Jonge C (2000) Failure of elimination of paternal mitochondrial DNA in abnormal embryos. Lancet 355:200

Sutovsky P, Moreno RD, Ramalho-Santos J, Domiko T, Simerly C, Schatten G (1999) Ubiquitin tag for sperm mitochondria. Nature 403:371–372

't Hart LM, Jansen JJ, Lemkes HHPJ, de Knijff P, Maassen JA (1996) Heteroplasmy levels of a mitochondrial gene mutation associated with diabetes mellitus decrease in leucocyte DNA upon aging. Hum Mutat 7:193–197

Torroni A, Carelli V, Petrozzi M, Terracina M, Barboni P, Malpassi P, Wallace DC, Scozzari R (1996) Detection of the mtDNA 14484 mutation on an African-specific haplotype: implications about its role in causing Leber hereditary optic neuropathy [letter]. Am J Hum Genet 59:248–252

Upholt WB, Dawid IB (1977) Mapping of mitochondrial DNA of individual sheep and goats: rapid evolution in the D loop region. Cell 11:571–583

Uusimaa J, Remes AM, Rantala H, Vianionpaa L, Herva R, Vuopala K, Nuutinen M, Majamaa K, Hassinen IE (2000) Childhood encepahlopathies and myopathies: a prospective study in a defined population to assess the frequency of mitochondrial diseases. Pediatrics 105:598–603

Vilkki J, Savontaus ML, Nikoskelainen EK (1990) Segregation of mitochondrial genomes in a heteroplasmic lineage with Leber hereditary optic neuroretinopathy. Am J Hum Genet 47:95–100

Wallace DC (1999) Mitochondrial diseases in mouse and man. Science 283:1482–1488

Wallace DC, Brown MD, Melov S, Graham B, Lott M (1998) Mitochondrial biology, degenerative diseases and aging. Biofactors 7:187–190

Wright S (1969) Evolution and the genetics of populations. University of Chicago Press, Chicago

Nucleic Acids and Molecular Biology, Vol. 18
Hans-Jürgen Bandelt, Vincent Macaulay, Martin Richards (Eds.)
Human Mitochondrial DNA and the Evolution of *Homo sapiens*

Numts Revisited

Claudio M. Bravi · Walther Parson · Hans-Jürgen Bandelt (✉)

Dept. of Mathematics, University of Hamburg, Bundesstr. 55, 20146 Hamburg, Germany
bandelt@math.uni-hamburg.de

1 Introduction

Loosely speaking, mitochondrial DNA (mtDNA) does not only thrive in the mitochondria of the cell but it also thrives in various bits and pieces within the nuclear DNA of the chromosomes. The first hints at similarity between nuclear DNA and mtDNA came from hybridization experiments involving mouse liver (du Buy and Riley 1967). The existence of contiguous DNA sequences in nuclear genomes that have extensive similarity to (noncontiguous) mtDNA sequences was subsequently confirmed in yeast, locust, fungus, sea urchin, human, maize, and rat. These findings were published in a series of seminal papers in 1983 (in exact chronological order by month: Farrelly et al. 1983; Gellissen et al. 1983; Wright and Cummings 1983; Jacobs et al. 1983; Tsuzuki et al. 1983a; Kemble et al. 1983; Tsuzuki et al. 1983b; Hadler et al. 1983). The transferred fragments from mtDNA are usually called nuclear inserts of mtDNA, or *numts*, for short (Lopez et al. 1994).

We now know that the transfer of genetic material from organelle to nucleus is a ubiquitous mechanism of evolutionary change in eukaryotes, which may have started soon after the arrival of the ancestral endosymbiont in the cytoplasm as an organelle (see Box 2 in Zhang and Hewitt 1996), and the integration of mitochondrial fragments in the nucleus is an ongoing process that shapes nuclear genomes (Richetti et al. 2004). Translocation of genetic material is also known to take place within and between the nuclear chromosomes, for example, by retrotransposition involving different types of elements in insertion mutagenesis or through unequal crossover in deletion mutagenesis (Chen et al. 2005). Then subsequent rearrangement and partial duplication within the nucleus can also affect earlier settled inserts as hitchhikers, which thus complicates the analysis of transfer events.

As to the precise mechanism by which mtDNA slips into the nucleus, one can only speculate: "Mitochondrial DNA released from sperm cells during fertilization could be an important source for mtDNA insertion into the germline nuclear genome in multicellular organisms" (see supplementary information in Heilig et al. 2003). Conversely, Zhao et al. (2004) even envision that "it is conceivable that Numts could be transferred back to the mtDNA".

The idea about the potential paternal input of mtDNA into the nuclear DNA or mtDNA of the fertilized egg seems to have lingered since the futile attempts to infer recombination of human mtDNA (see Bandelt et al. 2005 for critical reassessment and pertinent references).

Although the migration of mtDNA to the nucleus is certainly an interesting process, why should it matter when we are primarily interested in mtDNA as a marker for human migrations? Well, the first answer is that those numts which arrived in the human line, say, between 0.2 million and six million years ago could serve as outgroups to modern human mtDNA, and consequently would help to root the human mtDNA tree. Since mitochondrial pseudogenes residing in the nucleus evolve much more slowly than their functional counterparts in the mitochondrion, numts may be considered as snapshots of the mtDNA at the time of transfer, thus representing nuclear fossils of ancient mtDNA (Zischler et al. 1995a, 1998; Perna and Kocher 1996; Zischler 2000). Human numts could therefore supplement or in part even replace mtDNA sequences from common chimpanzee and bonobo as outgroups for estimating ancestral states at nucletotide positions that changed near the root in the human mtDNA phylogeny. Gorilla mtDNA, apart from being somewhat too distant from human mtDNA, enjoys a particularly rich spectrum of accompanying numts, thus making it difficult to determine whether purported mitochondrial sequences truly derive from that genome (Jensen-Seaman et al. 2004; Thalmann et al. 2004, 2005).

The second answer to the previous question is that authentic ancient mtDNA—from the mitochondrion—normally competes with a mixed pallet of numts in the course of the sequencing process, depending on where the employed sequencing primers bind. With modern mtDNA, at least under normal conditions (but see later), numts do not stand a chance of proliferating in the cycles of the polymerase chain reaction (PCR) because nuclear DNA is greatly outnumbered compared to mtDNA in copy number. The situation is drastically different with the meagre amounts of authentic short stretches of ancient DNA that could potentially be retrieved from old bones or teeth. In the absence of any authentic mtDNA, both mtDNA and numts from modern contaminant DNA stand a good chance of proliferating. Then, numt DNA may preferentially be generated and falsely get reported as the specimen's mtDNA because modern human mtDNA (stemming from the researcher himself or herself, say) would readily be identified and discarded as an obvious contaminant but the more deviant numt DNA could stay unrecognized.

The breaking news of mid-November 1994 was that small fragments of DNA might have survived 80 million years in the bone material from a dinosaur (Woodward et al. 1994). But, as indicated in the accompanying Research News of *Science* at the time, this 'dino' DNA find was greeted with some scepticism by the academic community (Gibbons 1994). In aiming at countering the emerging evidence that human numts might have been in play (Zischler et al. 1995b), one of the arguments brought forward by Woodward

(1995) expressed the idea that the similarity of dino DNA and human numts may be spurious because "if the analysis is based on a single gene or locus, any selective pressures that may be exerted on the resulting protein must be taken into consideration". This rather flexible general-purpose argument warning that, because of selection, one could never know what happens with the 'single locus' mtDNA lingers to the present day whenever one wants to downplay the role of mtDNA in elucidating phylogenetic relationships.

Willerslev and Cooper (2005) have reminded the reader of further spectacular claims of DNA sequences surviving for millions of years which either turned out to have originated from human or microbial contamination or were not replicable by independent experiments. It might seem easier to transform lead into gold than to obtain a stretch of authentic DNA from, say, Miocene amber inclusions of fossilized blood droplets. But, of course, it is the hope that dies last: "haemolymph droplets may serve as reservoirs for fossil DNA" (Penney 2005). Another example—of a somewhat more recent age than the Miocene, though—constitutes the short hypervariable segment I (HVS-I) sequence attributed to the 'Lake Mungo 3' sample by Adcock et al. (2001): besides the many problematic aspects of that study (Cooper et al. 2001), the variation reported comes suspiciously close to some numt sequence rather than what would be expected from real mtDNA of an early offshoot from the human mtDNA line (see Fig. 10.14 in Klein and Takahata 2002).

2
Numts in Humans

The numts that have been discovered in human chromosomes so far may considerably vary in size, from some oligonucleotides to large pieces of several kilobases. Woischnik and Moraes (2002) have estimated that there were 612 independent integrations of mtDNA sequences in the human nuclear genome, reflecting the continued colonization of the human genome by mtDNA; the total contribution of numts to the human genome was then estimated to be at least 0.016%. After arrival in the nucleus, numts may undergo considerable rearrangement (see Fig. 1 in Tourmen et al. 2002). Nuclear duplications of numts are not infrequent (Bensasson et al. 2003).

If a new insertion of a piece of mtDNA in the nuclear genome arrives at an unfortunate place, then it could modify the expression of a protein or inhibit the synthesis of a functional protein. This could then lead to a particular disease, as observed in a sporadic case of Pallister–Hall syndrome, where a 72-bp insertion (12243–12314) into exon 14 of the *GLI3* gene created a premature stop codon (Turner et al. 2003). Later, a case of mucolipidosis IV was reported by Goldin et al. (2004), where a 93-bp segment from mitochondrial NADH dehydrogenase 5 was inserted into exon 2 of *MCOLN1*, which abolished proper splicing of *MCOLN1*. Chen et al. (2005) have revealed the mitochondrial origin of a previously unrecognized 36-bp insertion in exon 9 of the *USH1C* gene. Inspection of the sequence features for these two fragments led the authors to propose a novel mutational mechanism, *trans*-replication slippage, for the generation of this sort of mitochondrial–nuclear sequence transfer.

There is a (not unexpected) twist to this: suppose that some mtDNA mutations are suspected to participate in the aetiology of a particular disease since they were found in a patient to be heteroplasmic at higher levels than in controls. Under unfortunate circumstances the patient's mtDNA fragment could have stemmed in part from a numt, so that the perceived evidence was spurious. A classic case in this respect is that of Davis et al. (1997), who claimed that normal individuals and—at a higher level—patients with sporadic Alzheimer's disease harbour a specific population of mutated mitochondrial cytochrome *c* oxidase (*COX*) genes that coexist with normal mtDNAs. Hirano et al. (1997) and Wallace et al. (1997), in back-to-back publications, then demonstrated that the observed heteroplasmy was an artefact induced by coamplification of a numt. Herrnstadt et al. (1999) subsequently determined the nuclear insertion responsible for those earlier results as a 5842-bp numt on chromosome 1 with more than 98% identity to the mtDNA region 3914–9755. Further it was shown that a fragment of this numt was easily amplified in human cell lines depleted of mtDNA by means of primers that are normally employed for amplifying the paralogous mtDNA fragment.

Such cases could not come as a surprise, as Zhang and Hewitt (1996) have already warned: "One should seriously consider the possibility of nuclear copies of mitochondrial sequences if (1) PCR amplification constantly produces more than one band or different bands, (2) sequence ambiguities or background bands persist, (3) unexpected deletions/insertions, frameshifts or stop codons occur, (4) nucleotide sequences obtained are radically different from those expected, or (5) phylogenetic analysis yields an unusual or contradictory tree topology." Unfortunately, these caveats are not always taken seriously enough, so that human numts in camouflage still constitute a prolific reservoir for 'exciting' findings, as we will see later.

3 A Numt from an Egyptian Mummy

"With the advances in science technology, there is now a new 'weapon' that can help Egyptologists in their quest to construct the definitive chronology of Egyptian kings, namely DNA testing" (http://www.egyptologyonline.com/using_dna.htm). In fact, there has been a long tradition in extracting mtDNA from Egyptian mummies, beginning with the pioneering paper of Pääbo (1985). The advance of the PCR technique in the 1990s led to enthusiasm for the potential of application of the new approach to ancient DNA, but the then-prevailing optimism about retrieving authentic short fragments of mtDNA from ancient samples of modern humans (e.g. Audic and Béraud-Colomb 1997) has rather given way to scepticism (Pääbo et al. 2004; Gilbert et al. 2005; Bandelt 2005; Chap. 9), notwithstanding that routine application of ancient DNA analysis to human remains still flourishes in a niche of molecular anthropology (e.g. Jones 2004).

A classic case from the PCR era is the finding of "an unusual mitochondrial DNA sequence variant from an Egyptian mummy" by Hänni et al. (1994). These authors claimed to have obtained authentic information from a portion (16057–16400) of the control region including most of the first hypervariable segment I (HVS-I) and from a short stretch (8221–8290) covering a tiny fragment of the *COII* gene and most of the intergenic NC7 region (for full names of mitochondrial genes, see http://www.mitomap.org/cgi-bin/mitomap/tbl1gen.pl). The changes detected in HVS-I were 16221-16281-16331T and in the coding region fragment 8251-8255-8260-8269-8273A-8276-8278-8279-8284+T, where unsuffixed numbers indicate transitions (and suffixed numbers transversions or insertion) at the corresponding mtDNA positions relative to the Cambridge reference sequence (CRS), which is identical to the revised Cambridge reference sequence (rCRS) for these two short fragments.

As to the HVS-I part, the 16281 transition is a prime candidate for a phantom mutation, that is, an artificial mutation generated by the electrophoresis (Chap. 6). Although the 16221 polymorphism is known to be real, it also occurred as a frequent artefact in a flawed mtDNA study from 1996 (see data analysis in Bandelt et al. 2002). The A to T transversion at 16331 is most unusual—actually, it has never been found anywhere else; it may very well be a phantom mutation, like the other two mutations, or it may constitute postmortem damage (Chap. 5). Therefore, most likely, we are seeing here the rCRS consensus motif of recent contamination spiced with phantom or postmortem mutations.

The variants found in the short *COII*/NC7 fragment are really unusual since no contemporary mtDNA would show so much variation there. The authors sequenced both strands (repeatedly from independent amplification) and displayed the electropherograms for the two strands between positions 8243 and 8287 (Fig. 1A in Hänni et al. 1994). A BLAST search (http://www.ncbi.nlm.nih.gov/BLAST/), however, now gives a near match for this 82-bp sequence (and an exact match for the displayed 45-bp portion) within a 9-kb numt (GenBank accession number NC_000005) located in chromosome 5 that is paralogous to the mtDNA region 6117–15183 (rCRS). Namely, this numt differs from Hänni's sequence only by a transition at position 8288, which, however, is just outside the pair of displayed electropherograms. The conclusion we can draw from this is that this piece of DNA deemed to be mummy mtDNA did not actually come from the mitochondrion but was amplified from a piece of chromosome 5, presumably of some contaminant DNA.

4
A Numt from the Sperm's Head?

There is a view that mutations in mtDNA (other than well-recognized pathogenic mutations) could be responsible for sperm dysfunction such as asthenozoospermia or oligozoospermia (St. John et al. 2005). Moreover, it is contended that "there is increasing evidence that mitochondrial DNA (mtDNA) anomalies in sperm may lead to infertility" (May-Panloup et al. 2003). Pereira et al. (2005) have recently investigated whether Portuguese men with oligozoospermia showed any significant association with mtDNA haplogroups; however, they "did not find any mtDNA lineage association with the phenotype of male infertility". In the presence of a strong expectation that mitochondrial abnormalities could play a role in sperm dysfunction, abnormal cases reporting an extreme amount of seeming somatic mutations are greeted with enthusiasm. This would parallel the situation with perceived mtDNA instabilities in tumours, where, however, artefacts of mtDNA handling and sequencing seem to be the rule rather than the exception (Salas et al. 2005).

Thangaraj et al. (2003) claimed to have found a case of tissue-specific mosaicism for mtDNA in spermatozoa. The authors had targeted mtDNA genes *COI*, *S(UCN)*, *COII*, *K*, *ATP8*, *ATP6*, and *ND3* of the sperm's mtDNA of a subject with oligoasthenoteratozoospermia and they found as many as nine missense and 27 silent mutations as well as a 2-bp deletion, clustering in the region 6241–9167. Most of the nucleotides listed for the reference sequence, however, do not correspond to the nucleotides of the rCRS; instead, a simple pattern emerges: almost all positions seem to have been shifted from the rCRS by − 1, presumably because the convention concerning the 'empty nucleotide' at position 3106 in the rCRS (which was established in order to retain the original numbering of the CRS) was not followed. We have therefore added + 1 to all position numbers except for three irregular instances where we added + 6, − 1 and 0, respectively, in order to accommodate the postulated reference nucleotide; then we compared this sequence with three different versions of the same fragment from a 5.8-kb numt located on chromosome 1 that we mentioned before (Table 1).

It is certainly an irony that from the large panel of numts a fragment of exactly the same numt which had already misled Davis et al. (1997) was amplified again.

5
A Bouquet of Numts?

Idiosyncratic laboratory results that have never been confirmed anywhere else should be treated with great scepticism, as they could have arisen through

Table 1 Variant nucleotides of nuclear inserts of mitochondrial DNA(*mtDNA*) (*numt*)-derived sequences relative to the revised Cambridge reference sequence (*rCRS*)

rCRS Position	Nucl.	Numt chrom. 1[a]	Numt Herrnstadt[b]	Heteropl. Davis[c]	ρ^0 cell Hirano[d]	ρ^0 cell Wallace[e]	Sperm Thangaraj[f]
Shared mutations							
6023	G	A	A		A	A	
6221	T	C	C			C	
6242	C	T	T		T	T	T
6266	A	C	C		C	C	C
6299	A	G	G		G	G	
6366	G	A	A	A	A	A	
6383	G	A	A		A	A	A
6410	C	T	T		T	T	
6452	C	T	T			T	
6483	C	T	T	T	T	T	
6512	T	C	C		C	C	
6542	C	T	T		T	T	
6569	C	A	A		A	A	
6641	T	C	C		C	C	
6935	C	T	T		T	T	
6938	C	T	T		T	T	
7146	A	G	G	G	G	G	
7232	C	T	T			T	
7256	C	T	T		T	T	
7316	G	A	A		A	A	A
7521	G	A	A			A	
7650	C	T	T	T	T	T	T
7705	T	C	C		C	C	C
7810	C	T	T		T	T	T
7868	C	T	T	T	T	T	T
7891	C	T	T		T	T	T
7912	G	A	A		A	A	A
8021	A	G	G	G	G	G	G
8065	G	A	A		A	A	A
8140	C	T	T		T	T	T
8152	G	A	A		A	A	A
8167	T	C	C		C	C	C
8195	C	Del	Del			Del	Del
8196	A	Del	Del			Del	Del
8203	C	T	T		T	T	T
8254	C	C	C			T	T
8392	G	A	A		NA	NA	A
8455	C	T	T		NA	NA	T
8461	C	T	T		NA	NA	T
8503	T	C	C		NA	NA	C

Table 1 (continued)

rCRS position	Nucl.	Numt chrom. 1[a]	Numt Herrnstadt[b]	Heteropl. Davis[c]	ρ^0 cell Hirano[d]	ρ^0 cell Wallace[e]	Sperm Thangaraj[f]
Shared mutations							
8545	G	A	A		NA	NA	A
8655	C	T	T		NA	NA	T
8677	A	C	C		NA	NA	C
8701	A	G	G		NA	NA	G
8718	A	G	G		NA	NA	G
8860	A	G	G		NA	NA	G
8943	C	T	T		NA	NA	T
9060	C	A	A		NA	NA	A
9075	C	T	T		NA	NA	T
9168	C	T	T		NA	NA	T
Private mutations							
5984	A				G		
5985	G				C		
6345	T						C
6691	G	GG					
6772	A						C
7267	C						G
7310	T						A
7616	G					A	
8080	C				T		
8674	A				NA	NA	T
9102	C		T		NA	NA	
9175	C		T		NA	NA	

NA not applicable

[a] Variant nucleotides from an aligned paralogous 5.8-kb numt on chromosome 1 (GenBank, accession number NC_000001:603515-610692) relative to the region 5904–9207 in the rCRS

[b] Sequence variation as listed in Table 1 of Herrnstadt et al. (1999) but with the + 1 shift corrected at 9102; the corresponding sequence from GenBank (accession number AF134583) differs in this range at position 6383 (+ 1 shifted) and at further positions where ambiguous nucleotides were reported instead ("N", "Y", or "S")

[c] Six variant positions detected in cloned and sequenced *COI* and *COII* genes (Davis et al. 1997)

[d] Unambiguous nucleotide differences found between the rCRS sequence and the sequence of uncloned PCR products (for the *COI* and *COII* genes) from total DNA isolated from ρ^0-negative cells, according to Table 1 of Hirano et al. (1997)

[e] Consensus sequences obtained by Wallace et al. (1997) from four clones (*COI*) and five clones (*COII*), respectively, amplified from a mtDNA-free cell line

[f] Sequence variation attributed to "sperm mtDNA" by Thangaraj et al. (2003); position numbers are adjusted by assuming that all positions were systematically shifted by – 1, except for positions 6266 (– 6 shift), 8674 (+ 1 shift) and 8677 (no shift), which probably underwent secondary shifts

large-scale missequencing (e.g. reflected by phantom mutations) or through the amplification of numts. An interesting case is the one presented by Khrapko et al. (1997), who analysed a very short (approximately 100-bp) fragment from the coding region (with a rather unusual methodology—mutational spectrometry) and claimed that "human organs such as colon, lung, and muscle, as well as their derived tumors, share nearly all mitochondrial hotspot point mutations" within the range 10031–10129. If the 17 perceived mutational hotspots were real, then they should routinely show up in one or other sequence from large coding-region data sets. However, they do not, except for the 10084 transition. As we have already commented "the real causes of this unexpected result remain obscure, ... so that one cannot exclude the possibility that it is strongly affected by artifacts" (Salas et al. 2005).

We queried the human genome at http://www.ensembl.org/Homo_sapiens/blastview employing the algorithm BLASTN with search sensitivity set at 'no optimization'. A 99-bp mtDNA segment (10031–10129 in rCRS) modified with As at 10068 and 10098 (as frequently detected by Khrapko et al. 1997) was employed as a query sequence. Then, a total of 30 nuclear matches were obtained, and the hits were recovered with 100 bp added in both the 5′ and 3′ directions. The resulting approximately 300 bp sequences were then aligned with the rCRS fragment spanning 9931–10229. In order to aid the visual inspection of the mutational pattern, one further artificial sequence was included in the alignment, viz. the 99-bp segment spanning 10031–10129 in rCRS with the 17 changes reported by Krapkho et al. (1997). After visual inspection of the alignment, we retained 20 numts bearing some reasonable similarity with that profile within that segment (percentage identity between 67 and 85).

These numts then all include a longer than 220-bp stretch at which they are quite similar to each other and paralogous to the mtDNA fragment 9912–10139. In particular, with respect to this alignment, 15 of them bear A at 10068 and 10070 as well as C at 10076 that also occur in the hotspot spectrum (Fig. 3 in Khraphko et al. 1997), and moreover, all but one share the full recognition site GTAC at 10009–10012 for the restriction enzyme *Rsa*I, which was used by Khraphko et al. (1997) "to liberate the desired fragment (base pairs 10009–10231)". It thus appears that there must have been ample opportunity to catch portions of nuclear DNA. Since mutant fractions were enriched in successive steps by the approach employed, it seems that a bouquet of numts were preferentially targeted that eventually contributed to the mutational spectrum. This would also explain why this pattern of 17 hotspots was repeatedly recovered in those experiments. It seems, however, impossible to repeat these experiments in order to confirm our suspicion, because important information is missing from the "Materials and Methods" section of that paper, such as the exact amount of DNA that was used—which definitely matters (see later). Nonetheless, we believe that the best explanation for those unusual findings is that mutations were selectively fished from a number of numts.

Unfortunately, this story continues with the most recent publication of Coller et al. (2005), where again the same short fragment of the coding region was targeted and the stance of the peculiar hotspot spectrum was retained and renewed—now focusing on 12 such 'hotspots' (transitions at 10034, 10042, 10057, 10058, 10060, 10063, 10066, 10071, 10075, 10076, 10081 and 10084), which overlap with the earlier 17-hotspot spectrum in ten mutations. Nowhere in this publication is the possibility of mistaking numt variation for mtDNA variation considered. No extensive direct sequencing of multiple largely overlapping fragments spanning a wide part of the coding region (if not the entire mtDNA) has been considered necessary by those authors.

6
Adverse Laboratory Conditions

As outlined earlier, numts can get a chance of successful amplification by winning over the authentic mtDNA target when PCR conditions are suboptimal. In this respect, primer-site mutations, the DNA concentration and the number of PCR cycles are the most important considerations. Using laser-activated fluorescent detection technology—which has been the common standard for analysing DNA polymorphisms over the past decade—34 PCR cycles are required to retrieve an interpretable electrophoretic signal from a few targeted DNA molecules (Gill et al. 2005). Routine laboratory protocols generally involve DNA concentrations that are more practical (usually around 1 ng, corresponding to approximately 167 copies of nuclear DNA) and can use fewer PCR cycles, normally between 28 and 30 in order to obtain a useful result. Here we refer to nuclear DNA copy numbers, as the majority of applied and cited DNA quantitation methods determine the total DNA concentration in the extract, to which nuclear DNA contributes more than 99.9% of the molar mass. Depending on the biological tissue from which the DNA was extracted, 1 ng of nuclear DNA would correspond to approximately 167 000 copies of mtDNA, more than enough for mtDNA amplification and sequencing.

This actually indicates that only 1/1000th of the DNA content would be necessary for successful mtDNA typing. While this is true for direct amplicon testing methods such as fragment length analysis, it has become good laboratory practice to start with a slightly increased mtDNA amount for PCR with subsequent sequencing, because the resulting raw data gain in peak height and signal-to-noise ratio results in an improved reading quality of the individual sequences. It is further desirable to reduce the number of PCR cycles and to refrain from repetitive, i.e. (semi-)nested PCR in order to reduce the risk of contamination by external DNA and to minimize the reduction of reading quality due to a decreased signal-to-noise ratio, as systematically tested by Brandstätter and Parson (2003). In this context, numts come into play, inasmuch as strongly deviating PCR conditions, e.g. elevated mtDNA template number and/or increased number of PCR cycles, can favour the amplification of numts, especially when mutations in primer binding sites reduce the PCR efficiency for the authentic mtDNA target.

PCR cocktails are usually designed to allow for a relatively wide range of applied (mt)DNA amount owing to an excess of dNTPs, primers, and $MgCl_2$. Moreover, there is usually enough molar amount of enzyme available in the cocktail to enable amplification of higher DNA concentrations; this is why, for example, the enormous amount of 100 ng of DNA can result in a successful amplicon. However, the specificity of such an assay is out of control and may result in amplification of artefacts, here in particular of numts, since both the mitochondrial as well as the nuclear target are present in highly elevated concentrations.

Unusual experimental findings based on such suboptimal reaction conditions should then be questioned prior to any speculative interpretations. A pertinent case constitutes the recent study by Wu et al. (2005), who claimed that "somatic mtDNA mutations and mtDNA depletion occur in gastric cancer and that mtDNA depletion is involved in carcinogenesis and/or cancer progression of gastric carcinoma". These effects were observed by sequencing cloned PCR fragments that were amplified from 100 ng of template DNA—Lee et al. (2005) even used 200 ng of DNA. Wu et al. (2005) showed in their Fig. 2 tandem duplication (Fig. 2b in Wu et al. 2005) and triplication events (Fig. 2c in Wu et al. 2005) to exemplify their findings. These events, however, rather seem methodological artefacts of excessive PCR amplicons that underwent self-ligation via the polycytosine stretches during the cloning process, which the authors used prior to sequence analysis. For a similar reason, the amplification strategy employed by Thangaraj et al. (2003) was somewhat unfortunate as it used 50 ng of (nuclear) DNA for amplification, even though sperms contain a greater nuclear DNA-to-mtDNA ratio compared with other cells. This possibly paved the way for the fragment of the 5.8-kb numt, from the sperm's head, to compete with and eventually win over the mtDNA from the sperm's mid-piece.

7
Conclusion

In general, numts are not measurably coamplified in routine applications of mtDNA sequencing because authentic mtDNA sequences will greatly predominate in copy number over any paralogous numts. The vast majority of insertions into nuclear DNA are too short anyway, not comprising more than 40 nucleotides (Chen et al. 2005). There are, however, two specific instances where a better match between primers and the nuclear targets can frustrate attempts to purify mtDNA, so that numts would get amplified preferentially

(Collura and Stewart 1995). First, a mutation in the authentic mtDNA sequence at the 3′-most primer position can inhibit proper primer binding and thus permit amplification of some numt instead. It has been demonstrated very clearly that primer-site mutations even prevent successful amplification at positions distant from the 3′ end of the primer, such as 4, 8, and 12 bases away, respectively (Heinrich et al. 2004). "The second instance involved an attempt to amplify mtDNA from a sperm head fraction of vaginal swabs. DNA preparations from the sperm head fraction are known to be highly deficient in mtDNA, and amplification of the pseudogene in this case can be traced to a higher representation of nuclear than mtDNA templates" (Morgan et al. 1997).

There may be other cell/tissue types with lowered mtDNA content such as hair, for example. A higher ratio of nuclear genomes to mitochondrial genomes in hair (compared with blood) was also seen as the possible cause for amplifying primarily numt DNA in hair DNA while obtaining mtDNA in blood DNA of Asian elephants with the same PCR primers (Greenwood and Pääbo 1999). May-Panloup et al. (2003) have also warned that, in studies of mtDNA anomalies in sperm, numts might have been erroneously recognized by the wide-ranging mtDNA probes such as those used in the Southern blot technique. A similar situation with low copy number may be encountered in the search for ancient mtDNA, especially with samples that contain virtually no intact authentic mtDNA fragments and are likely contaminated with endogenous or exogenous DNA (Willerslev and Cooper 2005). The chances of detecting unintentionally amplified numts can, however, be maximized by amplifying greatly overlapping fragments (obtained with different primer pairs).

On the other hand, there is also a clear advantage in having a variety of numts at one's disposal, inasmuch as they constitute molecular 'fossils' within the more slowly evolving nuclear DNA and are thus (relatively) 'frozen' snapshots of ancient mtDNA configuration. For phylogenetic analyses of human mtDNA the numts that have high similarity to their human mtDNA paralogues could greatly enhance the rooting of the mtDNA trees because chimp mtDNA is somewhat distant. For instance, Kivisild et al. (2006) employed a compound sequence that was composed of numts, giving priority to numt fragments with highest overall percentage identity, which altogether covered most of the coding region. This was of considerable help in diminishing the ambiguity (incurred because of the few available chimp mtDNA sequences) in rooting a tree of coding-region sequences sampled worldwide (Chap. 7).

References

Adcock G, Dennis E, Easteal S, Huttley G, Jermelin L, Peacock W, Thorne A (2001) Mitochondrial DNA sequences in ancient Australians: implications for modern human origins. Proc Natl Acad Sci USA 98:537–542

Audic S, Béraud-Colomb E (1997) Ancient DNA is thirteen years old. Nat Biotechnol 15:855–858

Bandelt H-J (2005) Mosaics of ancient mitochondrial DNA: positive indicators of non-authenticity. Eur J Hum Genet 13:1106–1112

Bandelt H-J, Kong Q-P, Parson W, Salas A (2005) More evidence for non-maternal inheritance of mitochondrial DNA? J Med Genet 42:957–960

Bandelt H-J, Quintana-Murci L, Salas A, Macaulay V (2002) The fingerprint of phantom mutations in mitochondrial DNA data. Am J Hum Genet 71:1150–1160

Bensasson D, Feldman MW, Petrov DA (2003) Rates of DNA duplication and mitochondrial DNA insertion in the human genome. J Mol Evol 57:343–354

Brandstätter A, Parson W (2003) Mitochondrial DNA heteroplasmy or artefacts—a matter of the amplification strategy? Int J Legal Med 117:180–184

du Buy HG, Riley FL (1967) Hybridization between the nuclear and kinetoplast DNA's of *Leishmania enriettii* and between nuclear and mitochondrial DNA's of mouse liver. Proc Natl Acad Sci USA 57:790–797

Chen J-M, Chuzhanova N, Stenson PD, Férec C, Cooper DN (2005) Meta-analysis of gross insertions causing human genetic disease: novel mutational mechanisms and the role of replication slippage. Hum Mutat 25:207–221

Coller HA, Khrapko K, Herrero-Jimenez P, Vatland JA, Li-Sucholeiki X-C, Thilly WG (2005) Clustering of mutant mitochondrial DNA copies suggests stem cells are common in human bronchial epithelium. Mutat Res 578:256-271

Collura RV, Stewart C-B (1995) Insertions and duplications of mtDNA in the nuclear genomes of Old World monkeys and hominoids. Nature 378:485–489

Cooper A, Rambaut A, Macaulay V, Willerslev E, Hansen AJ, Stringer C (2001) Human origins and ancient human DNA. Science 292:1655–1656

Davis RE, Miller S, Herrnstadt C, Ghosh SS, Fahy E, Shinobu LA, Galasko D, Thal LJ, Beal MF, Howell N, Parker WD Jr (1997) Mutations in mitochondrial cytochrome *c* oxidase genes segregate with late-onset Alzheimer disease. Proc Natl Acad Sci USA 94:4526–4531

Farrelly F, Butow RA (1983) Rearranged mitochondrial genes in the yeast nuclear genome. Nature 301:296–301

Gellissen G, Bradfield JY, White BN, Wyatt GR (1983) Mitochondrial DNA sequences in the nuclear genome of a locust. Nature 301:631–634

Gibbons A (1994) Possible dino DNA find is greeted with skepticism. Science 266:1159

Gilbert MTP, Bandelt H-J, Hofreiter M, Barnes I (2005) Assessing ancient DNA studies. Trends Ecol Evol 20:541–544

Gill P, Curran J, Elliot K (2005) A graphical simulation model of the entire DNA process associated with the analysis of short tandem repeat loci. Nucl Acids Res 33:632–643

Goldin E, Stahl S, Cooney AM, Kaneski CR, Gupta S, Brady RO, Ellis JR, Schiffmann R (2004) Transfer of a mitochondrial DNA fragment to *MCOLN1* causes an inherited case of mucolipidosis IV. Hum Mutat 24:460–465

Greenwood AD, Pääbo S (1999) Nuclear insertion sequences of mitochondrial DNA predominate in hair but not in blood of elephants. Mol Ecol 8:133–137

Hadler HI, Dimitrijevic B, Mahalingam R (1983) Mitochondrial DNA and nuclear DNA from normal rat liver have a common sequence. Proc Natl Acad Sci USA 80:6495–6499

Hänni C, Laudet V, Coll J, Stehelin D (1994) An unusual mitochondrial DNA sequence variant from an Egyptian mummy. Genomics 22:487–489

Heilig R et al (2003) The DNA sequence and analysis of human chromosome 14. Nature 421:601–607

Heinrich M, Müller M, Rand S, Brinkmann B, Hohoff C (2004) Allelic drop-out in the STR system ACTBP2 (SE33) as a result of mutations in the primer binding region. Int J Legal Med 118:361–363

Herrnstadt C, Clevenger W, Ghosh SS, Anderson C, Fahy E, Miller S, Howell N, Davis RE (1999) A novel mitochondrial DNA-like sequence in the human nuclear genome. Genomics 60:67–77

Hirano M, Shtilbans A, Mayeux R, Davidson MM, DiMauro S, Knowles J, Schon EA (1997) Apparent mtDNA heteroplasmy in Alzheimer's disease patients and in normals due to PCR amplification of nucleus-embedded mtDNA pseudogenes. Proc Natl Acad Sci USA 94:14894–14899

Jacobs HT, Posakony JW, Grula JW, Roberts JW, Xin JH, Britten RJ, Davidson EH (1983) Mitochondrial DNA sequences in the nuclear genome of *Strongylocentrotus purpuratus*. J Mol Biol 165:609–632

Jensen-Seaman MI, Sarmiento EE, Deinard AS, Kidd KK (2004) Nuclear integrations of mitochondrial DNA in gorillas. Am J Primatol 63:139–147

Jones PN (2004) American Indian mtDNA and Y chromosome genetic data: a comprehensive report of their use in migration and other anthropological studies. (http://www.iiirm.org/publications/Articles%20Reports%20Papers/Genetics%20and%20Biotechnology/Jones%20DNA.pdf)

Kemble RJ, Mans RJ, Gabay-Laughnan S, Laughnan JR (1983) Sequences homologous to episomal mitochondrial DNAs in the maize nuclear genome. Nature 304:744–747

Khrapko K, Coller HA, André PC, Li X-C, Hanekamp JS, Thilly WG (1997) Mitochondrial mutational spectra in human cells and tissues. Proc Natl Acad Sci USA 94:13798–13803

Kivisild T, Shen P, Wall D, Do B, Sung R, Davis K, Passarino G, Underhill PA, Scharfe C, Torroni A, Scozzari R, Modiano D, Coppa A, de Knijff P, Feldman M, Cavalli-Sforza LL, Oefner PJ (2006) The role of selection in the evolution of human mitochondrial genomes. Genetics 172:373–387

Klein J, Takahata N (2002) Where do we come from? The molecular evidence for human descent. Springer, Berlin Heidelberg New York

Lee H-C, Yin P-H, Lin J-C, Wu C-C, Chen C-Y, Wu C-W, Chi C-W, Tam T-N, Wie Y-H (2005) Mitochondrial genome instability and mtDNA depletion in human cancers. Ann N Y Acad Sci 1042:109–122

Lopez JV, Yuhki N, Masuda R, Modi W, O'Brien SJ (1994) *Numt*, a recent transfer and tandem amplification of mitochondrial DNA to the nuclear genome of the domestic cat. J Mol Evol 39:174–190

May-Panloup P, Chrétien M-F, Savagner F, Vasseur C, Jean M, Malthièry Y, Reynier P (2003) Increased sperm mitochondrial DNA content in male infertility. Hum Reprod 18:550–556

Morgan MA, Parsons TJ, Holland MM (1997) Amplification of human nuclear pseudogenes derived from mitochondrial DNA: a problem for mitochondrial DNA identity testing? Proceedings of the 8th International Symposium on Human Identification—1997. Promega, Maddison, p 128. http://www.promega.com/geneticidproc/ussymp8proc/34.html

Pääbo S (1985) Molecular cloning of ancient Egyptian mummy DNA. Nature 314:644–645

Pääbo S, Poinar H, Serre D, Jaenicke-Després V, Hebler J, Rohland N, Kuch M, Krause J, Vigilant L, Hofreiter M (2004) Genetic analyses from ancient DNA. Annu Rev Genet 38:645–679

Penney D (2005) Fossil blood droplets in Miocene Dominican amber yield clues to speed and direction of resin secretion. Palaeontology 48:925–927

Pereira L, Gonçalves J, Goios A, Rocha T, Amorim A (2005) Human mtDNA haplogroups and reduced male fertility: real association or hidden population substructuring. Int J Androl 28:241–247

Perna NT, Kocher TD (1996) Mitochondrial DNA: molecular fossils in the nucleus. Curr Biol 6:128–129

Ricchetti M, Tekaia F, Dujon B (2004) Continued colonization of the human genome by mitochondrial DNA. PLoS Biol 2:e273

Salas A, Yao Y-G, Macaulay V, Vega A, Carracedo Á, Bandelt H-J (2005) A critical reassessment of the role of mitochondria in tumorigenesis. PLoS Med 2:e296

St John JC, Jokhi RP, Barratt CLR (2005) The impact of mitochondrial genetics on male infertility. Int J Androl 28:65–73

Thalmann O, Hebler J, Poinar HN, Pääbo S, Vigilant L (2004) Unreliable mtDNA data due to nuclear insertions: a cautionary tale from analysis of humans and other great apes. Mol Ecol 13:321–335

Thalmann O, Serre D, Hofreiter M, Lukas D, Eriksson J, Vigilant L (2005) Nuclear insertions help and hinder inference of the evolutionary history of gorilla mtDNA. Mol Ecol 14:179–188

Thangaraj K, Joshi MB, Reddy AG, Rasalkar AA, Singh L (2003) Sperm mitochondrial mutations as a cause of low sperm motility. J Androl 24:388–392

Tourmen Y, Baris O, Dessen P, Jacques C, Malthièry Y, Reynier P (2002) Structure and chromosomal distribution of human mitochondrial pseudogenes. Genomics 80:71–77

Tsuzuki T, Nomiyama H, Setoyama C, Maeda S, Shimada K, Pestka S (1983a) The majority of cDNA clones with strong positive signals for the interferon-induction-specific sequences resemble mitochondrial ribosomal RNA genes. Biochem Biophys Res Commun 114:670–676

Tsuzuki T, Nomiyama H, Setoyama C, Maeda S, Shimada K (1983b) Presence of mitochondrial-DNA-like sequences in the human nuclear DNA. Gene 25:223–229

Turner C, Killoran C, Thomas NS, Rosenberg M, Chuzhanova NA, Johnston J, Kemel Y, Cooper DN, Biesecker LG (2003) Human genetic disease caused by de novo mitochondrial-nuclear DNA transfer. Hum Genet 112:303–309

Wallace DC, Stugard C, Murdock D, Schurr T, Brown MD (1997) Ancient mtDNA sequences in the human nuclear genome: a potential source of errors in identifying pathogenic mutations. Proc Natl Acad Sci USA 94:14900–14905

Willerslev E, Cooper A (2005) Ancient DNA. Proc R Soc Lond Ser B 272:3–16

Woischnik M, Moraes CT (2002) Pattern of organization of human mitochondrial pseudogenes in the nuclear genome. Genome Res 12:885–893

Woodward SR (1995) Detecting dinosaur DNA: response. Science 268:1194

Woodward SR, Weyand NJ, Bunell M (1994) DNA sequence from Cretaceous period bone fragments. Science 266:1229–1232

Wright RM, Cummings DJ (1983) Integration of mitochondrial gene sequences within the nuclear genome during senescence in a fungus. Nature 302:86–88

Wu C-W, Yin P-H, Hung W-Y, Li AF-Y, Li S-H, Chi C-W, Wei Y-H, Lee H-C (2005) Mitochondrial DNA mutations and mitochondrial DNA depletion in gastric cancer. Genes Chromosomes Cancer 44:19–28

Zhang D-X, Hewitt GM (1996) Nuclear integrations: challenges for mitochondrial DNA markers. Trends Ecol Evol 11:247–251

Zhao X, Li N, Guo W, Hu X, Liu Z, Gong G, Wang A, Feng J, Wu C (2004) Further evidence for paternal inheritance of mitochondrial DNA in the sheep (*Ovis aries*). Heredity 93:399–403

Zischler H (2000) Nuclear integrations of mitochondrial DNA in primates: inference of associated mutational events. Electrophoresis 21:531–536

Zischler H, Geisert H, Castresana J (1998) A hominoid-specific nuclear insertion of the mitochondrial D-loop: implications for reconstructing ancestral mitochondrial sequences. Mol Biol Evol 15:463–469

Zischler H, von Haeseler A, Pääbo S (1995a) A nuclear fossil of the mitochondrial D-loop and the origin of modern humans. Nature 378:489–492

Zischler H, Höss M, Handt O, von Haeseler A, van der Kuyl AC, Goudsmit J, Pääbo S (1995b) Detecting dinosaur DNA. Science 268:1192–1193

Nucleic Acids and Molecular Biology, Vol. 18
Hans-Jürgen Bandelt, Vincent Macaulay, Martin Richards (Eds.)
Human Mitochondrial DNA and the Evolution of *Homo sapiens*

Estimation of Mutation Rates and Coalescence Times: Some Caveats

Hans-Jürgen Bandelt · Qing-Peng Kong · Martin Richards ·
Vincent Macaulay (✉)

Department of Statistics, University of Glasgow,
University Avenue, Glasgow G12 8QQ, UK
vincent@stats.gla.ac.uk

1 Introduction

Reconstructing the spread of modern humans using genetic markers, just as for the phylogeography of any species in general, relies on multiple lines of evidence, and it is important to obtain as much information as possible from the distribution of the lineages. Genetic dating, especially using uniparental markers such as mitochondrial DNA (mtDNA) and the non-recombining portion of the Y chromosome, then plays a crucial role. One might expect (or at least hope) to be able to estimate the time of important events in the evolution and prehistory of modern humans—provided that the underlying phylogenies can be estimated with confidence, a molecular clock can be stipulated and well calibrated, and the interpretive gap between the evolution of the molecules and the prehistory of their carriers can be narrowed. None of these preconditions are without pitfalls (Bandelt et al. 2003b).

The phylogenetic aspect may seem rather trivial in this context, but past research on mtDNA has proven it to be far from straightforward. The data, methods, and arguments employed have often been inadequate for the particular task or even badly flawed. The choice of phylogenetic method, however, often seems to depend on preconception rather than rational understanding: "although the technical side of tree building may appear to be a matter of pure graph theory and combinatorial optimization, the fundamental issues that determine the validity of these methods are sometimes discussed in terms more suited for religion" (see p. 447 in Gusfield 1997). A deemed general superiority of maximum likelihood (ML) over maximum parsimony (MP), or vice versa, would ignore their close theoretical relationship and the related computational issues of these (and other) approaches (Steel 2002). The costs and benefits of the different methods have to be weighed against each other in every single case. To manage very large problem sizes, one can only resort to MP methods or hybrid distance/parsimony/ML methods anyway, rather than fall into the trap of some crude approximate ML heuristic. We need perhaps to worry not only over potential zones of inconsistency, such as the 'Felsenstein zone' for MP or the 'Farris zone' for ML (Steel and Penny 2000), but also to guard against a 'triviality zone': for some kinds of data almost nothing is gained from a computationally expensive ML approach compared with a simple (weighted) parsimony analysis.

By the late 1990s, more than 50 complete mtDNA lineages had been published in the field of medical genetics (Hedrick and Kumar 2001), but most of them suffered from missequencing and misreading (Macaulay et al. 1997), so these data necessitate very critical evaluation (for East Asian data, see Kivisild et al. 2002). Therefore the first (fairly) reliable set of complete mtDNA sequences (Ingman et al. 2000), sampled from around the world, quickly gained widespread acceptance. The messages conveyed with this study were, by way of hitchhiking, then accepted as equally reliable, namely, the displayed neighbour-joining tree as an estimate of the phylogeny, the asserted overall coalescence time of 171 500 years, and the claim that the control region did not conform to a molecular clock. The confusions have ramified since then with the claim that natural selection has profoundly shaped the phylogenetic tree of human mtDNA in East Asia (Mishmar et al. 2003). Unfortunately, in all these claims there are technical errors, biases, and pitfalls that deserve closer examination.

Calibrating the molecular clock has been the subject of a great deal of controversy, to the extent that in the mid-1990s concerns were raised that the clock for the first hypervariable segment of the mtDNA control region might have been misestimated by a factor of 10 or more (Howell et al. 1996; Pääbo 1996; Parsons et al. 1997). Subsequent discussions emphasized the numerous approaches to dating, such as calibration against the fossil record, calibration against and comparisons with the archaeological record, and comparison with other systems such as coding-region restriction fragment length polymorphisms (RFLPs) for which further approaches were available (Macaulay et al. 1997). These considerations suggested that those concerns, which were based on estimates from pedigree studies, were largely unwarranted. Furthermore, it became clear that there was some confusion in the way the arguments had been expressed (Sigurdarðóttir et al. 2000). We nevertheless still see the old arguments recycled about a tenfold higher 'pedigree rate' (Howell et al. 2003); see later.

One way to date the most recent common ancestor of a genetic locus in a population is simply to calculate the average number of mutations from the ancestor on the tree (Morral et al. 1994) and scale (divide) by the mutation rate for the whole locus. The statistic, the average number of mutations, is often referred to as rho (ρ) (Forster et al. 1996). In the same way, one could date any of the nodes within the tree on the way to the root. This straightforward approach to dating makes no assumptions about prehistoric demography, and is therefore sometimes referred to as 'model-free', although this only refers to the lack of a population-genetics model (Stumpf and Goldstein 2001). As a consequence this kind of approach is only able to estimate some of the parameters associated with a gene tree that population geneticists are usually interested in. For example, it can estimate the coalescence time, the time for the most recent common ancestor (TMRCA), of a clade or of a whole population sample, but does obviously not allow the estimation of other parameters in a diachronic model of the demographic process and population structure. It has become popular to make use of very simple models within the framework of coalescent theory that are mathematically tractable. This is then referred to as a 'model-based' approach. Such estimations can be performed using programs such as Genetree (http://www.stats.ox.ac.uk/~griff/software.html) and BATWING (http://www.maths.abdn.ac.uk/~ijw/downloads/batwing/batguide/). The problem then is that a naïve model may be of little help and may give results that can be positively misleading. Because of the extreme difficulties of adequately characterizing the demography for the distant past, particularly in what will surely appear as the prehistory of prehistory, it seems likely that 'model-free' approaches may be more robust and will continue to be useful in the estimation of time depth of a particular ancestral type, at least until more effective models of demography have evolved.

Before we can discuss these issues in detail, we will present a reanalysis of the mtDNA data set of Ingman et al. (2000), especially as the original analysis left a number of lacunae for justification (see also Chap. 7). A reanalysis of these 53 complete mtDNA sequences is needed because these data have already fallen victim to superficial approaches (e.g. Meyer and von Haeseler 2003) that have elicited most extraordinary claims (see later). Although the number of available complete mtDNA genomes has exploded since the year 2000, we can take this small data set, for the purpose of demonstration, as a nutshell of worldwide mtDNA variation. We can compensate for this restricted view by inserting more up-to-date information in the form of postulated intermediate ancestral types, which have crystallized since then from several continentally focused analyses of mtDNA (see Chaps. 7, 8 for a more thorough discussion).

2
Prerequisite: a Global mtDNA Tree

Any tests of the molecular clock, or the estimation of the mutational rate spectrum within the molecule, or the calculation of coalescence times should be based on some solid estimate of the mtDNA phylogeny. We shall illustrate this with a phylogenetic tree for the data of Ingman et al. (2000) that strives to incorporate additional information from complete mtDNA sequences published since then. It builds upon an ML analysis that clarifies the branching pattern near the root (Mishmar et al. 2003; Macaulay et al. 2005), using the two available complete mtDNA sequences from chimpanzees (Horai et al. 1995). Decisions about intermediate branching nodes, shown in Fig. 1, representing the postulated ancestral sequences of haplogroups, employ data from Chen et al. (1995) and Maca-Meyer et al. (2001) (concerning haplogroup L1c), Bandelt et al. (2001, 2003a), Finnilä et al. (2001), Torroni et al. (2001), Herrnstadt et al. (2002, 2003), Kivisild et al. (2002, 2006), Salas et al. (2002), Ingman and Gyllensten (2003), Kong et al. (2003), Mishmar et al. (2003), Achilli et al. (2004), Palanichamy et al. (2004), Tanaka et al. (2004), and Friedlaender et al. (2005). In devising this tree, parsimony was taken as a default option (whenever no extrinsic information could assist in the reconstruction of the mutational changes along the tree), but we have given higher priority to the coding region, because the control region is well known to harbour some extreme mutational hotspots (see later). The latter observation is now on solid ground (inferred from numerous closely related sequences and matrilineal pedigrees)—although some scholars still prefer to adduce rather mystical arguments for regarding mutational hotspots as old mutations in the human line that have been dispersed by recombination (Hagelberg 2003).

The tree displayed in Fig. 1 reproduces the full sequence information except for the extremely variable length polymorphisms of long C stretches in the first two hypervariable segments (HVS-I and HVS-II) around nucleotide positions 16189 and 310, respectively. The information for position 16519 and for the CA dinucleotide repeats scored at 522–523 is recorded (and highlighted in italics) in the tree for the sake of completeness, but it is impossible to trace their evolution reliably, so we decided to exclude these mutations from coalescence time estimation. Even with this caveat, the phylogeny estimate of Fig. 1 cannot be regarded as 'bulletproof'—in contrast to single nucleotide polymorphism (SNP) phylogenies for the Y chromosome (when based on sufficient mutation screening). In particular, not too much faith should be placed in the mutational pattern near the global (L) root ('mitochondrial Eve') since knowledge about the extensive variation within the African mtDNA haplogroup L0 is far too meagre, with its real complexity only beginning to emerge (Kivisild et al. 2004, 2006).

This uncertainty also concerns some reconstructed coding-region mutations in the vicinity of the roots of L0 and its potential sister haplogroup L1′2′3 (which could equally be denoted by L1′2 or L1′3 or L1′4 or L1′5 or L1′2′3′4′5, etc.). For instance, the tree of Fig. 1 and the corresponding tree of Kivisild et al. (2006), which is based on a new set of 277 coding-region sequences and rooted with chimp sequences and a few nuclear inserts of mtDNA (Chap. 3), differs in the reconstruction of the recurrent changes at position 9755. In the latter tree three mutations at 9755 are postulated (on the branches to L0d, L0f, and L1′2), whereas the alternative reconstruction would

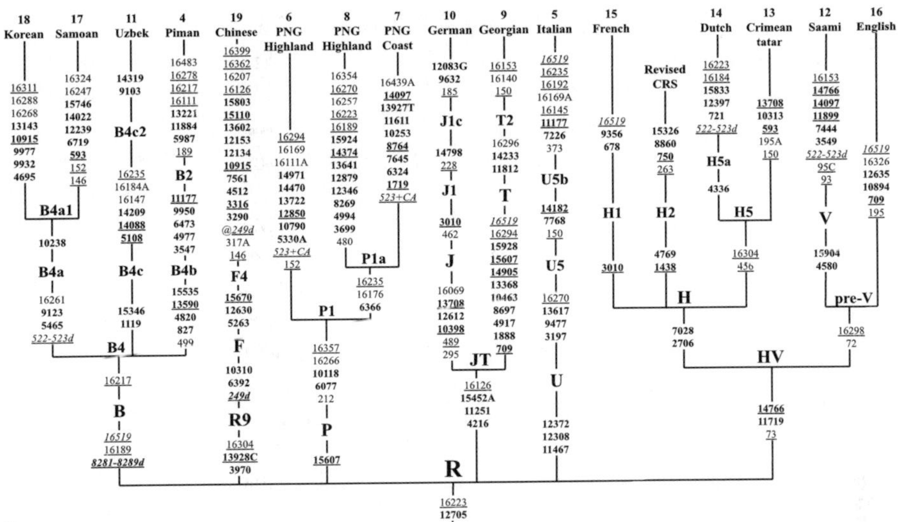

a

Fig. 1 **a** haplogroup R, **b** haplogroups M and N, **c** haplogroups L2 and L3, **d** haplogroups L0 and L1. Phylogenetic tree representing the 53 complete mitochondrial DNA sequences from Ingman et al. (2000). Sequences are numbered according to that publication, with ethnic or geographic affiliation added. *Capital letters with suffixes* (*numbers* and *lowercase letters* in alternation) indicate the position of intermediate sequences ancestral to the respective haplogroups. Mutations allocated to the links are numbered according to the revised Cambridge reference sequence (*rCRS*; Andrews et al. 1999), which is distinguished from the H node by transitions at positions 263, 750, 1438, 4769, 8860, and 15326. Mutations are transitions unless a suffix A, G, C, T, or d indicates a transversion or a deletion, respectively; insertions are specified by + and the inserted nucleotides; for transversions and indels, @ indicates a (back) mutation to the corresponding rCRS nucleotide(s). *Underlining* highlights recurrent mutations. Mutations hitting the coding region are in **bold**. Mutations in *italics* are disregarded for time estimations. *PNG* Papua New Guinea

posit changes on the branches to L0a and L0d and thereby interchange the reconstructed ancestral nucleotide at 9755. Although this reconstruction is more parsimonious for the unrooted tree (but equally parsimonious for the rooted tree), we will possibly never know which reconstruction is the likely correct one. The placement of certain control-region mutations is particularly problematic. In the case of the 16129 and 189 transitions this amounts to no more than educated guesses: we have opted for a parsimonious reconstruction (relative to this data set), which will certainly underestimate the total number of forward and backward changes. With the addition of more L0 sequences (Watson et al. 1997; Kivisild et al. 2004), it becomes more plausible that 16129 is part of the motif (compared with the revised Cambridge reference sequence, rCRS; Andrews et al. 1999) of the L root. Further upstream, in and near haplogroup L2′3, it is the recurrent 16311 transition that is difficult

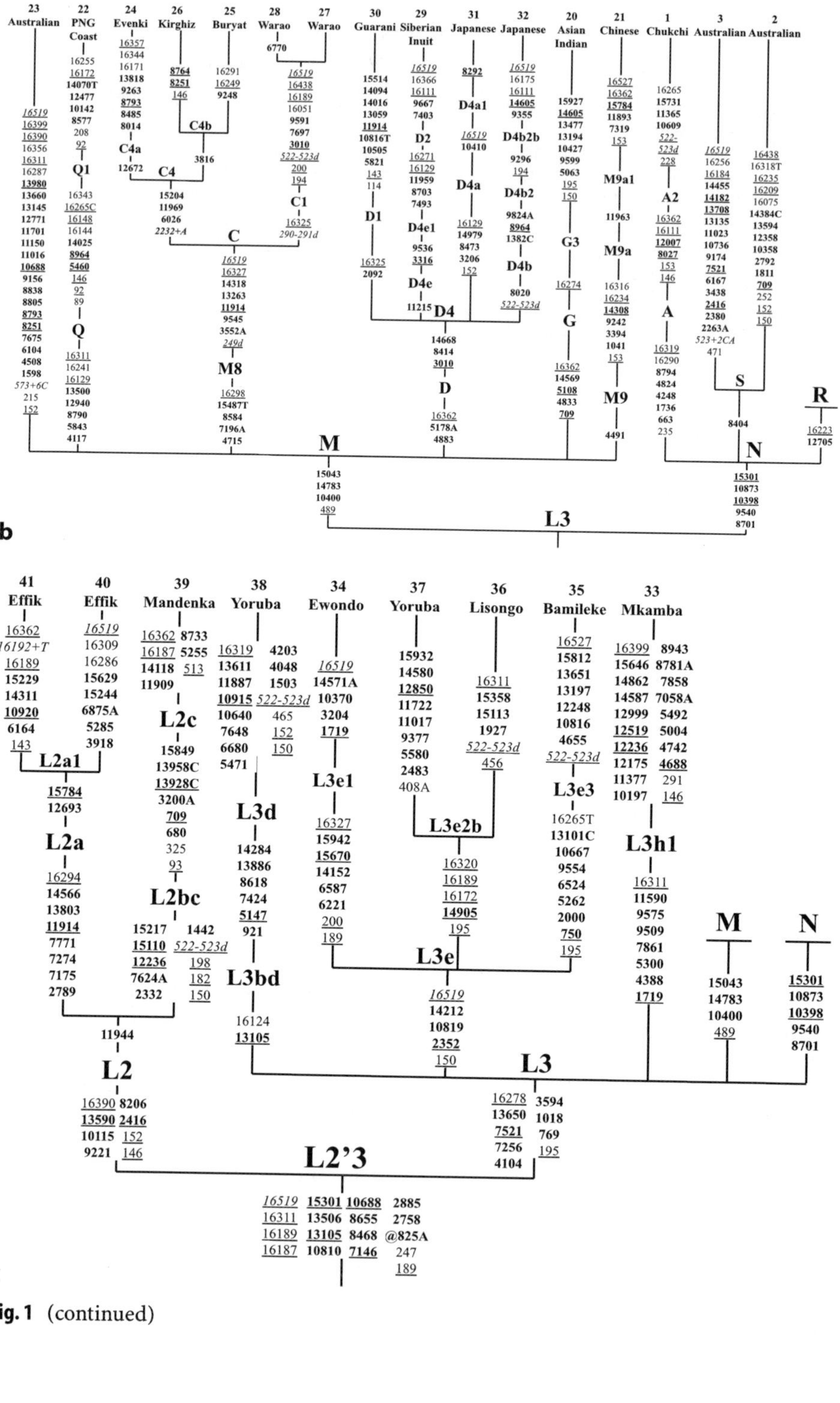

Fig. 1 (continued)

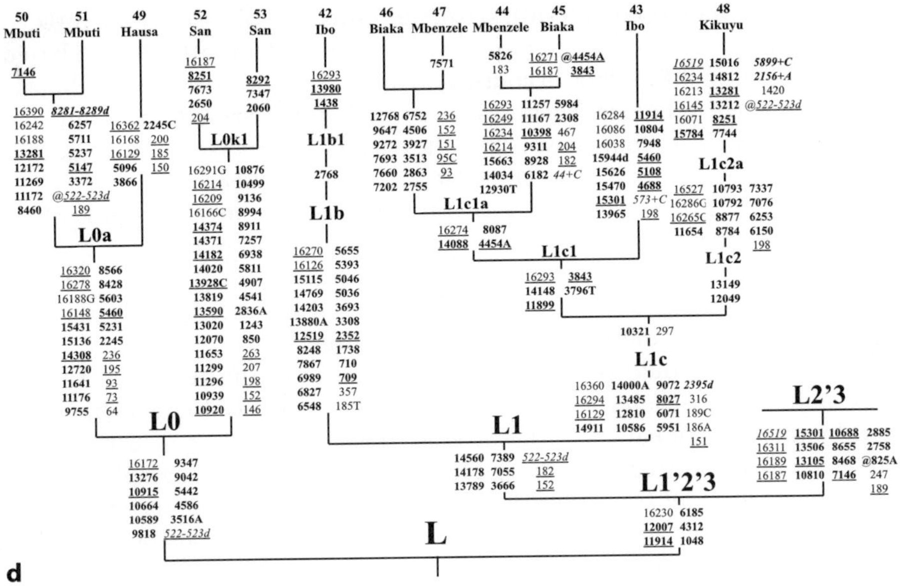

d

Fig. 1 (continued)

to assess (see Fig. 2A of Kivisild et al. 2004 for an alternative reconstruction). As we will see later, we have to resign ourselves to considerable underestimation (by MP) of nucleotide changes at mutational hotspots deep in the mtDNA tree of Fig. 1.

An estimate of the mtDNA phylogeny with mutations reconstructed, no matter how tentative it may be, is nevertheless of considerable help for judging more limited or technically flawed attempts at analysing portions of worldwide human mtDNA variation. The problems can be manifest in, for example, (1) the choice of too small a fragment of mtDNA, with insufficient characters that yield little phylogenetic resolution, or (2) a poorly tested or misapplied phylogenetic method, or (3) data analysed with no view to a coherent global picture, or (4) specific sequencing errors afflicting the data. As to this last point, one should bear in mind that most complete sequencing studies show some imperfections and may even be severely flawed (Yao et al. 2003; Bandelt et al. 2005a, b). Many control-region studies, of course, traditionally also have a high error rate, which to some extent disguises the real mutational process (Chap. 6).

To illustrate point 1, contrast the mtDNA tree based on the cytochrome *c* oxidase subunit III gene (*COIII*) of mtDNA (9207–9990) from Wilder et al. (2004) with the total mtDNA tree of Fig. 1. Not a single *COIII* mutation would highlight the evolutionary pathway from the L root up to the M root. Several independent Asian M and African L branches therefore emanate directly from the root, whereas haplogroup N, in contrast, is located on a single branch, supported by one mutation (at 9540). By chance, this bonsai form of the mtDNA phylogeny is quite star-like, resembling some of the earliest mtDNA trees based on low-resolution RFLP data, in contrast to the more highly resolved phylogeny of entire mtDNA genomes. These features should be borne in mind before entering into any direct comparison with the current bonsai Y-chromosome SNP tree, which—by chance—is basally somewhat less star-like.

As to point 2, Howell et al. (2004) employed quartet puzzling (QP), which is a majority consensus method aiming at ML (but rarely reaching it, since QP is rather a crude heuristic that operates on quartet trees). In the presence of considerable homoplasy this method is bound to return nearly star-like trees. If character evolution is reconstructed along such trees, then a single mutational event could be multiplied many times in the reconstruction as the inbuilt consensus may collapse a link carrying a whole branch characterized by this mutational event. It therefore does not come as a surprise that the control-region trees shown by Howell et al. show, for example, a large number of recurrent transitions for the pair 16189 and 16192, more than necessary, and rather at variance with the observation that position 16192 is hypermutable in the presence of a long C stretch caused by C at 16189 (Howell and Smejkal 2000). We are not aware of extremely recurrent events involving tandem mutations at 16189 and 16192 anywhere in the mtDNA phylogeny, despite the high polymorphism at 16189.

Exemplifying point 3, the study of Howell et al. (2004) again constitutes a good case in point where the exhibited mutational patterns do not admit a coherent solution. The three separate networks shown in their Figs. 1–3 (supposed to reflect the phylogenetic relationships among mtDNA lineages of haplogroup L2) are actually in conflict with each other and bear a number of further shortcomings. Firstly, a (too) distant outgroup, represented by some haplogroup L0a lineage, has been employed in an inconsistent way, without clarifying intermediate branching points from more extensive analyses. The tentative root or reference node thus remains obscure (coming somewhat close to the L1′2′3 node of our Fig. 1). A more reasonable decision would have been to sandwich haplogroup L2 between its nearest relatives, L1/L5 and L4/L6 (Chap. 7).

3 Transition/Transversion Rate Ratio

Basic models of molecular evolution, such as the Kimura model, in principle suitable for mtDNA sequences, employ at least two parameters, a rate α for transitions and another one, β, for transversions. The more realistic HKY85 model introduces additional parameters g_A, g_G, g_C, g_T to account for different nucleotide frequencies (Nei and Kumar 2000). The transition–transversion rate ratio (TTRR) $\kappa = \alpha/\beta$ then measures how much faster a transition to a specific nucleotide would occur compared with a transversion to the same nucleotide. Since there are eight kinds of transversions but only four kinds of transitions, the transition–transversion count ratio (abbreviated as TI : TV; Strandberg and Salter 2004) that compares the total number of transitions relative to the total number of transversions observed in a data set aims at estimating $\alpha/2\beta$ (i.e. half the TTRR) in the case of the Kimura model (where $g_A = g_G = g_C = g_T$). For the more general HKY85 model, the expected transition–transversion count ratio is $\alpha/2\beta$ times $2(g_A g_G + g_C g_T)/g_R g_Y$, where $g_R = g_A + g_G$ and $g_Y = g_C + g_T$.

Unfortunately, both concepts (i.e. rate versus count) of transition–transversion ratio are often intermingled, so in general one cannot be sure what exactly is meant in case studies by the transition–transversion ratio. Worse, in forensics, this concept is traditionally misunderstood as a mere polymorphism count relative to the (revised) CRS (e.g. Zupanič Pajnič et al. 2004); that is, each variant position of each sequence contributes one unit to the count of the corresponding mutation category. For instance, when Baasner et al. (1998) observed G at position 263 in 49 out of 50 cases, then this single ancestral nucleotide contributed a count of 49 to their A→G category. Yet another interpretation of the transition–transversion ratio would compare the number of positions showing a transition with the number of positions showing a transversion (Lutz-Bonengel et al. 2003).

Leaving the notational confusion aside, there are fundamental obstacles in estimating the transition–transversion ratio as well as additional parameters for more complex models, for example, the HKY85 model with gamma-distributed positional rates: (1) systematic bias introduced by the estimator when the total number of observed transversions is small, (2) partial saturation of transitions at certain hotspot positions, (3) the relatively high frequency of non-synonymous transversions, and (4) the low quality of the sequence data employed.

As for point 4, laboratory artefacts such as phantom mutations or documentation errors frequently lead to erroneous transversions (Chap. 6). The innocent use of flawed data would therefore bias the TTRR towards low values. Without taking particular measures to avoid phantom mutations, it would be hazardous to estimate the TTRR. It is instructive to distinguish the different kinds of transversions. When we take the tree of Fig. 1, for instance, we count a total of 35 transversions in the coding region. With respect to the L root, that is, scored forward in time, we see 18 transversions to A, eight to C, eight to T, and only one transversion to G. This remarkable imbalance between predominant transversions to A and a single transversion to G is not a sequencing artefact, but is also seen in other data sets. For instance, in the Eurasian mtDNA tree combined from Kong et al. (2003) and Palanichamy et al. (2004), there are 17 transversions to A, nine to C, seven to T, and three to G. Data sets that dramatically deviate from this pattern then certainly need particular attention.

For example, the original 560 coding-region sequences of Herrnstadt et al. (2002) had several odd features that were, for example, manifest in the number and kind of transversions recorded; these (uncorrected) raw data can still (as of 14 December 2005) be retrieved from the mtDB database maintained by Max Ingman (http://www.genpat.uu.se/mtDB/). Transversions to G were relatively frequent in these data of Herrnstadt et al.; see Table 1 for a small selection of the most frequent C to G transversions. Alarmingly, almost all transversions to G in those sequences can be allocated to the terminal branches of the global mtDNA phylogeny, thus constituting private mutations. More alarmingly, these private transversions to G were typically changes from C to G in a characteristic sequence environment with at least one G preceding, so that, for example, GGC often changed to GGG, as was the case, for example, with the transversion at 14974. Most alarmingly, some of these transversions even occurred in combination on different branches of the mtDNA phylogeny: for example 7927G and 7985G (Table 1). The basecalls at these positions were originally accepted without suspicion because the same mutations occurred in at least two overlapping PCR products (Corinna Herrnstadt, personal communication). This notwithstanding, most of those transversions (and other phantom mutations) disappeared in the revision of the data (Herrnstadt et al. 2003; http://mito546.securesites.net/science/560mtdnasrevision.php). There are also a suspicious number of transversions to G in the data of Mishmar et al. (2003), which, however, have not been revised to our knowledge. Namely, the old CRS errors 9559G and 14272G (Andrews et al. 1999) were reported there once, as well as the phantom variants 14974G (Table 1) and 317G (Brandstätter et al. 2005). Therefore, the ease with which phantom transversions can slip into data sets necessitates that particular measures be taken

Table 1 Fate of six frequent phantom transversions C to G

Mutation	Sample no. (Haplogroup) from Herrnstadt et al. (2002, 2003)[a] Original	Revised
7927G	66 (K1a1), 79 (K1a1), 171 (A2), 273 (H1)	—
7978G	166 (L3e2b), 172 (L2a), 273 (H1)	—
7985G	66 (K1a1), 67 (T2), 166 (L3e2b), 171 (A2), 173 (L1c), 174 (C), 175 (L2b), 238 (K1b), 274 (H6a), 277 (U5a)	—
13941G	41 (K2), 173 (L1c), 274 (H6a)	41
14460G	36 (H), 192 (L1b), 194 (L1c), 200 (A2), 202 (D1)	—
14974G	195 (L2a), 196 (L3e2a), 197 (B2), 257 (H1)	—

[a] Underlining highlights samples carrying at least two of these particular phantom mutations

before data are entered into calculations for the transition–transversion ratio or the inference of other features of the mutational process.

It is well understood that TTRR estimates appear to be time-dependent because highly recurrent transitions dominate the multiple hits at deep coalescences (Purvis and Bromham 1997). Corrections for multiple substitutions as effectively carried out in an ML approach (assuming gamma-distributed rates) cannot fully compensate for the resulting bias, mainly because the skewness of the positional rate spectrum is rather conservatively estimated at this point. That is, the hotspot transitions would have delivered more changes than inferred; on the other hand, transversions at sites (such as position 16311) where hotspot transitions occur could inadvertently be inferred as recurrent events. By definition, the TTRR depends on the imposed model of sequence evolution as well, albeit only marginally in practice. The factor $2(g_A g_G + g_C g_T)/g_R g_Y$ that is needed to translate half the TTRR into the transition–transversion count ratio is close to 1 for human mtDNA, viz. it is 0.956 for the entire molecule or the coding region alone (and 0.958 for the control region), based on the nucleotide distribution in the Afro-Eurasian consensus sequence (the ancestral type of haplogroup L3). Hence, conversely, when translating twice the count into the rate, a correction factor of 1.046 (or 1.044, respectively) is required.

For instance, we have recently estimated the TTRR in the coding region as 47.8 based on a small representative Eurasian mtDNA tree with a distant African outgroup lineage from haplogroup L0f (Macaulay et al. 2005). This value is only slightly larger than the double transition–transversion count ratio (44.8) obtained by parsimony. With the adoption of HKY85, this value translates into an estimated TTRR of 46.9. Thus, the correction for multiple hits relative to the hypothesized gamma-distributed positional rates effectively added to the latter value suspiciously little, namely only 0.9. Kivisild et al. (2006) had obtained a similar value (42.9) along their tree using Jukes–Cantor correction for multiple hits, which would have to be translated into 44.9 to fit into the HKY85 model. At face value, a TTRR between 45 and 48 might fit the general expectation, but it would be prudent to be somewhat sceptical because of the caveats already spelled out.

In order to approach a robust estimate of the ratio α/β, it is desirable to avoid deep coalescences and thus to keep the potential error in the tree topology low and maintain rather short branch lengths where most multiple hits could actually be reconstructed (by parsimony). Therefore, we would exclusively use mtDNAs sampled from Eurasia. We have put together the Eurasian trees displayed by Kong et al. (2003) and Palanichamy et al. (2004), dropping mtDNA sequences borrowed from other studies but retaining the rCRS. In the case of the control region we disregarded the seeming transversions from A to C at 16182 and 16183 as they may rather be regarded as compensatory length polymorphisms. Then we estimated the overall TTRR as 63.0, simply by dividing twice the number of reconstructed transitions by the corresponding number of transversions in the combined tree and then multiplying this by the factor 1.046 necessary when referring to the HKY85 model. The shared portion of the tree alone contributes 62.2 to this value, whereas the private mutations (along the terminal branches) altogether contribute 63.5. There is hence no alarming discrepancy between the shared and private patterns of transitions versus transversions, which might have pointed to the occurrence of phantom mutations that would accumulate along the terminal branches. Also the difference between the separate contributions of coding region and control region to the TTRR is not large: the TTRR estimated from the coding region is 59.3, whereas the TTRR for the control region is 71.1. The somewhat lower value for the coding region can be explained by constraints operating on non-synonymous changes.

The message to be drawn from this preliminary estimation is that the average TTRR is at least 60—in any case higher than previously estimated. Moreover, the control region does not seem to behave differently in this respect. We may contrast, however, synonymous changes with non-synonymous ones (that lead to a replacement of an amino acid): non-synonymous changes yield a TTRR smaller than 30 according to Kivisild et al. (2006), which amounts to merely two thirds of the TTRR for the total data as estimated in that paper. Hence, with appropriate multiple hit corrections applied, the TTRR for all other kinds of changes in the entire molecule might be about 70 (or higher). Saturation of some synonymous transitions and, in particular, of transitions in the control region should be the major concern, however. For example, the TTRR value estimated from the global mtDNA tree of Fig. 1 stays below 33 (without correction for multiple hits). In particular, the low TTRR values (in the order of 20–30) for the control region of African mtDNAs as proposed by Howell et al. (2004) are rather unrealistic and are likely incurred by methodological shortcomings (due to TREE-PUZZLE), insufficient multiple hit correction, possible inclusion of the transversions to C scored in 16182–16183, and perhaps also the suboptimal quality of the data.

In view of these findings, the claim of Strandberg and Salter (2004) in their simulation study (for estimating the ratio of the rates of transitions to transversions) that "a TI : TV ratio as high as 29.21 is rarely seen in real data sets" (which, in our case, would correspond to a TTRR of 61) is surprising. The parameter setting in their experiments that comes closest to human HVS-I sequences (TI : TV = 29.21 and $a = 0.2335$) returned a transition–transversion count ratio of 0.3754 ± 0.0316 in the case of parsimony, albeit with a simulated data set comprising only 14 sequences of length 231 bp showing extremely different nucleotide frequencies. This value would [in view of $2(g_A g_G + g_C g_T)/g_R g_Y \approx 1.37$ in this case] translate into a TTRR of 1.03, compatible with the expected value 1 for totally randomized data (relative to the prescribed nucleotide frequencies); therefore, this experimental design can hardly be regarded as useful. In order to compare a number of approaches, including ML and Bayesian methods, for the estimation of TTRRs, simulated data sets can only have bonsai form, because of the computational burden, which might evoke the impression that the estimation procedure can be carried out only for tiny real-world data sets. Unfortunately, similar ideas had been followed in estimating positional mutation rates, too, as we shall see next.

4
Spectrum of Relative Mutational Rates Along the Molecule

A robust estimation of the mtDNA phylogeny demands a good characterization of the variability of mutation rates among different nucleotide positions. In particular, it needs the identification of mutational hotspots since the latter could mislead phylogeny reconstruction and bias mutation counts. With such information available, it becomes easier to distinguish, to a greater or lesser extent, when mtDNA molecules share nucleotides by descent or only by state, a difference crucial to phylogeny estimation. On the other hand, this variability can only be well characterized with a reliable phylogeny. This then seems like a vicious circle. One strategy to deal with this problem has been to pursue inferential methods of (presumed) good repute and to execute them on an aligned mtDNA data set of medium size, without imposing any prior knowledge about the mutational process. Such a strategy might then be deemed to be objective; however, it nevertheless implicitly assumes equal rates at the outset, and in the specific case of human mtDNA, it ignores the enormous amount of published information that does not quite conform to the sequencing range of the targeted database. Thus, it is reasonable to incorporate all of the (reliable) information that is available. Then, before any fine-tuned work can begin, one would need a clear view on problematic positions that would better be disregarded entirely or at least be downweighted, provided that numerous small-scale studies would unanimously suggest this. In other words, a mild weighting procedure based on solid evidence seems superior to an approach that ignores any prior information.

Stochastic models of sequence evolution that stipulate parameters to account for different nucleotide frequencies, such as the HKY85 model, effectively weight changes because pyrimidine transitions (changes between C and T) will be slightly preferred over purine transitions (changes between G and A) in number when $g_C g_T > g_A g_G$. The real trend in human mtDNA appears to be even stronger than the 1.9-fold bias towards pyrimidine transitions would predict: for example, nine of the ten extreme hotspot mutations in the control region are pyrimidine transitions (see later). So, in the case of a direct conflict between a purine site versus a pyrimidine site, the former will win and force the pyrimidine transition to be recurrent as this increases the likelihood. But, in the case of complete mtDNA genomes, apart from that and a mild effect of correcting for multiple hits, ML cannot be expected to deliver trees and lengths that are considerably different from trees and lengths estimated with parsimony (see later). When it comes to the estimation of positional rates along a tree (which may well be an ML tree or a tree that is based on additional information), claims such as "the parsimony method is known to be biased" (Meyer et al. 1999) are too bold in regard to the specifics of the data. In fact, Deng and Fu (2000) have performed extensive studies (for small sample sizes, though) that could not detect a general bias in the parsimony approach of reconstructing and counting mutations along a tree – whether it be a most parsimonious tree, a UPGMA (unweighted pair-group method using arithmetic averages) tree, or an ML tree – in regard to the gamma shape parameter, i.e. the heterogeneity parameter a for the positional rates.

The first attempts at elucidating the mutational spectrum of human mtDNA had to rely on a limited number of HVS-I sequences (of short length) that were subjected to a method standard at the time; for example, Wakeley (1993) performed a most parsimonious reconstruction of mutations on a neighbour-joining tree of slightly more than 300 partial HVS-I sequences (scored from positions 16130–16379), whereas Hasegawa et al. (1993) used a putative most parsimonious (MP) tree for a set of nearly 400 HVS-I sequences. Later, Excoffier and Yang (1999) improved upon these early attempts by using more than 800 HVS-I sequences and by determining ML estimates of the parameters in a mutation model incorporating rate heterogeneity, on a number of (putative) MP tree topologies. What these approaches and more recent ones (Allard et al. 2002, 2004, 2005) have in common is their reliance on control-region information alone. The large amount of homoplasy in such data makes the task of inferring the relative rates of different nucleotide positions rather a tough one. In anticipating these problems, Malyarchuk and Rogozin (2004) chose to subdivide a large collection of HVS-I (or HVS-I plus HVS-II) sequences into 90 well-defined haplogroups and to score for each haplogroup whether the position in that haplogroup is polymorphic (contributing count 1) or not (count 0). In this way, however, the estimation of the mutational hotspots is biased downwards since the counts can never exceed 1 per haplogroup. Another (smaller) bias may be introduced by incorrect haplogroup affiliation, so instances of back mutations at haplogroup-defining sites may not be identified.

A better strategy for characterizing the mutational spectrum within HVS-I is to use more informative data. By supplementing HVS-I sequence information with information from the stabler coding region of mtDNA in the same samples, power is gained to distinguish recurrent mutation within HVS-I. Ideally, one would use complete sequence information, but in 2001 the complete mtDNA database was still too small, so at the time one had to resort to investigating combined HVS-I and high-resolution (14-enzyme) RFLP data (Bandelt et al. 2002). In the latter case, it seemed realistic to focus on fairly recent and well-characterized parts of the mtDNA phylogeny, in order to avoid missing recurrent mutations on the long internal branches near the root.

Thus, haplotypes were classified into haplogroups on the basis of the presence of certain compound motifs, and the mutations that have occurred on the branches connecting the haplogroups were ignored. Phylogenetic analysis then proceeded haplogroupwise, employing the reduced-median (RM) network method with the standard parameter setting (Bandelt et al. 1995, 2000). With some further moderate weighting, some of the reticulation present could be removed. In the few cases where a unique tree could not be inferred but only multiple best trees, an average count over most parsimonious reconstructions was performed, so the final numbers may be fractional. A list of the resulting mutational scores and more detailed information can be found on the website http://www.stats.gla.ac.uk/~vincent/fingerprint/table.html. Figure 2 displays the strong heterogeneity of the positional rates in HVS-I (of range 16051–16365 here) and shows the corresponding curve of the best-fit gamma distribution with parameter $a = 0.205$.

To give a brief summary here, we have divided the insertions and deletions (indels), transversions, and transitions into groups with rates exceeding the average transitional rate in HVS-I (calculated as 1.62 scored mutations per position within 16051–16365) by powers of 2. Table 2 then displays the top four groups. Note that none of the indels or transversions in HVS-I received scores above 3. Six RFLP sites fell into the class with scores from 3.5 to 6, whereas a single RFLP site, 1715*Dde*I, received a count of 9. Losses of 1715*Dde*I are typically induced by a mutation at 1719, which is indeed a frequent event: for example,

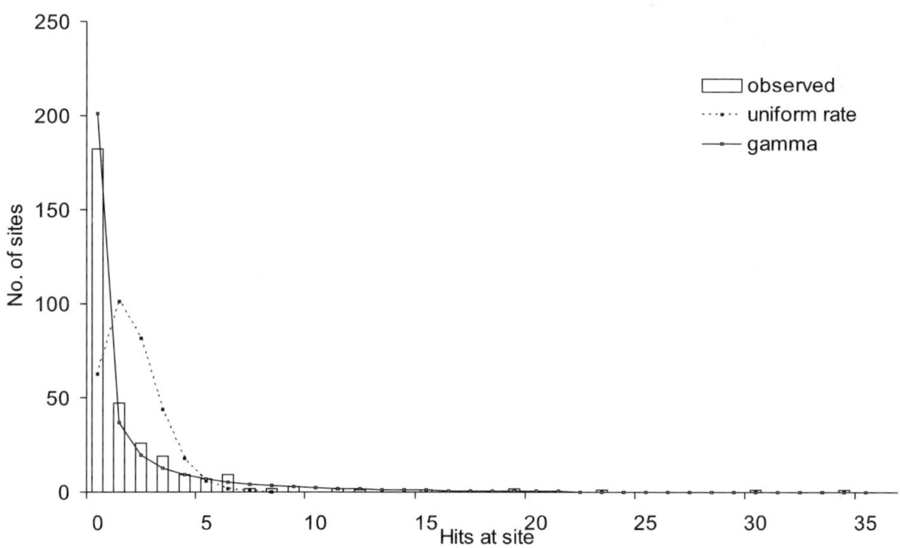

Fig. 2 Frequency spectrum of mutational hits in HVS-I (16051–16365) as inferred from a study of 873 combined HVS-I sequences and high-resolution restriction fragment length polymorphisms (V. Macaulay, H.-J. Bandelt, M. Richards, A. Torroni, unpublished data)

Table 2 The number of transitions reconstructed at position 16051–16365 derived from 873 combined high-resolution restriction fragment length polymorphism/HVS-I haplotypes

Hits	Transition
29–34	16093 16311
15–23	16129 16189 16192 16362
7–12	16051 16145 16172 16223 16256 16261 16274 16278 16291 16292 16293 16320
3.5–6	16092 16111 16126 16148 16193 16209 16213 16214 16234 16242 16249 16265
	16266 16290 16295 16296 16298 16300 16304 16319 16325 16335 16355

in the tree of Fig. 1 three mutations occur at 1719. The top five transitions are at 16093, 16129, 16189, 16311, and 16362. These are also the top five positions that unanimously receive the (cumulative) highest mutation scores from both Allard et al. (2002, 2004, 2005) and Malyarchuk and Rogozin (2004). From this preliminary analysis of HVS-I variation we conclude that there is solid evidence for identifying the top five transitional hotspots in HVS-I and that one has to assume a more than tenfold rate for the top two mutations compared with the average transitional rate. This indicates that the range for the different rate categories handled by Excoffier and Yang (1999) was a bit too narrow.

For the coding region, the study of Kivisild et al. (2006) listed the recurrent mutations reconstructed along their worldwide tree. Coble (2004) analysed 646 coding-region sequences from the literature (Ingman et al. 2000; Maca-Meyer et al. 2001; Herrnstadt et al. 2002, 2003) and recorded the mutational hits as well. Table 3 displays the extreme hotspots according to both lists. The topmost variable position appears to be 709 in both studies, but a handful of other transitions come close. The fastest transversion in the coding region (with 5 and 6 hits, respectively) is 13928C, which happens to be among the defining mutations for several major haplogroups (L0k, L2c, R9, and B6). These findings do not quite support the view of Ingman (2001) that "the numbers of back- and parallel mutations found outside of the D-loop were practically zero" and "the rate of evolution of the rest of the genome was surprisingly even between different sites, different genes and also between the different gene complexes". Many positions in the coding region appear to be prone to recurrent mutation, although the overall amount of homoplasy in the coding region is much lower compared with the amount of homoplasy in the hypervariable segments of the control region.

For practical purposes, phylogenetic analyses would be enhanced when the variation of positional rates is appreciated, at least tentatively in the form of a rough weighting scheme. Such a scheme would need to be adapted to the particular method employed and its parameter settings. For the parsimony-oriented network methods RM and median-joining (Bandelt et al. 1999), we

Table 3 Mutational hotspots in the coding region as derived from (A) 277 coding-region sequences (Kivisild et al. 2006) and (B) 646 coding-region sequences reanalysed by Coble (2004)

Hits	Transition	
	A	B
15	709	709
14	–	–
13	–	5460 11914
12	–	13708
11	13708	–
10	–	1719 15924
9	1888 8251	3010 10398
8	11914 10398	8251 14470 15784

would envision a scheme that assigns weights dependent on both the type of change (transition, transversion, and indel) and the affected position. Table 4 offers a provisional scheme mainly based on the HVS-I results mentioned before, on Allard et al. (2002, 2004, 2005) and on Malyarchuk and Rogozin (2004) for the other parts of the control region, and on Kivisild et al. (2006) for the coding region. The progression of weights is a bit fluid as it would still require specific testing; for example, instead of weights 0, 1/2, 1, 2, and 3, one could also handle 0, 1, 2, 3, and 4. Depending on the focus of a phylogenetic analysis, one might even wish to disregard (i.e. give weight 0 to) the ten extreme hotspot transitions at positions 146, 150, 152, 195, 16093, 16129, 16189, 16311, 16362, and 16519.

It is clear that the sorting of mutations into several weight classes needs further fine-tuning; in particular, more comprehensive studies on the positional rate spectrum are needed in the future when a few thousand (reliable) complete mtDNA sequences become available. It has, however, become apparent that a single set of parameters for an ML approach does not quite make the best of the data. Although for the vast majority of sites one would envision a standard 'default' model (such as HKY85) as being robust enough, particular oligonucleotide changes scattered across the molecule bearing idiosyncratic features might warrant tailored modelling.

5
Pitfalls with Estimation of Positional Rate Variability

To escape the vicious circle of estimating an mtDNA tree without imposing the information, for example about mutational hotspots, that would be neces-

Table 4 Weighting scheme for parsimony and network analyses

Weight	Region	Type of mutation[a]	Mutation/site/fragment
0	HVS-I	C run length polymorphism	16182C, 16183C, C indels scored at 16193
	HVS-II	C run length polymorphism	C Indels scored at 309 and 315
	HVS-III	Dinucleotide repeat	AC Indels in 515–524 (alias CA indels in 514–523)
		C run length polymorphism	C Indels scored at 573
1/2	HVS-I & 16519	Transition	16051 16078 16086 16092 16093 16111 16114 16124 16126 16129 16140 16145 16147 16148 16150 16163 16172 16173 16176 16186 16187 16189 16192 16193 16209 16212 16213 16214 16216 16217 16223 16227 16231 16232 16234 16235 16239 16240 16241 16242 16245 16249 16255 16256 16257 16258 16260 16261 16263 16264 16265 16266 16270 16274 16278 16284 16287 16288 16290 16291 16292 16293 16294 16295 16296 16298 16300 16301 16304 16309 16311 16316 16319 16320 16325 16327 16335 16352 16354 16355 16356 16357 16360 16362 16390 16519
		Transversion	16111A 16188A 16265C
		Indel	16166del

[a]Complex contiguous deletions and insertions (*indels*) are scored as one character (nucleotide) change

Table 4 (continued)

Weight	Region	Type of mutation[a]	Mutation/site/fragment
1/2	HVS-II	Transition	93 146 150 151 152 182 183 185 189 194 195 198 199 200 204 207 228
	HVS-III	Transition	499
	Coding	Transition, indel	709, C indels scored at 965
1	Control	Any	All remaining mutations (not listed above)
	Coding	Transition	1438 1598 1719 1888 3010 3394 5147 5231 5460 5821 6182 6221 7055 8251 8790 9545 9554 9950 10398 11914 12007 12172 12501 13105 13359 13368 13708 13966 14110 15110 15217 15514 15924 15930
		Transversion	13928C
2	Coding	Transition	All remaining transitions (not listed above)
		Spacer indel	Indels within 3305–3306 4401 5577–5586 5656 5892–5903 7517 8270–8294 8365 14743–14746 15954
3	Coding	Transversion/indel	All remaining transversions/indels (not listed above)

[a]Complex contiguous deletions and insertions (*indels*) are scored as one character (nucleotide) change

sary to arrive at a reliable estimate, pairwise methods have been advertised by some researchers. For instance, Pesole and Saccone (2001) have regarded the reliance on a phylogeny as a "major drawback" for the estimation of site-specific rates, and it has been maintained that "knowledge of the phylogeny is not necessary to infer the rates" (Meyer and von Haeseler 2003). The former authors observed, correctly, that for two sequences a mutation at a position with rate $2r$ will have a double chance to contribute to the distance between the two sequences compared with a mutation at a position with rate r. Apparently, they took it for granted that this proportionality principle extends to sample sizes larger than 3; however, it does not. For an instructive example, take four sequences connected by a tree that is not a star, such as the subtree connecting the haplogroup B4 sequences nos. 4, 11, 17, and 18 in Fig. 1. There are six ways to choose a pair from these four sequences and connect them by the corresponding subpath of this subtree: the middle link (separating the sequence pair 4/11 from pair 17/18) occurs in four of these six paths, whereas any of the four terminal links occur on three of the six paths. Then by the naïve pairwise argument, one would infer that the positions that changed on the middle link (here represented by positions 522–523, 5465, 9123, 10238, and 16261) had mutated at four thirds of the rate of the positions that had changed on the terminal links. The fallacy we encounter here is actually notorious in the field (see also later) when dealing with stochastic branching processes: as long as the underlying tree is not a perfect star, the interior links induce dependencies that cannot be ignored without introducing a bias for estimators operating on pairs.

Pesole and Saccone (2001) employed the mtDNA database Hvrbase of Handt et al. (1998) in order to estimate the mutational rate spectrum of HVS-I and HVS-II. Although drawing sequences from a Web database may appear convenient, it represents a risky strategy. First, the automatic inclusion of published mtDNA population data may incorporate error-ridden sequences suffering from all kinds of artefacts (such as those from Nasidze and Stoneking 2001; Chap. 6). Second, the database itself may represent imperfect images of the real data tables when the sequences have been for the most part manually transferred in the construction of the database (Bandelt and Parson 2004; Chap. 6). This was, unfortunately, the case with the Hvrbase data collection (Árnason 2003). After this news broke (Dennis 2003), Hvrbase was immediately taken offline for a couple of weeks to have the mistakes that were discovered by Árnason repaired, but other errors and inconsistencies remain, presumably because of the time and effort it would have taken to revise the database thoroughly. Databases derived from manual input can hardly be expected to be error-free (Forster 2003; Bandelt and Parson 2004).

To analyse the 'pairwise' approach to positional rate estimation, consider a (clean) hypothetical database of $n > 3$ sequences (with n even, for the sake of the exercise). Then the most extreme polymorphism count for a position showing no more than two distinct nucleotides in the database is attained when $n/2$

sequences show one nucleotide (say, C) at that position and the other $n/2$ sequences share the other nucleotide (say, T). Then $n^2/4$ pairs of sequences are distinguished at this position, compared with just the $n - 1$ pairs that would be distinguished at a position displaying a unique deviant nucleotide. The ratio of counts is thus approximately $n/4$, which the naïve 'pairwise' approach would take as the estimator for the relative difference in positional rates. Specifically, Pesole and Saccone (2001) took 1308 HVS-I sequences and 458 HVS-II sequences from the Hvrbase database. For $n = 1308$, the factor is thus more than 320, and even for $n = 458$, the factor is still more than 110. Figure 3 of Pesole and Saccone (2001) is very compatible with these numbers, in particular, in part B of the figure, one can tentatively identify the minimum positive score and then read off from the diagram that the maximum rate is a factor of 100–200 larger than that. It is then not surprising that the heterogeneity parameter a (sometimes called alpha) of the gamma distribution that best fitted those scores was as low as 0.09 and 0.05 for HVS-I and HVS-II, respectively.

The approach of Pesole and Saccone (2001) has been misleadingly labelled 'empirical' by Meyer and von Haeseler (2003), possibly because it is without the blessing of ML. But in reality, it is simply flawed as it evaluates products of polymorphism counts instead of unbiased estimates of the relative rates. With sufficiently large data sets, the deviation from the true relative rates can be enormous. For instance, the statement by Pesole and Saccone (2001) that their "data suggest that the nucleotide changes T146C, T152C, and T195C are so fast that they can occur several times during the life span of an individual" constitutes an extreme exaggeration. But even without an analysis of the logic behind the pairwise method, their assertion can quickly be refuted by taking all published HVS-I and HVS-II sequences into consideration. For example, the HVS-II transition motif 73, 146, 152, 195, 263, 315+C relative to the rCRS is basal to the African mtDNA haplogroup L2. Although the coalescence of one of its subclades, L2a, certainly dates back several tens of thousands of years (Salas et al. 2002), this consensus motif of L2a is still present in the majority of its members. If these mutations were so superfast that they occurred every handful of generations on average, then such patterns would be completely destroyed in fewer than a thousand years.

A sort of ML approach for the problem of differential positional rates has been suggested that uses QP (Strimmer and von Haeseler 1996). This QP method performs exact ML calculations only on quartets of DNA sequences and then aggregates the estimated quartet trees in a random order that mimics the analogous agglomerative distance method of Farris (1970). This agglomerative procedure is repeated a prescribed number of times with respect to randomly generated orders (puzzling steps). Eventually, the multitude of resulting trees is aggregated into a majority consensus of all candidate trees obtained. It is therefore not a method comparable to 'genuine' ML algorithms or parsimony searches or minimum-evolution distance approaches, since the optimization component of the method occurs only at the level of quartets. QP rather falls into the category of consensus procedures that aim at elucidating the support for certain splits in an unrooted tree or clades in a rooted tree, for example by returning those splits/clades with majority support. Consensus approaches, however, can only deliver a meagre view of the structure of a data set when the amount of homoplasy is high (as is definitely the case with the control region).

To estimate relative rates in the two hypervariable segments Meyer et al. (1999) employed the (flawed) database Hvrbase and used an approach which averaged positional rate estimates (derived from successive runs of the QP procedure) over numerous subsamples of 50 haplotypes each taken from a data set of 1229 distinct sequences of HVS-I (region 16024–16382), conditioning on point estimates of the parameters in a mutation model (similarly obtained by averaging over subsamples) that incorporated rate variation via a discrete gamma distribution with eight rates. The same sort of approach was then also applied to 385 distinct HVS-II sequences (region 57–371). Although both HVS-I and HVS-II information is available for several populations in Hvrbase (Handt et al. 1998), no attempt was made to compare the trees estimated from analysing HVS-I and HVS-II separately. The discrepancies that this might have revealed would have called into question the method or the data. Indeed, although some mutations in HVS-II are known to be phylogenetically quite informative, it has been demonstrated that HVS-II sequences alone could normally produce only nonsense trees by whatever method (Bandelt et al. 2000).

The approach by Meyer et al. (1999) for estimating relative positional rates is biased for the same reason as the pairwise method of Pesole and Saccone (2001). Since $50 \ll 1229$, the resampling scheme employing only 50 sequences in each go would lead to a strong bias even in the absence of homoplasy. To see this, assume that the total database has n sequences and that s, the resampling size, satisfies $1 < s < n$ (where n is even, say). Take an extreme case where the data are homoplasy-free and the tree has one central link associated with a single mutation at site p separating $n/2$ from $n/2$ sequences. The mutation on this central link may be compared to a peripheral link supported by a single mutation at site q (which thus separates 1 from $n-1$ sequences). Then the probability that q is polymorphic in a subsample of size s is

$$1 - \binom{n-1}{s} \Big/ \binom{n}{s} = \frac{s}{n} .$$

In comparison, the site p is polymorphic in the subsample with probability

$$1 - 2\binom{n/2}{s} \Big/ \binom{n}{s} \approx 1 - 2^{1-s} \approx 1 , \quad \text{when } 1 \ll s \ll n .$$

Hence, the ratio of polymorphism probabilities for q over p is approximately s/n. For $n = 1229$ and $s = 50$, this means that the ratio is approximately 0.04, yielding a difference in estimated rates of more than 1 order of magnitude.

A remarkable discrepancy between the Meyer et al. rates and our estimates of relative rates in HVS-I can, for instance, be seen for the transitions at 16093 and 16294. We estimate that the (transitional) rate of 16294 is 1 order of magnitude smaller than the rate of 16093. There is, however, hardly any major clade of the worldwide phylogeny that seems to be supported by the latter position. In contrast, the 16294 transition participates in motifs of the major African haplogroups L1c and L2a and the West Eurasian haplogroup T. Therefore 'rate' estimation based on polymorphism count would place 16294 faster than 16093. For example, approximately 12.1% of the HVS-I and HVS-II sequences in the SWGDAM database (Monson et al. 2002) have a T at 16294, but only approximately 5.5% have a C at 16093. This is rather well reflected by the relative rates of approximately 5 : 3 displayed in Fig. 2 of Meyer et al. (1999). A similar argument can be made in the case of HVS-II. For example, the transition at 73 supports haplogroups HV and L0a, whereas the 247 transition separates the African haplogroups L0, L1, and L5 from haplogroup L2′3 (nearly without exception). In East Asia, where these haplogroups are virtually absent (there is only one potential member of haplogroup HV, CHN.ASN.000281, in the Chinese fraction of the SWGDAM database), not a single independent mutation at these two positions can be found in the SWGDAM database. In the whole database, however, an A at 247 occurs at a frequency of approximately 6.7% and the rCRS nucleotide A at 73 at approximately 18.5%. It therefore does not come as a surprise that both positions rank in the highest two rate categories of Meyer et al. (1999). Although their method thus evidently falls far short of the goal, these 'rate' estimates are still being cited routinely and uncritically; see, for example, Pakendorf and Stoneking (2005) or Thalmann et al. (2005).

No matter how absurd or flawed an approach or how biased a method for estimating the positional rate spectrum, real mutational hotspots will always show up in the high-rank mutation category because any such position receives a relatively high polymorphism count in whatever subset of the database. Indeed, mutational hotspots will show up on many links of the mtDNA phylogeny—some being more peripheral, others more central. Therefore all approaches are bound to correlate highly because inferred relative rates will positively correlate with polymorphism counts. This explains why Meyer et al. (1999) found that their relative rate "estimates agree by and large with results from other sequences of human mtDNA".

Meyer and von Haeseler (2003) made yet another attempt to estimate site-specific mutation rates, now for the whole mitochondrial genome. In order to provide "a complete picture of the site-specific substitution rates in human mitochondrial DNA", they chose the 53 complete mtDNA sequences from Ingman et al. (2000) and subjected them to QP. The resulting QP tree underlying their estimates of positional rates was not revealed, but let us (optimistically) assume that it is not much worse than the neighbour-joining tree of Ingman et al. (although certainly less resolved). It is, however, an enigma how such a small tree with 50–100 links at which mutations can be scored could give any meaningful picture of the whole mutational spectrum for hundreds of polymorphic sites: most sites that are polymorphic in this data set are so because of just a single mutation in the phylogeny, so any estimate of the rate at that site will have an uncertainty as large as the estimate itself.

The first, trivial, step in dealing with complete mtDNA sequences is to align the sequences, which is usually done with some experience and insight into the intricacies of the few stretches showing natural length polymorphism. According to Meyer and von Haeseler (2003), "the sequences were aligned by eye". As our reanalysis of their raw data (kindly provided by Arndt von Haeseler) shows, this must have gone wrong in many cases. In regions affected by deletions and insertions their alignment regularly went out of phase (e.g. in regions 310–317, 522–523, 2157–2232, and 15944–16193). The problem in region 2157–2232 was apparently induced by insertions at position 2156 in sample 48 and at 2232 in samples 24–26 (Fig. 1), which were not correctly treated. In the region 15944–16193, ranging from an indel position until the end of a C stretch, rate scores for single positions are 'smeared' across the corresponding neighbourhood. Moreover, several single nucleotides are misallocated, too. For example, positions 248–249 were not identified as being hit by a deletion of one A (in samples 24–28), but were scored as if this had been an ordinary nucleotide substitution. These errors were not detected because of the lack of any reference mtDNA phylogeny to serve as an external control.

To judge the rate estimates obtained by Meyer and von Haeseler (2003), let us first consider the pair of sequences 46 and 47 displayed in Fig. 1, which are distinguished by a single mutation (at the parsimoniously uninformative position 7571). The sister relationship of this pair is supported by as many as 12 coding-region mutations, none of which are observed anywhere else in the tree. Nonetheless, the mutational rate at each of these positions was inferred to be about twice the rate at position 7571, although all of the 13 mutations in question would trivially be reconstructed as unique events by any phylogenetic method (even by TREE-PUZZLE). At the other extreme, the highest polymorphism count (23) is attained for position 15301. Consequently, it is exactly this position that received the highest rate in the whole mitochondrial genome according to Meyer and von Haeseler (2003). In contrast, the topmost mutational hotspot (leaving aside length polymorphisms of C stretches), position 152 (according to Allard et al. 2002, 2004, 2005; Malyarchuk and Rogozin 2004), is reported with a score that is 31% smaller than that for 15301, which well reflects the 26% lower polymorphism count of this position. Familiarity with the many published mtDNA population data sets would have highlighted the implausibility of these relative rate estimates: mutational hotspots are nearly always polymorphic in small samples of restricted geographic or phylogenetic scope, whereas conservative positions typically appear unvaried when the haplogroup which they support is absent from the sample. For example, the mtDNA data set of Coble et al. (2004) comprising

241 complete mtDNA sequences from the common European mtDNA haplogroups (H, V, K, J, and T) does not record any single mutation at 15301.

Figure 3 (compare with a similar diagram obtained by Coble 2004) reveals that the positional rate estimated by Meyer and von Haeseler (2003) linearly scales with the polymorphism count, that is, the minor allele frequency at the respective position. This count equals the number of mutations at the position when a perfect star with 53 tips representing the distinct sequences is imposed (instead of any realistic phylogenctic tree), so the variation at a particular position is accommodated most parsimoniously by assigning the majority nucleotide to the ancestor and counting each occurrence of the minority nucleotide(s) observed at the position as an independent change. This result, of course, could have been obtained more directly by hand rather than via the intensive computation of TREE-PUZZLE! Although the number of positions for which the rate was overestimated by orders of magnitude is rather small, the phylogenetic signal for the deeper branches of the phylogeny is to be sought exactly among these positions. Therefore any weighting of sites that aims to conform to the Meyer and von Haeseler estimates would introduce a systematic bias and thus be more harmful than uniform or even random weighting.

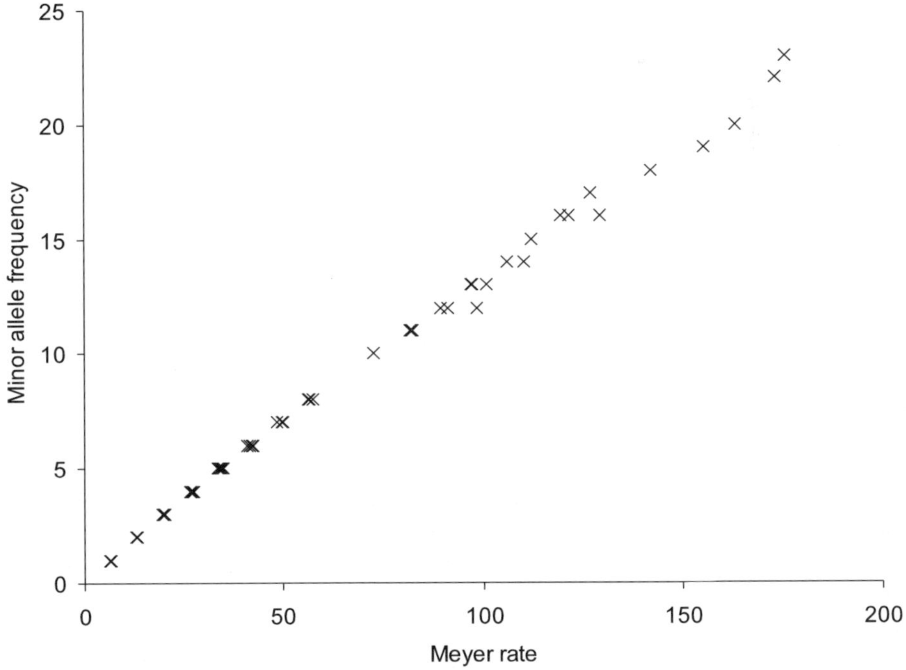

Fig. 3 Comparison of the positional rates as postulated by Meyer and von Haeseler (2003) and the polymorphism counts for the data set of Ingman et al. (2000)

Although the linear relationship of these rate estimates with polymorphism counts can be fully explained by assuming that the tree deriving from TREE-PUZZLE was nearly a perfect star, one could also envision other causes for linearity. Suppose that the pairwise method AP of Meyer and von Haeseler (2003) that compares all sequence pairs from the Ingman data set had been implemented with a shortcut, namely to compare each single sequence with a consensus sequence (which would essentially coincide with the ancestral sequence of haplogroup L3). Then the polymorphism counts would be approximately realized by the total number of changes at a specific site read off from the collection of all paths connecting the consensus sequence to the 53 sequences of the data set. The usual application of the AP method would rather yield a quadratic relationship by realizing products of polymorphism counts, exactly as in the pairwise approach of Pesole and Saccone (2001). One can firmly rule out that the alternative IP method of Meyer and von Haeseler (2003) had been used for the analysis of the Ingman et al. data: this approach would employ a (single?) packing of 26 paths that have no link in common and connect all but one of the tips (represented by the 53 sequences). In fact, in this situation some inner links and one terminal link would not be covered at all, so the mutations unique to those links would be completely missing; however, this would be at odds with Fig. 3.

Meyer and von Haeseler (2003) sought support for their method(s) by carrying out some simulations and contended that "simulations show that the evolutionary rate of fast-evolving sites can be reliably inferred". Simulation studies alone, however, will not necessarily identify a particular bias in the estimation of relative rates since the number of positions with medium to slow rate that mutated deep in the model tree can be quite restricted. The shape of the model tree and its reconstruction (by TREE-PUZZLE) can also matter a lot. But, when the specific outcomes of each single run do not come under individual scrutiny, an interpretation of the associated summary statistics is more akin to reading tea leaves than to understanding causal relationships. Simulations that employ summary statistics, quite popular in biology as a general-purpose tool, should not be a substitute for common sense and an investigation of the logic behind a method (plus the little mathematics that might be involved).

There are two take-home messages from this painful case: (1) freelance inference of major properties of a genetic marker without basic knowledge of the marker system itself may go very badly wrong; (2) misguided work from the far side of bioinformatics can hit molecular evolution journals—as long as the framework looks impressive and fashionable. Since ML computations are usually not well documented and cannot practically be followed by hand or via inspection, there is a real danger of falling into the pit of ill-designed experiments and statistical biases, without reviewers or editors ever being alerted.

6 Calibration of the Mutational Clock

The concept of a mutational clock, as an abstraction of early observations in protein evolution, uses stochastic models of DNA sequence evolution (Gillespie 1991). The standard framework is that of a continuous-time Markov chain governing the mutational changes at each site along the molecule (that is to say, the appearance of a new mutation is completely independent of previous mutations in the system). Special features of Markov chains employed in the modelling include reversibility, which has some technical advantages (Semple and Steel 2003). There has been much discussion in the past about the validity of the molecular clock; however, as Bromham and Penny (2003) have pointed out, "even an approximate clock allows time estimates of events in evolutionary history".

There have been attempts to question the validity of a molecular clock, in particular, for human mtDNA, either for the control region (Ingman et al. 2000) or from some specific branch (haplogroup L2) of the mtDNA phylogeny (Torroni et al. 2001). Whereas the former claim for the non-clock-like nature of control-region evolution could be dismissed because the test employed is not very informative and the tool TREE-PUZZLE is unreliable, the latter phenomenon (further elaborated on by Howell et al. 2004) may constitute a one-off deviation (not seen elsewhere in the mtDNA tree), which would rather point to (minor) deficiencies of the HKY85 model in satisfactorily explaining mtDNA sequence evolution. In Macaulay et al. (2005) we investigated the clock-like behaviour of the coding region of human mtDNA in the case of a small sample representative of basal Eurasian mtDNA variation. Assuming the HKY85 mutation model (with indels ignored, as usual) with gamma-distributed rates (approximated by a discrete distribution with 32 categories), we then used PAML 3.13 (Yang 1997) to check the molecular clock hypothesis by maximizing the likelihood over the branch lengths and the parameters κ (the TTRR) and a (the heterogeneity parameter in the gamma distribution), both with and without the clock enforced and using a likelihood ratio test to compare the maximized likelihoods. We could not reject the null hypothesis, viz. that the clock adequately describes the data.

Most recently, molecular dating has been questioned in a more fundamental way: there has been the concern that "the estimated times of various evolutionary events require a rethink" (Penny 2005) in view of the "firm evidence" provided by Ho et al. (2005) that evolution seems to tick faster on shorter timescales compared with longer ones. The argument put forward is that (slightly) deleterious mutations which can survive in the short term eventually get weeded out in the long run. In other words, purifying selection would offer "the most convincing explanation for the observed relationship between the estimated rate and the depth of the calibration", according to Ho et al. (2005). But what are the observations, and what data are they based on? The basic observation is that the inferred rate of change seems to decay exponentially with the age of the calibration point, leading to a contrast between the fast 'pedigree rate' and the slow 'phylogenetic rate' (see later). The data sets comprise either short mtDNA gene sequences or segments of control-region sequences.

In the latter case, the decay is most extreme and nearly reaches its asymptote after a million years. This is by no means unexpected: saturation is the most straightforward explanation and as we will see later, we have to take into account partial saturation at hotspot sites in the control region even after, say, 60 000 years! So, what could one expect at a time of 25 million years (the oldest calibration point handled by Ho et al. 2005) other than a near total loss of phylogenetic signal? Ho et al. (2005) seem to believe that partial saturation could be eliminated as a cause for their observations because (1) their method of analysis uses a sophisticated substitution model that could potentially "remove the impact of mutational saturation", (2) "variation at possible hot spots only appears to form a small proportion of the total variation among d-loop sequences", (3) there is no clearly non-uniform distribution of polymorphisms along the respective mtDNA stretches, and (4) relatively low sequence variation indicates the absence of mutational saturation.

There is no hard evidence provided for the first claim (which would run counter to the phylogenetic wisdom that highly recurrent characters, in great numbers, would swamp phylogenetic signals and destroy meaningful inferences); and the second claim is clearly wrong in the light of estimates for the rate heterogeneity in HVS-I, for example (Fig. 2). The arguments given for the other two assertions are equally unconvincing: some mutational hotspots cluster in particular stretches, while others appear in a more erratic way isolated amidst more conservative neighbourhoods—the hypervariable position 16519 may serve as a paradigmatic example in this latter respect. After all, one cannot judge the amount of rate heterogeneity by contemplating the spectrum of polymorphism along the molecule, as was aimed at with Ho et al.'s Fig. 2. Whilst functional constraints may leave many positions virtually invariable, some of the few remaining ones may nonetheless change almost freely. In the absence of focused phylogenetic analyses such claims remain merely speculative.

The coalescence time for human HVS-I sequences obtained under the exponential decay law of Ho et al. (2005) would lead to some puzzling consequences. Namely, the TMRCA for present-day human mtDNA they favour is 76 000 years ago, whereas most other scholars would prefer an estimate of about 200 000 years ago. Then, using Eq. 7 of Ho et al. (2005), the time for the peopling of Eurasia (equated with the ages of the Eurasian founder mtDNAs, at about one third of the global TMRCA) would be driven close to the Late Glacial Maximum (22 000 years ago), and Native American beginnings (at about one ninth of the global TMRCA) would then be contemporaneous with the first farming communities on the lowlands of north-central Europe

(7000 years ago). With respect to the function for the slightly faster [sic!] 'protein-coding' rate as proposed by Ho et al. (2005), these two pioneer settlement events (Eurasia and America) would be dated at 24 000 and 8000 years ago, respectively, relative to the asserted global TMRCA of 76 000 years ago. Needless to say, such 'express train' scenarios for pioneer Eurasian and American settlements do not sit easily with the archaeological and anthropological record. Thus, this time-dependent calibration seems to be nothing but a tempest in a teacup. Such episodes come and go—an earlier one (Pääbo 1996) heralded the rejuvenation of the peaks of bell-shaped mtDNA mismatch distributions (traditionally dated at approximately 40 000 years ago) and their association with the 'agricultural revolution' 5000–10 000 years ago, for exactly the same reason—a decay principle for rate calibration. The proposed timing and interpretation equally led to absurd consequences (Bandelt and Forster 1997; Macaulay et al. 1997) and seems to have been abandoned since then.

For the time frame from the TMRCA of present-day humans until now, though, the decay in the rate calibration is marginal, according to Eq. 6 and Table 1 of Ho et al. (2005): the rate at the TMRCA still amounts to 63% of the fastest rate from the pedigree horizon. Since the claimed huge (factor of approximately 10) disproportion between 'pedigree rate' and 'phylogenetic rate' (Howell et al. 1996, 2003) must be manifest within the human mtDNA phylogeny, the factor could never be more than 1.6 by Eq. 6. A factor of 10 is, however, achieved at a time depth no more recent than approximately 410 000 years ago. Hence the decay law (Eq. 6) of Ho et al. (2005) cannot explain what it aims to explain, at least in the case of human HVS-I variation. Then the 'rethought' TMRCA of 76 000 years ago cannot be justified by purifying selection, but only by drastically miscalibrated maximum-minimum mutation rates in connection with other factors (such as partial saturation) that deflate the TMRCA estimate. In fact, it is well accepted in the field that the mutation rate in the human mtDNA control region is much higher (by a factor of more than 5) than in the coding region, whereas Ho et al. (2005) posit that it is even lower within the human mtDNA phylogeny. With HVS-I sequences, as innocently employed by Ho et al. (2005), partial saturation is indeed an issue, as we shall demonstrate next.

In order to contrast coding region and control region at different time scales, we have split the mtDNA tree of Fig. 1 into two parts, the upper 'Eurasian/Oceanian' part and the lower 'African' part, which have only a node (the L3 root) but no link in common. We disregard indels (as is usually also done with standard stochastic models of sequence evolution), the transversions 16182C and 16183C (because they are highly correlated with the nucleotide at 16189), and the 16519 transition (which is so frequent that often no realistic reconstruction is possible). In the Eurasian/Oceanian part we then score 279 mutations for the coding region and 155 for the control region, whereas in the African part 313 mutations are scored for the coding region but only 123 for the control region. Thus, 64.3% of the total variation seen in the Eurasian/Oceanian tree is attributed to the coding region, whereas the corresponding contribution is 71.8% for the African tree. This most likely indicates that, as we would expect, the count of control-region mutations is deflated in the deeper African parts of the mtDNA tree (where branches are typically longer). Among the control-region mutations we find 86 'Eurasian/Oceanian' and just 49 'African' transitions in the stretch 16090–16365, which was considered by Forster et al. (1996) for a calibration of the HVS-I clock, so the transitions in region 16090–16365 are particularly underrepresented in the African part of the tree. If we were to calibrate the mutation rate in the control region against the coding-region rate, expressed by the above estimate of 5140 years per scored mutation, then one would obtain a rate of one control-region mutation per 9250 years for the Eurasian/Oceanian tree, but 13 080 years in the case of the African tree. The difference becomes even greater when one restricts the scoring to the transitions in region 16090–16365: the Eurasian/Oceanian part would suggest 16 680 years but the African part 32 830 years per transition within this stretch. Although these numbers have a large standard error (SE) because of extremely small sample sizes, there is a very clear trend: the deeper parts of the mtDNA tree would be significantly impoverished in transitions in the hypervariable segments as reconstructed by parsimony.

Purifying selection, however, does seem to have left its mark in the mtDNA phylogeny. According to Elson et al. (2004) and Kivisild et al. (2006), the deeper parts of the mtDNA phylogeny are relatively impoverished in non-synonymous substitutions compared with synonymous ones. The kind and quantity of decay of the non-synonymous mutations with time is not yet clear. Exponential decay is not very plausible, but perhaps a sigmoid-like curve could come into play as the cleansing of slightly deleterious mutations may be rather rapid so that most of them disappear in less than 10 000 years. The decay may easily be overestimated because of partial saturation for some non-synonymous changes (relative to the bulk of synonymous changes) and, in addition, because of a thin coating of phantom mutations, innocently interpreted as private mutations.

Occasional concerns that the molecular clock might be elusive and not tick regularly for human mtDNA should not, then, hinder us from attempting a calibration. Given an outgroup to human mtDNA, such as chimpanzee mtDNA, one would first have to infer an approximate coalescence time from the palaeoanthropological/palaeontological record (and possibly backed up by independent calibrations for other genetic markers). Most estimates employed so far range between five and seven million years. To come up with a very precise estimate, however, seems quite unrealistic, inasmuch as the interspecies coalescence time of the mtDNA lineages for humans and chimps will predate the corresponding species split by tens or hundreds of thousand years ('lineage sorting'). In Mishmar et al. (2003) and Macaulay et al. (2005)

the human–chimpanzee species split was assumed to be 6.0 million years ago, with 0.5 million years added to account for lineage sorting, so the human–chimpanzee mtDNA split was taken as 6.5 million years ago. With the ML approach described before, the complete mtDNA genomes analysed by Mishmar et al. (2003) together with two chimpanzee mtDNAs led to an estimated average mutation rate for the human mtDNA coding region of 1.26×10^{-8} base substitutions per nucleotide per year, which corresponds to 5140 years per substitution in the whole coding region. This coding-region rate estimate agrees well with rule-of-thumb estimates for the HVS-I rate (e.g. Forster et al. 1996) and hence earlier proposed control-region rates, so it can be considered as a 'conventional' rate, at least for Eurasian mtDNAs (Kong et al. 2003). We envision that a future recalibration of the mutation rate might discard non-synonymous substitutions altogether (as done by Kivisild et al. 2006) but embrace mutations in the control region at slowly evolving sites, so that the two spectra of positional rates are more or less comparable.

In principle, one way to assess the mutation rate is through studying pedigrees. New mutations that appear in a matrilineal genealogy can be observed directly (as differences in the sequences of mother and daughter), rather than inferred via phylogenetic reconstruction. Although the pedigree approach might seem promising at first (or even second) sight, in reality it is fraught with problems that seem insurmountable: (1) ascertainment bias; (2) the impact of artefacts; and (3) heteroplasmy and somatic mutations. In fact, early pedigree studies claimed a much higher mutation rate than the conventional rate that is still in use; see Sigurdarðóttir et al. (2000) for a critical assessment.

As for point 1, once a fast 'pedigree rate' has attracted attention in the field and become the expectation, studies that yield no new record high or remain inconclusive become less likely to be pursued and instead remain unpublished. To break out of this bias, one would have to generate a very large data set de novo without incorporating previous pedigree results. The study has to be large since the chances of observing a fresh mutation are quite low. Suppose that the control region is to be targeted and, for the sake of the exercise, assume an expected reciprocal rate of one mutation per 9250 years, say. With a (matrilineal) generation time of 27 years, this rate implies that a new mutation gets established in a matriline once every 343 mother–offspring transmissions, on average. Leaving aside any other problems, a sample of size 10 000 pairs, but certainly no less, might be adequate for the purpose of calibration. Some economy can be achieved by using deeper matrilines (where the older nodes may be unobserved, but the pedigree is known), although care needs to be taken with unrecorded adoptions and mutations that may occur in transmissions that are not directly observed (Sigurdarðóttir et al. 2000).

As for point 2, one has to bear in mind that the rate is being measured by single-generation events but the typical application involves time frames of 1000 generations or more—in other words, extreme extrapolation is employed. Ho et al. (2005) are right—although for another reason—when they contend that "it is invalid to extrapolate molecular rates of change across different evolutionary timescales". Among the reasons that we would envision in the first place are sequencing artefacts, which are deemed to be unavoidable (according to Herrnstadt et al. 2003; Chap. 6). Suppose we had a new sample of 560 pairs of control-region sequences from mother–offspring pairs. In a worst-case scenario, we could have, say, 1.6 phantom mutational events per pair on average (extrapolating the conservative error estimate from the data of Nasidze and Stoneking 2001; Chap. 6). If we diluted and dispersed such error-loaded sequences among ideal sequences, so that only approximately 1% of the sequences were not ideal and carried that specific error load, then we would infer a 'pedigree rate' that is more than sixfold higher than the conventional rate. Even if the chance of generating such erroneous sequences was only 0.1% (and thus amounted to roughly one wrong nucleotide in more than 700 000 bp), the 'pedigree rate' would still appear to be increased by more than 50% compared with the 'phylogenetic rate'. Thus, a tiny number of artefacts can have an enormous impact on rate estimation via extrapolation. It is almost impossible to arrive at the necessary precision without an experimental design with strongly overlapping fragments and employing numerous primer pairs (Brandstätter et al. 2004). Exclusively sequencing one strand with hardly any overlap of fragments, as apparently done by Howell et al. (2003), cannot ensure good enough quality.

As for point 3, inherited and de novo arisen heteroplasmy and (age-dependent) somatic mutations can hardly be distinguished. Although somatic mutations receive particular attention in medicine, there are only a few results that can be trusted since sample mix-up and contamination, phantom mutation processes, and documentation errors all contribute their share to the category 'somatic mutation' (Salas et al. 2005). Few studies have attempted to sequence the entire genome and have focused on cohorts of patients rather than healthy individuals (e.g. He et al. 2003). However, to appreciate adequately the real level of inherited heteroplasmy and acquired somatic mutations in general, one would need many cell/tissue samples from each individual of a large control group. Then, the haematopoietic system deserves particular attention, by investigating mtDNA from single colonies of stem cells, which would avoid cloning of mtDNA from single cells (Gattermann 2004).

Once we have agreed upon a mutation rate, be it conventional or merely a working hypothesis, we can set out to estimate the coalescence times of various clades (haplogroups) of the mtDNA phylogeny. The most unspectacular way to estimate the age of a particular ancestral ('root') haplotype, given the mutation rate, is to consider all available descendant individual sequences and take the arithmetic mean over all distances to the root haplotype (thereby requiring standard assumptions about a molecular clock). In doing so, it would be preferable to calculate the distances along an estimated tree—provided that realistic tree estimates are available. In a situation where any links in this tree bear a large number of mutations (such that unresolvable recurrent mutations may be an issue) one might employ some correction for multiple hits along those links. No matter how the calculation is eventually executed, we shall refer to this folkloric method as the 'rho' estimation, since the resulting age of the haplogroup (determined by the root haplotype in question) is denoted by ρ in the context of human mtDNAs (Forster et al. 1996).

One of the seeming disadvantages of the simple rho method is that it does not automatically provide the user with an estimate for the SE: the quantities that are averaged in the definition of ρ have a complicated and, more importantly, unknown, correlation structure, induced by the underlying genealogy, which makes any exact estimation of the SE impossible. Population geneticists therefore often employ coalescent model, which, however, are based on assumptions, for example, about population structure and the effective size of the population, so the resulting TMRCA and SE are typically both on very shaky grounds (Chap. 10). On the other hand, there would, of course, be a way to estimate the SE of ρ directly if we had the real genealogy at hand. Since the genealogy is unknown but some of its branches are spotlighted by mutations, one could take the estimated rooted tree as a rough proxy for the genealogy. Then one may assume that all links are governed by independent Poisson-distributed random variables with parameters equal to the numbers of mutations associated with the links (possibly corrected for multiple hits, if deemed necessary). Then ρ can be represented as the expected value of a weighted sum of these Poisson variables, where each weight equals the relative number of times the corresponding link occurs on the paths between individual sequences and the root; the SE associated with ρ calculated for this weighted sum of random variables is then denoted by σ (Saillard et al. 2000). Exactly the same approach was proposed by Britton et al. (2002) in a different context, where it was called the mean path length method.

The ML estimator has the advantage that it balances the age estimates for sister branches and does not tolerate reversion of ages for nested haplogroups, as for example happens with the ρ estimator in the case of haplogroups R9 and R (Kong et al. 2004). Since the ML estimate of any single clade depends on the global tree, its SE is in general smaller than the corresponding local estimate σ. This ML default scenario has, however, also a disadvantage when it comes to estimating the time of a pioneer settlement of a continent or subcontinent. Normally, a number of haplotypes rather that a single one would have been involved, so one cannot simply take the coalescence time of all sampled sequences from that region as the time for the settlement event; instead, some founder analysis would have to be applied first. If founder status receives sufficient support from the phylogeographic analysis of the descendant lineages in each case and, in addition, a worldwide phylogenetic analysis (based on ML or MP) does not yield significantly different ages of the ancestral founder types, then it seems reasonable to hypothesize equal ages for the founder types when dealing with mtDNA variation in the geographical region under scrutiny. In principle, one could adjust to this new hypothesis with an additional constraint in ML estimation (that the founder haplotypes are assigned the same age), or in the case of the rho method, some weighted average of the ρ values of the founder haplotypes can be constructed, in order to make an estimate of the single age that is based on all the data.

7 Pitfalls with Age Estimations

For a single locus, there has been a quite spectacular breadth of age estimates for the TMRCA of mtDNA. In some cases, the explanation for these divergent estimates is easy to spot. There has been an almost perennial confusion between *substitution rates* and *divergence rates* that seems to remain no matter how many times it is pointed out. The former measures the rate of change from an ancestral sequence to a contemporary descendant sequence, whereas the latter measures the rate of change between two sequences that diverged from a common ancestor. So, reading the substitution rate as a divergence rate would slow down the evolution by a factor of a half, whereas the other way round the evolutionary rate would get doubled; see Bandelt et al. (2003b) for a discussion of a pertinent example. This can cause particular confusion when it arises in a work that aims to compare different estimates of the mutation rate in the control region (Santos et al. 2005).

Perhaps the overestimates have caused the greatest difficulties. The brave multiregionalist, confronted with the shallow coalescence time usually estimated for human mtDNA, is easily tempted to refer to some of the more extreme age estimates yielding up to half a million years or so, obtained in the mid-1990s and based on meagre data and dubious methods. Such outlier estimates are then labelled 'conservative' and are still being woven into multiregional models to this day (Eswaran et al. 2005). There are, however, more recent and younger age estimates which equally lack any solid basis.

A particularly obscure time estimation, yielding a coalescence time of 240 000 years for the mtDNA of present-day humans, was executed on mtDNA data by Templeton (2002), who selected several different small data sets, based on HVS-I and HVS-II sequences, or high-resolution RFLPs, or complete sequences. It is unclear how these different data were aggregated and accommodated for time estimation—since they were all assigned the same age. However, those mtDNA samples coalesce on different most recent common ancestors: for instance, one pooled RFLP data set from East Asia and America comprises only mtDNA lineages from haplogroups M and N and therefore coalesces on the root of haplogroup L3, whereas another one, the Ingman et al. mtDNA sample (Fig. 1), coalesces on the root of all present-day human mtDNAs. In reality, these coalescence times are expected to differ by a factor of about 2/5. It rather looks as if Templeton (probably misled by the earliest five-enzyme RFLP data published in the early 1990s) may have erroneously assumed the multiregional stance of a common coalescence of any continental portion of mtDNA diversity. No analysis of complete mtDNA data (Fig. 1) would justify such a preconception, however.

The date of 240 000 years derived from a wrongly rooted tree was based on an estimator—nucleotide diversity π—that is known to be biased unless the population conforms to the idealized constant-size coalescent model of a panmictic population. The latter assumption is, however, rather absurd, given the worldwide sampling, and would effectively impose a most extreme version of the current multiregional model of human evolution, namely of unlimited gene flow. The data themselves hint at a violation of this model assumption: under the standard constant-size model, both π (which is the average distance in the sample) and the average distance ρ to the most recent common ancestor would equally constitute estimators of the parameter θ (that governs the model). For a panmictic but rapidly expanding population (Rosenberg and Hirsch 2003) or under a sudden expansion and wide geographical spread leading to non-panmixia (Forster et al. 1996), π would come close to 2ρ, the ultimate upper bound for π that is realized in perfect star phylogenies. For the worldwide coding-region data (Ingman et al. 2000) displayed in Fig. 1, π gives a larger value than ρ ($\rho = 38.0$ and $\pi = 43.6$), so employing the π estimator (rather than the ρ estimator) would yield a 15% overestimate of the coalescence time in this case. Note that the contrast between ρ (which depends on the assumed root) and π would have been larger if an erroneous root (say, the ancestral type of haplogroup L3) had been employed, for example by following Templeton's stipulation.

Ingman et al. (2000) have come up with another approach to calibrate the molecular clock for the coding region of human mtDNA. First a human–chimp species split was assumed at five million years ago, which was then apparently equated with the corresponding split between the mtDNAs in question. We do not regard either assumption as completely realistic, but let that pass for a moment. In any case, under these hypotheses averaged genetic distances (adjusted for multiple hits) between humans and chimps were scaled to five million years ago, yielding an estimate of the substitution rate in the coding region of 1.70×10^{-8} substitutions per position per year. This amounts, in the form of the reciprocal substitution rate, to 3810 years per mutation. In this count indels are disregarded. So far, so good. Then the coalescence time for haplogroup L3 was estimated via ρ as 8.85×10^{-4} substitutions per position or an average of 13.67 mutations across the whole coding region from the root of L3. However, this count, whether corrected for multiple hits or not, is too low and does not correspond to the reconstructed root of L3. There is hardly any ambiguity in the estimate of the sequence of this root since there are several independent branches emanating from it. We counted a total of 585 mutations from 38 members of haplogroup L3, yielding an average of 15.39 mutations to the root of L3. This translates into a TMRCA for

L3 of 58 700 years ago when scaled by Ingman et al.'s rate—rather than the 52 000 years ago given in the article. This discrepancy may be due either to an erroneous root or to the use of the distance estimate along the neighbour-joining tree—which does not constitute a reasonable estimate as a branching point in a neighbour-joining tree does not necessarily correspond to any reconstructed sequence.

Then, surprisingly, the total coalescence time, the TMRCA of the entire sample of 53 coding-region sequences, was calculated in a different way. Instead of attempting to reconstruct the sequence of the most recent common ancestor and taking the average distance to this sequence, Ingman et al. selected the two most divergent sequences and scaled their distance to time. In other words, the longest path in the phylogenetic tree was taken to reflect twice the coalescence time. The idea that a longest path in an estimated phylogeny or simply a pair of sequences with the largest mutational distance could directly be converted into the coalescence time crops up in biology many times: for instance, Merilä et al. (1997) used this approach to estimate the coalescence time of greenfinch mtDNA in Europe. This approach, however, is biased: the length of a longest mutational path (under an infinite-sites coalescent model) will (very slowly) diverge to infinity as the sample size rises to infinity. One cannot even guarantee that the most recent common ancestor lies on such a path! A more correct approach along these lines would have been to consider all paths passing through the reconstructed most recent common ancestor and then to take the average of their lengths.

Ingman et al. (2000) scored about 90 mutations (presumably resulting from correction for multiple hits) along the longest path, which they turned into $1/2 \times 90 \times 3810$ years, yielding 171 500 years ago as the TMRCA of human mtDNA. In Fig. 1 we count 87 mutations on this longest path. Our estimate of ρ, however, amounts to 38 mutations—less than half of 87. This would translate to just 144 800 years ago. The average distance to the reconstructed root is 34.60 for the L0 sequences but is 38.35 for the other sequences (from L1′2′3). Therefore an evolutionary path through the root has length 72.95 mutations on average. Then half of this, 36.48 mutations, would scale to an even slightly lower value of 139 000 years ago. A reasonably estimated age would then certainly stay below 150 000 years ago, even after correction for multiple hits. This age, and not the reported 171 500 years, would then have been the TMRCA associated with this sample, given the calibration of the mutation rate. Yet this flawed age estimate has been taken at face value by numerous authors citing this paper.

The mtDNA haplogroups M and N, completely covering the Eurasian/Oceanian mtDNA pool, have subsequently been redated in a similar spirit, using an additional batch of complete mtDNA sequences (Ingman and Gyllensten 2003), in order to determine the time of the out-of-Africa dispersal of modern humans. To estimate the ages—the coalescence times—of haplogroups M and N, the authors again took a single pair from either haplogroup, namely the most distant pair, for which the connecting pathway passes through the respective root—instead of, say, computing the average of the distances between all such pairs. This maximum choice principle then strongly biases upwards the estimates of the corresponding TMRCAs—which, on the other hand, was compensated for by the opposite trend incurred by the extremely low age (five million years ago) assumed for the human–chimp mtDNA split.

Our estimates of 139 000 or 145 000 years ago for the human TMRCA derived from the data and hypotheses of Ingman et al. (2000) (adopting their assumptions regarding the calibration) would at face value virtually coincide with the estimated time of 143 000 years ago published earlier by Horai et al. (1995). The latter authors pioneered the use of complete mtDNA sequences for the calibration issue and were careful to distinguish synonymous from non-synonymous changes when comparing three human mtDNA sequences (viz. a partly corrected CRS, a sequence from the East Asian haplogroup M10, and one from the African haplogroup L0a2) and four great ape mtDNA sequences (from orangutan, gorilla, common chimpanzee and bonobo). The haplogroup M10 sequence was selected from ten available complete mtDNA sequences from Japanese patients with neuromuscular disorders (Ozawa et al. 1991) as "the one ... that is most distantly related to the European ... in terms of nucleotide sequence and is not associated with any disease-specific mutations". This further example of the maximum choice principle resulted in a bias qualitatively similar to that produced by Ingman et al. (2000), although the bias was less severe in the case of Horai et al. (1995) because the pool of sequences for selection was much smaller and effectively led to a maximal choice from just six haplogroup M sequences not showing the pathogenic 3243 transition, viz. from the mtDNAs of patients PD,P-2 (from haplogroup D5a), DCM,P-1 (M10), HCM,P-1 (M7a), DCM,P-2 (M7a), FICM (D5c), and HCM,P-2 (D4b).

This kind of choice, however, was unfortunate for the calibration of the mutation rate of the control region relative to that of the coding region. Since coding-region variation clearly dominates control-region variation, a maximum distance choice will tend to select an mtDNA with an excess of coding-region substitutions over control-region substitutions. This is indeed the case here: the M10 sequence is separated from the M root by 19 coding-region substitutions and only six control-region substitutions. By chance, the choice of the partly corrected CRS further exacerbates the bias in that ten substitutions in the coding region are opposed to two in the control region relative to the respective haplogroup roots. The numbers of substitutions from the roots of haplogroups M and R we would expect are, rather, 12 (coding) and 6 (control), respectively. Thus, instead of the expected ratio of 24 : 12, the particular choice of the two mtDNA lineages yielded a ratio of 29 : 8. Further, the path to the African mtDNA lineage would necessarily underrepresent control-region mutations because of partial saturation. This is clearly reflected in the seemingly low share of the total variation for the control region, which therefore yielded a low reciprocal substitution rate of one mutation in 12 700 years as estimated by Horai et al. (1995). Consequently, the separate rates for the two hypervariable segments were also too low. Several authors (e.g. Bonatto and Salzano 1997) have since then employed these biased 'slow' rates.

The age estimates from Horai et al. (1995) for the roots of L3 and the entire haplogroup L of all present-day human mtDNAs rescaled to a human–chimp mtDNA split of 6.5 million years ago (as assumed by Mishmar et al. 2003, allowing for recent fossil discoveries and some lineage sorting) would yield 93 000 and 190 000 years ago. Using the reciprocal rate of 5140 years per coding-region substitution from Mishmar et al. (2003), we calculate 79 000 and 195 000 years ago, respectively, when basing the calculation on the sequence data of Ingman et al. (2000) and employing the ρ estimator (without correction for multiple hits). For the data set of Macaulay et al. (2005) the corresponding values are slightly larger, viz. 84 000 and 202 000 years ago; the ML estimates in the latter case gave almost the same point estimates: 84 000 and 205 000 years ago.

8 Conclusion

A satisfactory specification of the parameters governing the stochastic process that describes the evolution of mtDNA in humans has not yet been achieved. In particular, there is a need for more specific investigation that determines the amount of asymmetry in positional mutation rates and violations of independency of mutations at different positions (Howell et al. 2003). For the time being, this process is perhaps best described in rather simple terms, which require the estimation of only a few parameters, with stretches of sequence (such as long C stretches) that seem to evolve in a mode different from most of the rest of the mtDNA genome discarded. One certainly cannot claim that this reflects the real evolution adequately, but the current hope is that the basic assumptions are rather robust against minor violations. It has not yet been convincingly demonstrated that the molecular clock for human mtDNA, embedded in such a simplistic framework, would be violated drastically, so that age estimation would go astray. The extreme form of weighting that only accepts the coding region but rejects the entire control region is at best provisional and certainly not recommendable in the long run. An informed strategy would use rules to decide on a site-by-site basis and contrast synonymous with non-synonymous mutations.

Finally, to investigate to what extent selection (in particular, purifying selection) plays a role, one would need to screen a very large number of coding-region sequences; for a beginning, see Moilanen and Majamaa (2003), Elson et al. (2004), Vilmi et al. (2005), Kivisild et al. (2006), and Chap. 7. Earlier claims by Mishmar et al. (2003) that selection has left a geographic/climatic imprint on human mtDNA variation are unfounded, because the approach draws upon pairs of mtDNA lineages that are not independent, so significance cannot be maintained. The problems of the molecular clock and selection, however, do not seem to be the most acute ones for mtDNA research in regard to human evolution. Rather, the use of simplistic models from population genetics in combination with a continuous stream of technical flaws and biases renders published age estimates quite hazardous, allowing 'end-users' from anthropology, for example, simply to pick out the ages that serve the story they wish to tell, no matter how technically wrong the dating method might be. Calibration of the molecular clock using distant outgroup information, such as that derived from the human–chimp split, is not fully satisfactory. Improved calibration may come from calibrated radiocarbon dates in favourable pioneer-settlement situations with a well-defined founder mtDNA scenario and a rich archaeological record.

Acknowledgements We thank Toomas Kivisild, Antonio Salas, and Yong-Gang Yao for critical advice. We are grateful to Chiara Rengo, Rosaria Scozzari, and Antonio Torroni for information about unpublished RFLP data.

References

Achilli A, Rengo C, Magri C, Battaglia V, Olivieri A, Scozzari R, Cruciani F, Zeviani M, Briem E, Carelli V, Moral P, Dugoujon J-M, Roostalu U, Loogväli E-L, Kivisild T, Bandelt H-J, Richards M, Villems R, Santachiara-Benerecetti AS, Semino O, Torroni A (2004) The molecular dissection of mtDNA haplogroup H confirms that the Franco-Cantabrian glacial refuge was a major source for the European gene pool. Am J Hum Genet 75:910–918

Allard MW, Miller K, Wilson MR, Monson KL, Budowle B (2002) Characterization of the Caucasian haplogroups present in the SWGDAM forensic mtDNA data set for 1771 human control region sequences. J Forensic Sci 47:1215–1223

Allard MW, Wilson MR, Monson KL, Budowle B (2004) Control region sequences for East Asian individuals in the Scientific Working Group on DNA analysis methods forensic mtDNA data set. Legal Med 6:11–24

Allard MW, Polanskey D, Miller K, Wilson MR, Monson KL, Budowle B (2005) Characterization of human control region sequences of the African American SWGDAM forensic mtDNA data set. Forensic Sci Int 148:169–179

Andrews RM, Kubacka I, Chinnery PF, Lightowlers RN, Turnbull DM, Howell N (1999) Reanalysis and revision of the Cambridge reference sequence for human mitochondrial DNA. Nat Genet 23:147

Árnason E (2003) Genetic heterogeneity of Icelanders. Ann Hum Genet 67:5–16

Baasner A, Schäfer C, Junge A, Madea B (1998) Polymorphic sites in human mitochondrial DNA control region sequences: population data and maternal inheritance. Forensic Sci Int 98:169–178

Bandelt H-J, Forster P (1997) The myth of bumpy hunter-gatherer mismatch distributions. Am J Hum Genet 61:980–983

Bandelt H-J, Parson W (2004) Fehlerquellen mitochondrialer DNS-Datensätze und Evalution der mtDNS-Datenbank "D-Loop-BASE". Rechtsmedizin 14:251–257

Bandelt H-J, Forster P, Sykes BC, Richards MB (1995) Mitochondrial portraits of human populations using median networks. Genetics 141:743–753

Bandelt H-J, Forster P, Röhl A (1999) Median-joining networks for inferring intraspecific phylogenies. Mol Biol Evol 16:37–48

Bandelt H-J, Macaulay V, Richards M (2000) Median networks: speedy construction and greedy reduction, one simulation, and two case studies from human mtDNA. Mol Phylogenet Evol 16:8–28

Bandelt H-J, Alves-Silva J, Guimarães P, Santos M, Brehm A, Pereira L, Coppa A, Larruga JM, Rengo C, Scozzari R, Torroni A, Prata MJ, Amorim A, Prado VF, Pena SDJ (2001) Phylogeography of the human mitochondrial haplogroup L3e: a snapshot of African prehistory and Atlantic slave trade. Ann Hum Genet 65:549–563

Bandelt H-J, Quintana-Murci L, Salas A, Macaulay V (2002) The fingerprint of phantom mutations in mtDNA data. Am J Hum Genet 71:1150–1160

Bandelt H-J, Herrnstadt C, Yao Y-G, Kong Q-P, Kivisild T, Rengo C, Scozzari R, Richards M, Villems R, Macaulay V, Howell N, Torroni A, Zhang Y-P (2003a) Identification of Native American founder mtDNAs through the analysis of complete mtDNA sequences: some caveats. Ann Hum Genet 67:512–524

Bandelt H-J, Macaulay V, Richards M (2003b) What molecules can't tell us about the spread of languages and the Neolithic. In: Bellwood P, Renfrew C (eds) Examining the farming/language dispersal hypothesis. McDonald Institute for Archaeological Research, Cambridge, pp 99–111

Bandelt H-J, Achilli A, Kong Q-P, Salas A, Lutz-Bonengel S, Sun C, Zhang Y-P, Torroni A, Yao Y-G (2005a) Low "penetrance" of phylogenetic knowledge in mitochondrial disease studies. Biochem Biophys Res Commun 333:122–130

Bandelt H-J, Kong Q-P, Parson W, Salas A (2005b) More evidence for non-maternal inheritance of mitochondrial DNA? J Med Genet 42:957–960

Bonatto SL, Salzano FM (1997) Diversity and age of the four major mtDNA haplogroups, and their implications for the peopling of the New World. Am J Hum Genet 61:1413–1423

Brandstätter A, Peterson CT, Irwin JA, Mpoke S, Koech DK, Parson W, Parsons TJ (2004) Mitochondrial DNA control region sequences from Nairobi (Kenya): inferring phylogenetic parameters for the establishment of a forensic database. Int J Legal Med 118:294–306

Brandstätter A, Sänger T, Lutz-Bonengel S, Parson W, Béraud-Colomb E, Wen B, Kong Q-P, Bravi CM, Bandelt H-J (2005) Phantom mutation hotspots in human mitochondrial DNA. Electrophoresis 26:3414–3429

Britton T, Oxelman B, Vinnersten A, Bremer K (2002) Phylogenetic dating with confidence intervals using mean path-lenghts. Mol Phylogenet Evol 24:58–65

Bromham L, Penny D (2003) The modern molecular clock. Nat Rev Genet 4:216–224

Chen YS, Torroni A, Excoffier L, Santachiara-Benerecetti AS, Wallace DC (1995) Analysis of mtDNA variation in African populations reveals the most ancient of all human continent-specific haplogroups. Am J Hum Genet 57:133–149

Coble MD (2004) The identification of single nucleotide polymorphisms in the entire mitochondrial genome to increase the forensic discrimination of common HV1/HV2 types in the Caucasian population. PhD thesis, George Washington University

Coble MD, Just RS, O'Callaghan JE, Letmanyi IH, Peterson CT, Irwin JA, Parsons T (2004) Single nucleotide polymorphisms over the entire mtDNA genome that increase the power of forensic testing in Caucasians. Int J Legal Med 118:137–146

Deng H-W, Fu Y-X (2000) Counting mutations by parsimony and estimation of mutation rate variation across nucleotide sites—a simulation study. Math Comput Model 32:83–95

Dennis C (2003) Error reports threaten to unravel databases of mitochondrial DNA. Nature 421:773–774

Elson JL, Turnbull DM, Howell N (2004) Comparative genomics and the evolution of human mitochondrial DNA: assessing the effects of selection. Am J Hum Genet 74:229–238

Eswaran V, Harpending H, Rogers AR (2005) Genomics refutes an exclusively African origin of humans. J Hum Evol 49:1–18

Excoffier L, Yang Z (1999) Substitution rate variation among sites in mitochondrial hypervariable region I of humans and chimpanzees. Mol Biol Evol 16:1357–1368

Farris JS (1970) Methods for computing Wagner trees. Syst Zool 19:83–92

Finnilä S, Lehtonen MS, Majamaa K (2001) Phylogenetic network for European mtDNA. Am J Hum Genet 68:1475–1484

Forster P (2003) To err is human. Ann Hum Genet 67:2–4

Forster P, Harding R, Torroni A, Bandelt H-J (1996) Origin and evolution of Native American mtDNA variation: a reappraisal. Am J Hum Genet 59:935–945

Friedlaender J, Schurr T, Gentz F, Koki G, Friedlaender F, Horvat G, Babb P, Cerchio S, Kaestle F, Schanfield M, Deka R, Yanagihara R, Merriwether DA (2005) Expanding Southwest Pacific mitochondrial haplogroups P and Q. Mol Biol Evol 22:1506–1517

Gattermann N (2004) Mitochondrial DNA mutations in the hematopoietic system. Leukemia 18:18–22

Gillespie JH (1991) The causes of molecular evolution. Oxford University Press, New York

Gusfield D (1997) Algorithms on strings, trees, and sequences. Cambridge University Press, Cambridge

Hagelberg E (2003) Recombination or mutation rate heterogeneity? Implications for Mitochondrial Eve. Trends Genet 19:84–90

Handt O, Meyer S, von Haeseler A (1998) Compilation of human mtDNA control region sequences. Nucleic Acids Res 26:126–129

Hasegawa M, Di Rienzo A, Kocher TD, Wilson AC (1993) Toward a more accurate time scale for the human mitochondrial DNA tree. J Mol Evol 37:347–354

He L, Luo L, Proctor SJ, Middleton PG, Blakely EL, Taylor RW, Turnbull DM (2003) Somatic mitochondrial DNA mutations in adult-onset leukaemia. Leukemia 17:2487–2491

Hedrick P, Kumar S (2001) Mutation and linkage disequilibrium in human mtDNA. Eur J Hum Genet 9:969–972

Herrnstadt C, Elson JL, Fahy E, Preston G, Turnbull DM, Anderson C, Ghosh SS, Olefsky JM, Beal MF, Davis RE, Howell N (2002) Reduced-median-network analysis of complete mitochondrial DNA coding-region sequences for the major African, Asian, and European haplogroups. Am J Hum Genet 70:1152–1171 (erratum 71:448–449)

Herrnstadt C, Preston G, Howell N (2003) Errors, phantom and otherwise, in human mtDNA sequences. Am J Hum Genet 72:1585–1586

Ho SY, Phillips MJ, Cooper A, Drummond AJ (2005) Time dependency of molecular rate estimates and systematic overestimation of recent divergence times. Mol Biol Evol 22:1561–1568

Horai S, Hayasaka K, Kondo R, Tsugane K, Takahata N (1995) Recent African origin of modern humans revealed by complete sequences of hominoid mitochondrial DNAs. Proc Nat Acad Sci USA 92:532–536

Howell N, Smejkal CB (2000) Persistent heteroplasmy of a mutation in the human mtDNA control region: hypermutation as an apparent consequence of simple-repeat expansion/contraction. Am J Hum Genet 66:1589–1598

Howell N, Kubacka I, Mackey DA (1996) How rapidly does the human mitochondrial genome evolve? Am J Hum Genet 59:501–509

Howell N, Smejkal CB, Mackey DA, Chinnery PF, Turnbull DM, Herrnstadt C (2003) The pedigree rate of sequence divergence in the human mitochondrial genome: there is a difference between phylogenetic and pedigree rates. Am J Hum Genet 72:659–670

Howell N, Elson JL, Turnbull DM, Herrnstadt C (2004) African haplogroup L mtDNA sequences show violations of clock-like evolution. Mol Biol Evol 21:1843–1854

Ingman M (2001) Mitochondrial DNA clarifies human evolution. American Institute of Biological Sciences. http://www.actionbioscience.org/evolution/ingman.html

Ingman M, Gyllensten U (2003) Mitochondrial genome variation and evolutionary history of Australian and New Guinean aborigines. Genome Res 13:1600–1606

Ingman M, Kaessmann H, Pääbo S, Gyllensten U (2000) Mitochondrial genome variation and the origin of modern humans. Nature 408:708–713

Kivisild T, Tolk H-V, Parik J, Wang Y, Papiha SS, Bandelt H-J, Villems R (2002) The emerging limbs and twigs of the East Asian mtDNA tree. Mol Biol Evol 19:1737–1751 (erratum 20:162)

Kivisild T, Reidla M, Metspalu E, Rosa A, Brehm A, Pennarun E, Parik J, Geberhiwot T, Usanga E, Villems R (2004) Ethiopian mitochondrial DNA heritage: tracking gene flows across and around the Strait of Tears. Am J Hum Genet 75:752–770

Kivisild T, Shen P, Wall D, Do B, Sung R, Davis K, Passarino G, Underhill PA, Scharfe C, Torroni A, Scozzari R, Modiano D, Coppa A, de Knijff P, Feldman M, Cavalli-Sforza LL, Oefner PJ (2006) The role of selection in the evolution of human mitochondrial genomes. Genetics 172:373–387

Kong Q-P, Yao Y-G, Sun C, Bandelt H-J, Zhu C-L, Zhang Y-P (2003) Phylogeny of East Asian mitochondrial DNA lineages inferred from complete sequences. Am J Hum Genet 73:671–676 (erratum 75:157)

Lutz-Bonengel S, Schmidt U, Schmitt T, Pollak S (2003) Sequence polymorphisms within the human mitochondrial genes MTATP6, MTATP8 and MTND4. Int J Legal Med 117:133–142

Maca-Meyer N, González AM, Larruga JM, Flores C, Cabrera VC (2001) Major genomic mitochondrial lineages delineate early human expansions. BMC Genet 2:13

Macaulay VA, Richards MB, Forster P, Bendall KE, Watson E, Sykes BC, Bandelt H-J (1997) mtDNA mutation rates—no need to panic. Am J Hum Genet 61:983–986

Macaulay V, Hill C, Achilli A, Rengo C, Clarke D, Meehan W, Blackburn J, Semino O, Scozzari R, Cruciani F et al. (2005) Single, rapid coastal settlement of Asia revealed by analysis of complete mitochondrial genomes. Science 308:1034–1036

Malyarchuk BA, Rogozin IB (2004) Mutagenesis by transient misalignment in the human mitochondrial DNA control region. Ann Hum Genet 68:324–339

Merilä J, Björklund M, Baker AJ (1997) Historical demography and present day population structure of the Greenfinch, Carduelis chloris—an analysis of mtDNA control-region sequences. Evolution 51:946–956

Meyer S, von Haeseler A (2003) Identifying site-specific substitution rates. Mol Biol Evol 20:182–189

Meyer S, Weiss G, von Haeseler A (1999) Pattern of nucleotide substitution and rate heterogeneity in the hypervariable regions I and II of human mtDNA. Genetics 152:1103–1110

Mishmar D, Ruiz-Pesini E, Golik P, Macaulay V, Clark AG, Hosseini S, Brandon M, Easley K, Chen E, Brown MD, Sukernik RI, Olckers A, Wallace DC (2003) Natural selection shaped regional mtDNA variation in humans. Proc Nat Acad Sci USA 100:171–176

Moilanen JS, Majamaa K (2003) Phylogenetic network and physicochemical properties of nonsynonymous mutations in the protein-coding genes of human mitochondrial DNA. Mol Biol Evol 20:1195–1210

Monson KL, Miller KWP, Wilson MR, DiZinno JA, Budowle B (2002) The mtDNA population database: an integrated software and database resource for forensic comparison. Forensic Sci Commun 4(2). http://www.fbi.gov/hq/lab/fsc/backissu/april2002/miller1.htm

Morral N, Bertranpetit J, Estivill X, Nunes V, Casals T, Giménez J, Reis A et al. (1994) The origin of the major cystic fibrosis mutation (ΔF508) in European populations. Nat Genet 7:169–175

Nasidze I, Stoneking M (2001) Mitochondrial DNA variation and language replacements in the Caucasus. Proc R Soc Lond Ser B 268:1197–1206

Nei M, Kumar S (2000) Molecular evolution and phylogenetics. Oxford University Press, Oxford

Ozawa T, Tanaka M, Sugiyama S, Ino H, Ohno K, Hattori K, Ohbayashi T, Ito T, Deguchi H, Kawamura K, Nakane Y, Hashiba K (1991) Patients with idiopathic cardiomyopathy belong to the same mitochondrial DNA gene family of Parkinson's disease and mitochondrial encephalomyopathy. Biochem Biophys Res Commun 177:518–525

Pääbo S (1996) Mutational hot spots in the mitochondrial microcosm. Am J Hum Genet 59:493–496

Pakendorf B, Stoneking M (2005) Mitochondrial DNA and human evolution. Annu Rev Genomics Hum Genet 6:165–183

Palanichamy Mg, Sun C, Agrawal S et al. (2004) Phylogeny of mtDNA macrohaplogroup N in India based on complete sequencing: implications for the peopling of South Asia. Am J Hum Genet 75:966–978

Parsons TJ, Muniec DS, Sullivan K, Woodyatt N, Alliston-Grenier R, Wilson MR, Berry DL, Holland KA, Weedn VW, Gill P, Holland MM (1997) A high observed substitution rate in the human mitochondrial DNA control region. Nat Genet 15:363–368

Penny D (2005) Evolutionary biology: relativity for molecular clocks. Nature 436:183–184

Pesole G, Saccone C (2001) A novel method for estimating substitution rate variation among sites in a large dataset of homologous DNA sequences. Genetics 157:859–865

Purvis A, Bromham L (1997) Estimating the transition/transversion ratio from independent pairwise comparisons with an assumed phylogeny. J Mol Evol 44:112–119

Rosenberg NA, Hirsh AE (2003) On the use of star-shaped genealogies in inference of coalescence times. Genetics 164:1677–1682

Saillard J, Forster P, Lynnerup N, Bandelt H-J, Nørby S (2000) mtDNA variation among Greenland Eskimos: the edge of the Beringian expansion. Am J Hum Genet 67:718–726

Salas A, Richards M, De la Fe T, Lareu M-V, Sobrino B, Sánchez-Diz P, Macaulay V, Carracedo Á (2002) The making of the African mtDNA landscape. Am J Hum Genet 71:1082–1111

Salas A, Yao Y-G, Macaulay V, Vega A, Carracedo Á, Bandelt H-J (2005) A critical reassessment of the role of mitochondria in tumorigenesis. PLoS Med 2:e296

Santos C, Montiel R, Sierra B, Bettencourt C, Fernandez E, Alvarez L, Lima M, Abade A, Aluja MP (2005) Understanding differences between phylogenetic and pedigree-derived mtDNA mutation rate: a model using families from the Azores Islands (Portugal). Mol Biol Evol 22:1490–1505

Semple C, Steel M (2003) Phylogenetics. Oxford University Press, Oxford

Sigurdarðóttir S, Helgason A, Gulcher JR, Stefansson K, Donnelly P (2000) The mutation rate in the human mtDNA control region. Am J Hum Genet 66:1599–1609

Steel M (2002) Some statistical aspects of the maximum parsimony method. In: DeSalle R, Giribet G, Wheeler W (eds) Molecular systematics and evolution: theory and practice. Birkhäuser, Basel, pp 125–139

Steel M, Penny D (2000) Parsimony, likelihood, and the role of models in molecular phylogenetics. Mol Biol Evol 17:839–850

Strandberg AKK, Salter LA (2004) A comparison of methods for estimating the transition:transversion ratio from DNA sequences. Mol Phylogen Evol 32:495–503

Strimmer K, von Haeseler A (1996) Quartet puzzling: a quartet maximum-likelihood for reconstructing tree topologies. Mol Biol Evol 13:964–969

Stumpf M, Goldstein DB (2001) Genealogical and evolutionary inference with the human Y chromosome. Science 291:1738–1742

Tanaka M, Cabrera VM, González AM et al (2004) Mitochondrial genome variation in eastern Asia and the peopling of Japan. Genome Res 14:1832–1850

Templeton AR (1997) Testing the Out of Africa replacement hypothesis with mitochondrial DNA data. In: Clark GA, Willermet CM (eds) Conceptual issues in modern human origins research. de Gruyter, New York, pp 329–360

Templeton AR (2002) Out of Africa again and again. Nature 416:45–51

Thalmann O, Serre D, Hofreiter M, Lukas D, Eriksson J, Vigilant L (2005) Nuclear insertions help and hinder inference of the evolutionary history of gorilla mtDNA. Mol Ecol 14:179–188

Torroni A, Rengo C, Guida V, Cruciani F, Sellitto D, Coppa A, Calderon FL, Simionati B, Valle G, Richards M, Macaulay V, Scozzari R (2001) Do the four clades of the mtDNA haplogroup L2 evolve at different rates? Am J Hum Genet 69:1348–1356

Vilmi T, Moilanen JS, Finnilä S, Majamaa K (2005) Sequence variation in the tRNA genes of human mitochondrial DNA. J Mol Evol 60:587–597

Wakeley J (1993) Substitution rate variation among sites in hypervariable region 1 of human mitochondrial DNA. J Mol Evol 37:613–623

Watson E, Forster P, Richards M, Bandelt H-J (1997) Mitochondrial footprints of human expansions in Africa. Am J Hum Genet 61:691–704

Wilder JA, Mobasher Z, Hammer MF (2004) Genetic evidence for unequal effective population sizes of human females and males. Mol Biol Evol 21:2047–2057

Yang Z (1997) PAML: a program package for phylogenetic analysis by maximum likelihood. CABIOS 13:555–556

Yao Y-G, Macaulay V, Kivisild T, Zhang Y-P, Bandelt H-J (2003b) To trust or not to trust an idiosyncratic mitochondrial data set. Am J Hum Genet 72:1341–1346; 1348–1349

Zupanič Pajnič I, Balažic J, Komel R (2004) Sequence polymorphism of the mitochondrial DNA control region in the Slovenian population. Int J Legal Med 118:1–4

Nucleic Acids and Molecular Biology, Vol. 18
Hans-Jürgen Bandelt, Vincent Macaulay, Martin Richards (Eds.)
Human Mitochondrial DNA and the Evolution of *Homo sapiens*

Postmortem Damage of Mitochondrial DNA

M. Thomas P. Gilbert

Ancient DNA and Evolution, Niels Bohr Institute, University of Copenhagen,
Juliane Maries vej 30, 2100 Copenhagen, Denmark
mtpg@gfy.ku.dk

1 Introduction

It may come as a surprise to many readers that a book whose contents focus on mitochondrial evolution contains a chapter on the postmortem damage that mitochondrial DNA (mtDNA) undergoes. The mitochondria of deceased organisms are not subject to natural selection; thus, they do not evolve, and have not undergone any changes of interest to the population geneticist. So how is this chapter justified?

An insight into, and an appreciation of, postmortem damage may be viewed as useful for two main reasons. The more obvious of the two is the growing use of mtDNA as an archaeogenetic tool, through the study of ancient DNA (aDNA). Researchers in this field pit their technical wits against degraded archaeological and museum specimens in an attempt to tease out the last residues of DNA. This enables them to address a variety of otherwise unanswerable anthropological, archaeological, evolutionary, historical, and other biological questions. More recently, the study of postmortem damage has also provided novel insights into mutational processes within the mitochondria that are normally masked by their innate repair processes. In this chapter I hope to bring to the reader's attention an understanding of the problems damage brings to the aDNA researcher and the molecular phylogeneticist, the general biochemical processes involved in postmortem damage, how they are not so different from many of those involved in the in vitro process of mutation, and finally some of the findings that these insights have given us.

2 Ancient DNA

The study of postmortem damage and its effects on mtDNA first became of interest to scientists with the rise of the field of aDNA in the late 1980s.

Early successes on relatively young samples, including museum-preserved specimens of extinct mammals such as the quagga (Higuchi et al. 1984) and marsupial wolf (Thomas et al. 1989) were quickly eclipsed as an onset of techniques enabling both the retrieval of tiny quantities of DNA and its rapid and relatively accurate sequencing led the field to expand rapidly. In the early days of the discipline, reports as incredible as the successful retrieval of DNA from specimens that were millions of years old were commonplace. However this was not to last—unfortunately all such studies from the early 1990s dating to over one million years old have since been shown either to be impossible to replicate or to derive from identifiable sources of contamination (Austin et al. 1997a, 1997b; Sidow et al. 1991; Stankiewicz et al. 1998; Zischler et al. 1995). It is not the place of this chapter to recount the history of the field, but those readers who are interested are recommended to see Hofreiter et al. (2001b) for a very balanced review of the highs and lows that it has gone through.

As the field has matured it has become apparent that many of the early dreams were overoptimistic, and there is a natural limitation for aDNA studies owing to the postmortem degradation of DNA. This rate of decay is so fast that, based on calculated deamination and depurination kinetics for the four nucleotides, it has been estimated that under physiological salt conditions, neutral pH, and an ambient temperature of 15 °C, 100 000 years is a likely estimate of the time beyond which DNA will be degraded beyond retrieval (Lindahl 1993). However, it is possible that optimal conditions, such as preservation in ice or permafrost, may extend this period by 2 or 3 times. Conversely a good example of a much shorter time frame for survival is demonstrated by Marota et al. (2002). These authors calculate, using the decay rate of chloroplast DNA in papyri (and by inference human DNA in similar conditions), that DNA at Egyptian archaeological sites has a half-life of 19–24 years, which suggests an upper limit of 672 years on the survival of authentic DNA. What is quite clear from this and other studies is that an assessment of the likelihood of DNA retrieval from ancient specimens must take into account the degradation processes that they have undergone.

3
DNA Damage in Ancient Samples

3.1
DNA Degradation Immediately Following Cell Death

Exceptional circumstances apart, the degradation of endogenous DNA (e.g. that belonging to a sample of interest) commences shortly following its death. In humans, within 4–5 min of the death, cell autolysis commences

(Vass 2001). As the cells of the body are deprived of oxygen, carbon dioxide in the blood increases, pH decreases, and wastes accumulate which poison the cells. Concomitantly, unchecked cellular enzymes, including lipases, proteases, amylases, and nucleases, begin to dissolve the cell from the inside out. Soon the cells rupture, releasing nutrient-rich fluids that encourage the growth of internal and environmental microorganisms (mainly bacteria, fungi, and protozoa) involved in the further putrefaction of the remains. These contribute to further degradation of the DNA as they spread through the corpse. From the molecular biologist's point of view, despite the fact that most human diploid cells contain several billion bases of nuclear DNA, and thousands of copies of mtDNA, its decay is often so fast that within months, if not weeks, no PCR-amplifiable template remains. Nevertheless, in exceptional circumstances, this degradation can be significantly reduced or halted altogether. Such conditions, believed to destroy/inactivate the nucleases and/or inhibit the action of microorganisms, include rapid desiccation (e.g. natural mummies), low temperatures (e.g. ice mummies or samples buried in permafrost), and high salt concentrations (Hofreiter et al. 2001b) as well as those in which the flesh is quickly removed from the host, thus limiting the substrate that microorganisms thrive on and subsequent putrefaction (e.g. through mild cooking, through being eaten or through very rapid degradation of the flesh).

3.2 Long-term DNA Degradation

In such scenarios, slower processes of DNA degradation become important. While the histone proteins incorporated into nuclear DNA can be expected to offer some protection from damage, their absence in mitochondria renders it very susceptible to a range of biochemical attack (Poinar 2002). These biochemical modifications are believed to be analogous to those seen in vivo, and act via both the cross-linking and fragmentation of the molecule's chemical backbone and the alteration of individual nucleotide bases. Although the exact contributions of individual processes to the damage will vary with the direct environment surrounding specimens (Gilbert et al. 2003a) a brief summary of those believed to be important is presented in the following subsections. The reader should bear in mind that various situations exist that are of interest to the aDNA researcher but that are not discussed here, in which relatively unusual or artificial chemicals are present, and as a result other forms of damage may be predominant and present. For those interested, Rogers et al. (2000) provide a good discussion on the effect of preservation in amber on aDNA, while Vachot et al. (1996) and Douglas and Rogers (1998) address the issue of the effects of common formalin-derived museum fixatives.

3.2.1 DNA Fragmentation

A seminal study by one of the field's most respected names, Svante Pääbo (1989), still provides much of what is known today about DNA degradation. Although his analysis of the DNA retrieved from various desiccated mummy samples demonstrated that they contained not inconsiderable yields of double-stranded DNA, attempts at PCR amplifications on the DNA resulted in only small fragments of product (if any). This phenomenon is commonly found in extracts from ancient samples, and is believed to arise following cross-linkage of the DNA backbone (inhibiting enzymatic amplification), or complete double-strand breakage due to exposure of the sample to various forms of radiation.

Further modifications precluding amplification of DNA arise as a result of hydrolytic and oxidative damage (Fig. 1). Diesters, such as the bonds in the phosphate sugar backbone, are subject to quick hydrolytic cleavage, resulting in single-strand nicks, and are estimated as the most common form of hydrolytic damage that the DNA molecule must cope with (Greer and Zamenhof 1962; Shapiro 1981). Other structures within the molecule are also at risk from hydrolytic attack. Deoxyribose sugar lacks the 2′-OH bond that is present in ribose, which in turn weakens the glycosidic bond joining the bases to the sugars. This susceptibility to hydrolytic attack (in particular of the purines, adenine and guanine) may result in depurination into a baseless (apurinic) site via base protonation (Lindahl and Anderson 1972; Lindahl and Nyberg 1972). Such sites rapidly undergo cleavage, producing additional sources of single-strand nicks. Pyrimidines (cytosine and thymine) are also susceptible to such reactions, forming apyrimidinic sites, though at rates in vivo as much as 100–500 times slower than purines (Lindahl and Nyberg 1972; Lindahl and Karlström 1973; Shaaper et al. 1983).

The notion that hydrolytic damage relies to a degree on the presence of water may suggest that archaeological samples that have remained fairly dry are more protected from such damage; however, in such situations it is likely that oxidative modifications play a significant role (Poinar 2002). As in living systems, a large proportion of oxidative damage is believed to occur through the action of free radicals, such as the hydroxyl radical ($OH^{\bullet}$), peroxide radicals ($O_2^{\bullet}$), and hydrogen peroxide (H_2O_2) (Rogan and Salvo 1992; Lindahl 1993). These may arise from exogenous sources such as ionising radiation, UV light, as well as cellular processes that arise during bacterial and fungal degradation (Poinar 2002). Free radicals attack the integrity of the DNA molecule by adding to either C5 or C6 of pyrimidines via their shared double bond, or to C4, C5, and C8 of purines (Fig. 1). The radicals thus created are unstable and often go through a series of reactions in the presence of O_2 to form other reactive radicals. One such example is thymine, which once attacked by $OH^{\bullet}$ produces thymine peroxyl radicals. As with the hydrolytic

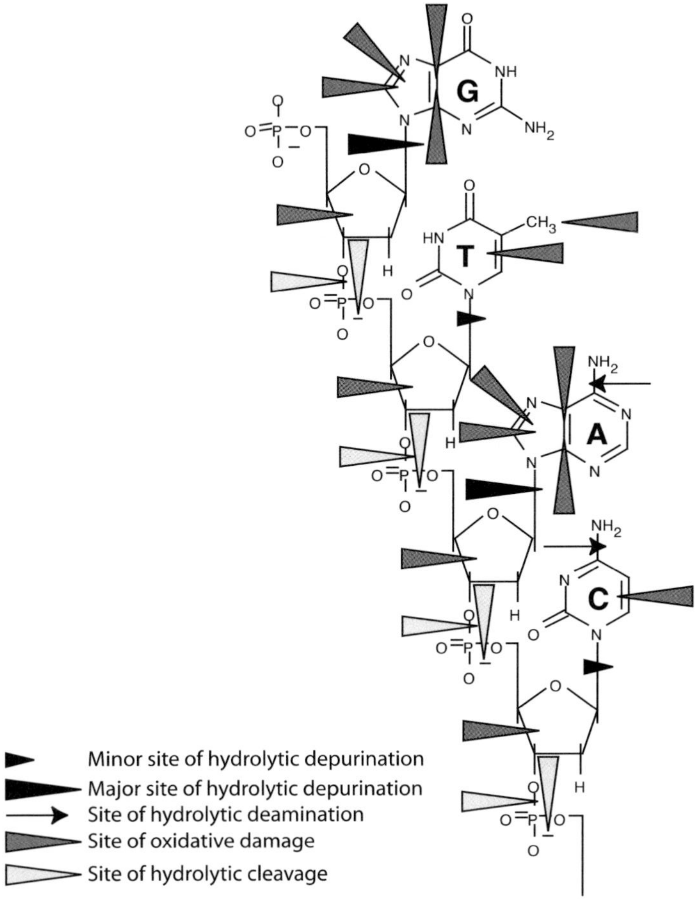

Fig. 1 Potential sites of biochemical degradation on the DNA molecule. (Reprinted and modified with permission from Lindahl 1993, copyright 1993 Nature Publishing Group; Hofreiter et al. 2001b, copyright 2001 Nature Publishing Group; Poinar 2002, copyright 2002 American Chemical Society)

damage discussed, the net result of the presence of such radicals is further fragmentation of the DNA molecule (Halliwell 1991). The sugar backbone can also be attacked by $OH^{\cdot}$ through removal of hydrogen atoms from any of the five carbons. Oxidative damage is also believed to affect the DNA molecule through the modification of nucleotides to hydantoins. Using gas chromatography/mass spectrometry (GC/MS), Höss et al. (1996) detected the occurrence of two predominant forms: the oxidised pyrimidines 5-hydroxy-5-methylhydantoin (5-OH-5-MeHyd), a major derivative of thymine following exposure to γ-radiation, and 5-hydroxyhydantoin (5-OH-Hyd). Unlike other forms of oxidative and hydrolytic damage, the authors suggested that these do not prevent DNA amplification through the fragmentation of the DNA, but simply block the polymerase during PCR.

3.2.2
Implications of DNA Fragmentation

To summarise, amplifiable aDNA is in short supply. It would thus seem logical (even if not always practised) that the first question any self-respecting aDNA researcher should ask themselves is: Does the sample of interest contain any DNA? Unfortunately, although the opportunities for a sample to undergo DNA damage are substantial, the extent of each form (and thus the resulting modifications) is closely linked to the diagenetic environment of the sample. Without this knowledge it becomes very difficult to predict exactly what damage DNA within a sample has undergone, or whether DNA is even present in the sample at all. However, several general rules of thumb can be outlined.

In the first place, DNA preservation correlates with archaeological site. A second rule is that an estimate of DNA preservation between any two unrelated samples cannot be predicted using a simple age correlate. It has been repeatedly demonstrated that unless two samples from an individual archaeological site are considered (and even then a correlation is not always guaranteed) age of sample does not correlate with state of DNA preservation (Pääbo 1989; Höss et al. 1996; Gilbert et al. 2003a). These rules are quite clear, and have prompted various methods of indirectly assessing the preservation of samples. One such, which has been demonstrated as providing reasonable predictions as to DNA survival, directly takes into account the biochemical limitation of temperature. This is the so-called thermal age of the sample, an age calculation taking into account the preservation temperature of the sample (Smith et al. 2001). Other, more indirect methods that have been used with some degree of success to estimate DNA preservation involve correlates such as the frequency of water change (Nielsen-Marsh 2000), and microbial content (Burger et al. 1999) or biochemical data such as amino acid racemisation (Poinar et al. 1996; Poinar and Stankiewicz 1999), composition (Bada et al. 1999), and levels of GC/MS-measured hydantoin bases (Höss et al. 1996).

Such DNA degradation presents three primary problems to the field of aDNA. Firstly, researchers are limited as to what they can recover. mtDNA, present at many thousand times the copy number of nuclear DNA, is thus a much easier template to recover (relatively). In many of the tissues that survive (e.g. bones, teeth, hair), pathogen DNA will be present at even lower copy numbers than nuclear DNA; thus, it will be even harder to retrieve. Secondly, owing to the previously described fragmentation and cross-linking, researchers are rarely able to amplify templates of more than several hundred base pairs in size (Pääbo 1989; Cooper and Poinar 2000; Hofreiter et al. 2001b). This naturally places limitations on what can be achieved in the field—to recover the complete genome of even recently extinct and well-preserved animals, such as permafrost-preserved mammoths, would take an awful lot of amplifications!

The third, and arguably the most serious, problem is that of sample contamination. The low quantity of aDNA extracted from a sample can easily be 'swamped' by the relatively much higher amount of DNA in external contaminants of the same or similar DNA sequence. For example, an archaeologist handling an old human femur without wearing protective gloves will coat the specimen in a layer of skin cells and bathe the specimen in a small quantity of sweat from his or her hands. Alternatively, an ancient sample exposed to a laboratory containing previously amplified human DNA will also quickly become contaminated. Without proper treatment to remove the foreign DNA, it is a very simple (and unfortunately common) matter to coamplify both sources of DNA. Depending on the relative concentrations, the host DNA, and hence the sequence produced, can be either completely swamped or modified so as to result in erroneous data.

3.3
Damage-Driven DNA Miscoding Lesions

Unfortunately the inconveniences that arise for those hunting aDNA as a result of DNA damage are not simply those preventing the preservation of an amplifiable template. Observed initially by Pääbo (1989), via the molecular cloning of amplified aDNA templates, further forms of DNA degradation (termed miscoding lesions) manifest themselves as modifications to the actual DNA sequence itself. Although Pääbo first demonstrated the presence of these modifications using enzymatic treatment of his aDNA samples, it is only recently that the full extent of the modifications has been realised, and the biochemistry behind the changes studied in detail. This is partially a result of the innate nature of PCR enzymes to incorporate a low frequency of 'incorrect' base pairings during amplification, ranging from approximately 2×10^{-4} to less than 1×10^{-5} per nucleotide cycle (Hansen et al. 2001). Recently, the advent of so-called hi-fidelity enzymes, which amplify with much lower misincorporation rates than 'standard' polymerases (even in the environment of ancient extracted samples) (Gilbert et al. 2003a, b), has enabled the identification and characterisation of two very common complementary pairs of base modification.

The most common damage-driven base changes observed are the four transitions cytosine to thymine (C→T), guanine to adenine (G→A), thymine to cytosine (T→C), and adenine to guanine (A→G) (Hansen et al. 2001; Gilbert et al. 2003a). However, because of the complementary nature of DNA, each of these observations can be explained by two possible causative events (Hofreiter et al. 2001a). Figure 2 demonstrates this for an observed C→T transition on the L strand. Owing to an original damage event on

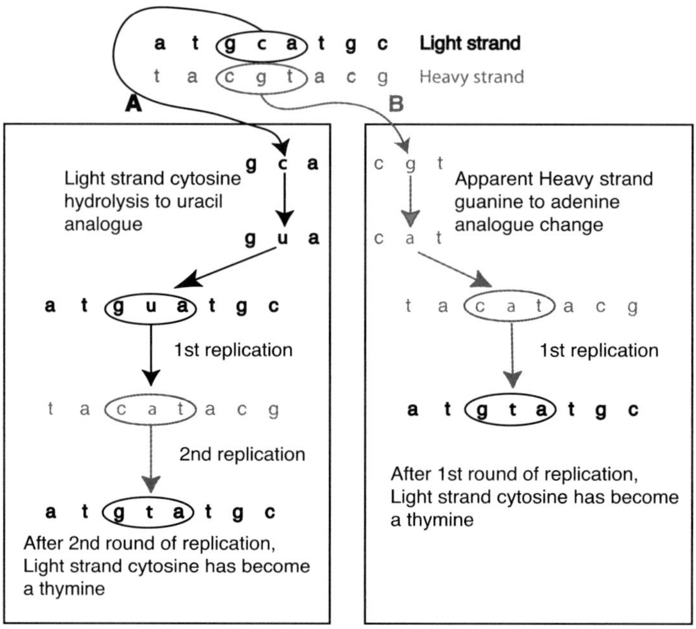

Fig. 2 Determination of a strand of origin for postmortem DNA damage events by using type 2 (C→T/G→A) transitions as an example. **A** L-strand C→T transitions after two cycles of amplifications, resulting in a permanent L-strand change. **B** A theoretical H-strand G→A change, producing the L-strand phenotype of C→T change following one cycle of amplification. However, since a direct G→A postmortem modification is chemically impossible, the example depicted is not possible. Thus, all C→T changes observed on the L strand must have occurred as L-strand C→T postmortem damage, and all G→A changes on the L strand must have occurred as H-strand C→T postmortem damage. (Reprinted and modified with permission from Gilbert et al. 2003a, copyright 2003 University of Chicago Press)

the L strand causing a C→U (uracil, a thymine analogue) transition, after two stages of replication we observe the C→T transition on the L strand (Fig. 2a). However, this phenotype can also occur via an H-strand G→A transition, which after one PCR cycle is observed as the C→T transition on the L strand (Fig. 2b). The same problem applies to each of the transitions, so that each postmortem biochemical change can result in two observed outcomes, depending on which strand is sequenced. Owing to this complementarity, Hansen et al. (2001) have termed A→G and T→C changes as type 1 transitions (A→G/T→C) and C→T and G→A changes as type 2 transitions (C→T/G→A), to indicate the uncertainty about which base was originally damaged.

While this situation appears intractable, a solution is offered by the possible biochemical pathways through which nucleotide damage can occur.

Hofreiter et al. (2001a) and Gilbert et al. (2003a) demonstrated that the postmortem damage-driven modification of guanine to an adenine analogue is highly unlikely, if not impossible. This excludes the possibility of the events shown in Fig. 2b. It can therefore be argued that any G→A transition observed on the L strand due to damage must have originated as an H-strand C→T modification event (actually arising as a hydrolytic deamination of cytosine to uracil), because G→A modification on the L strand is impossible. Conversely, any C→T transition observed on the L strand will actually have originated as an L-strand C→T modification event. A similar argument can be applied to type 1 damage using the discovery that in the postmortem environment the modification of T→C analogues is biochemically unlikely (Gilbert et al. 2003a). In this situation, any L-strand T→C modification will actually be due to an H-strand A→G event (actually arising following the hydrolytic deamination of adenine to hypoxanthine, a guanine analogue), while L-strand A→G events can be attributed to an original A→G damage event on the L strand.

These discoveries that the majority of postmortem transitions arise owing to only two biochemical modifications are rather surprising, as in vivo and in vitro studies of several polymerases have shown that a major oxidative derivative of thymine, 5-formyluracil (fU), has the capacity to pair with adenine, thymine, guanine, or cytosine (Yoshida et al. 1997; Zhang et al. 1997; Fujikawa et al. 1998; Zhang et al. 1999). The pairing of fU : G, producing a T→C modification, has also been demonstrated as the most common of these mispairings (Ånensen et al. 2001). However, experiments on ancient samples examined so far have demonstrated that, if present at all, this oxidative reaction is responsible for a negligible proportion of modifications (Gilbert et al. 2003a).

In vivo and in vitro studies have demonstrated that in comparable conditions the rate of deamination of cytosine is over 20 times that of guanine (Lindahl 1979; Karran and Lindahl 1980), and thus it is intuitive to expect that cytosine-dependent type 2 transitions will accumulate much faster. However, analyses of postmortem samples have demonstrated that the rate of accumulation of the breakdown products, and hence miscoding lesions, depends very much on the direct environment that the sample has experienced (Gilbert et al. 2003a). For example, the common environmental mutagens nitrous acid and bisulfite will preferentially deaminate cytosine over guanine (Schuster 1960; Lindahl and Nyberg 1974; Lindahl 1979). Cytosine is also more susceptible to heat-induced deamination than guanine (Shapiro and Klein 1966; Notari 1967), though using the assumption that each cytosine and adenine deamination is associated with similar activation energies, it is predicted that adenine deamination will persist at lower temperatures than cytosine deamination (Karran and Lindahl 1980). The pH of the immediate environment also affects each base differently (Jones 1966) and adds a further variable to the equations. This environmental variability has contributed to the observed phenomenon that with this (and other relevant forms of damage) little correlation can be seen with the ages and preservation of samples (Pääbo 1989; Höss et al. 1996; Gilbert et al. 2003a). So without information on the complete background of the sample and its environment, can anything be determined? It appears that, in general, for low amounts of damage, it is difficult to predict the predominance of either reaction, but as total deamination damage increases, an overall bias to type 2 modifications is observed (Hansen et al. 2001; Gilbert et al. 2003a).

3.3.1 Miscoding Lesions and Sequence Modification

The presence of modified nucleotides in samples containing low template numbers can be manifested as the serious problem of sequence misrepresentation. As demonstrated in Fig. 3, a single miscoding lesion on one template molecule in a PCR reaction can, once amplified, result in a sequence that is easily misinterpreted. These erroneous sequences may, in turn, lead to the overestimation of heterogeneity and other parameters (Lundstrom et al. 1992; Aris-Brisou and Excoffier 1996) in ancient populations. It is also possible that such modifications may help explain sequence 'anomalies' observed in various modern mtDNA data sets. Bandelt et al. (2001) provide a detailed list of many flawed data sets that contain sequences that have been suggested or shown to contain errors, with explanations for how they could have arisen. Many errors arise following human or computing error, and others are suspected to arise owing to biochemical parameters in the amplification and sequencing reactions (Chap. 6).

While it may at first seem that postmortem damage provides an excellent explanation for sequences containing spurious transitions (Stenico et al. 1996; Seo et al. 1998), it is unlikely that it is responsible in most cases for the previously described reason of template copy number. Most modern studies are based on DNA extractions that contain high enough numbers of templates to 'swamp' any transitions that might have arisen by damage. However, there are exceptions, including studies on tissues that both inherently contain very low levels of DNA and contain no DNA repair mechanisms, such as hair shafts or nails. In these cases it is quite plausible that heteroplasmy may be observed, derived from what is, in effect, postmortem DNA damage, combined with the low template numbers associated with such extractions.

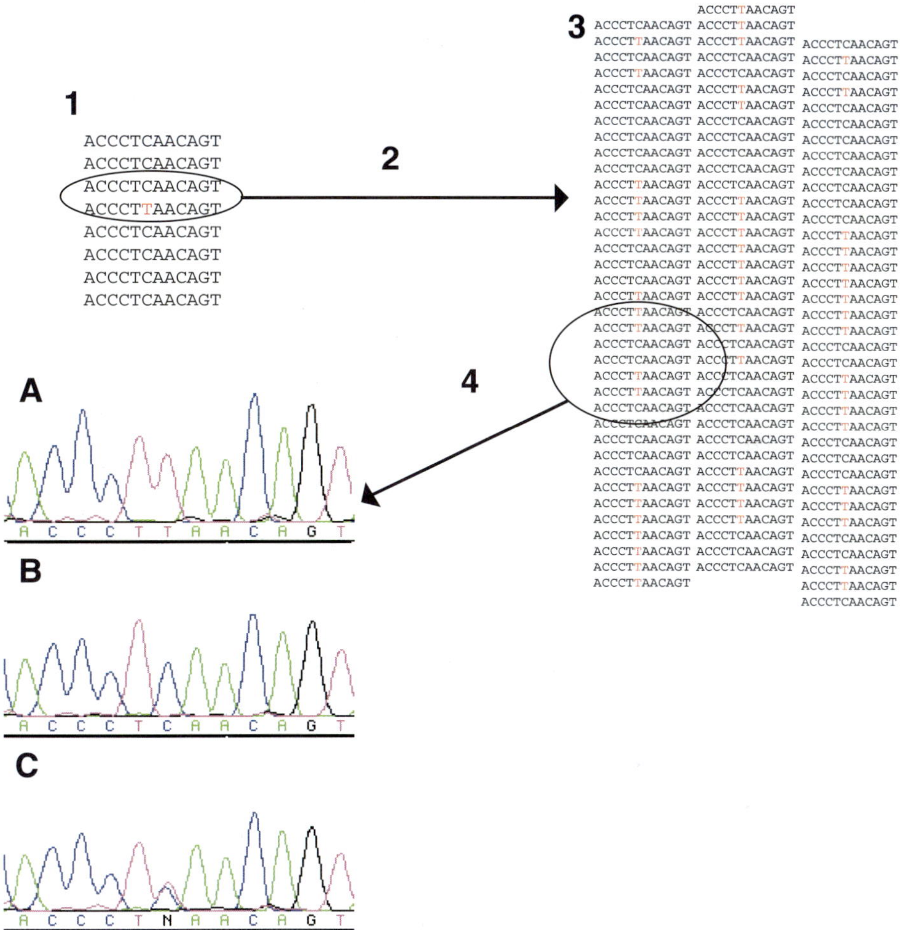

Fig. 3 Modification of sequences through miscoding lesions. *1* Ancient sample yields a low number of template molecules, one of which has undergone a C→T postmortem sequence modification. *2* Low template number results in a PCR starting from only two amplifiable molecules, one of which is the damaged sequence. *3* All amplicons derive from one of the two founder templates. *4* Sequencing reaction on a subset of this amplified product may result in three possible scenarios: *A* the number of 'original' templates in the reaction are significantly outnumbered by the 'modified' templates, yielding the incorrect sequence; *B* the number of 'original' templates in the reaction significantly outnumber the 'modified' templates, yielding the correct sequence; *C* a heterozygous phenotype is observed. Base may be called at the analyst's discretion or reported as heteroplasmy finding

4 Insights from Miscoding Lesions into in Vivo mtDNA Mutation

4.1 Mutational Hotspots and mtDNA Recombination

Miscoding lesion findings have not only raised an awareness of possible flaws in mtDNA sequence data sets, but they have also provided us with insights into the mutational processes that are important within the living mitochondrial genome. As described in Chap. 4, certain nucleotide positions within the mitochondrial hypervariable segment I (HVS-I), termed here as 'sites', appear to mutate in vivo at significantly higher rates than others. These sites, identified using a range of analytical techniques and data sets, have been described in the literature numerous times (e.g. Vigilant 1989; Hasegawa and Horai 1991; Hasegawa et al. 1993; Wakeley 1993; Aris-Brisou and Excoffier 1996; Macaulay et al. 1997; Excoffier and Yang 1999; Meyer et al. 1999; Stoneking 2000; Finnilä et al. 2001; Heyer et al. 2001; Allard et al. 2002, 2004, 2005; Forster et al. 2002; Malyarchuk et al. 2002, 2004) and are termed mutational 'hotspots'. Considerable debate exists as to exactly which sites are hotspots or not (Bandelt et al. 2002), or whether they even exist (Hagelberg 2003), and the importance of their existence has also been of fundamental interest in the intellectual battle over the presence or absence of recombination in the mitochondrial genome (Stoneking et al. 2001). At the root of this argument lies the fact that the majority of such hotspots have been identified on the basis of their occurrence as homoplasies on phylogenetic trees (Bandelt et al. 2002). As a result, critics argue that if the topology of the tree changes, then so do the hotspots. For example, although the base at Cambridge reference sequence (CRS) site 16325 is in some studies deemed as conservative (Excoffier and Yang 1999; Meyer et al. 1999), other authors suggest that more acceptable phylogenies demonstrate that it in fact mutates at above-average rates (Bandelt et al. 2002). These authors also remark that some studies are also biased in that mutations that happened early in the mtDNA phylogeny are all shifted into the high-rate categories (Chap. 4).

One simple solution to this debate is offered through 'real-time' studies of mitochondrial mutation, for example, using familial lines where offspring can be directly compared with living parents, or other such methods (Bendall et al. 1996; Mumm et al. 1997; Parsons et al. 1997; Sigurðardóttir et al. 2000; Heyer et al. 2001; Forster et al. 2002; Howell et al. 2003). Unfortunately, however, the mutation rate of mtDNA is still too slow to provide enough data points to provide a satisfactory model of mitochondrial hotspots. For example, in the study of Forster et al. (2002) on background-radiation-induced mutations in over 980 human samples, only 22 mutations were observed over the HVS-I and HVS-II regions. In an attempt to provide the larger data sets necessary to resolve this debate, researchers have turned to the analysis of the distribution of postmortem damage in ancient samples. While this may at first seem a bizarre step to take, this method has several advantages over the conventional studies.

Firstly, as already discussed, the hydrolytic deamination processes that lead to the generation of type 1 and type 2 transitions after death are fundamentally the same as some of the most common processes that generate point mutations in mtDNA in vivo. However, the lack of DNA repair capability and protection afforded to the DNA in vivo, such as base-excision repair and possibly mismatch repair (Bogenhagen 1999; Dianov et al. 2001; Mason et al. 2003), coupled with the long time spans over which the DNA is exposed to potentially damaging agents quickly enable researchers to develop a large data base of damage.

4.2 Effects of mtDNA Secondary and Tertiary Structure

Secondly, variation in rates of damage has been suggested to relate to DNA secondary and tertiary structure (Heyer et al. 2001; Gilbert et al. 2003b). For example, it is plausible that DNA secondary structural conformation predisposes particular sites to hydrolytic attack. While in vivo the process would normally be under the influence of selection, with two or three areas of exception, the HVS-I is not believed to contain any sequences of functional importance, and hence is not selected for. If this is the case, then it seems logical that such sites may appear both hypermutable in modern populations and common hotspots for postmortem damage.

4.3 Hotspots for Postmortem mtDNA Damage

Is there any evidence for such hotspots of postmortem damage? An initial study on the phenomena in ancient cave bear (*Ursus spelaeus*) sequences by Hofreiter et al. (2001a) found no statistically supported evidence for their existence. However in a much larger data set of the HVS-I (bases 16055–16410) of ancient human remains, strong support was found for their existence (Gilbert et al. 2003b). Hotspots have also been found in a further, as yet unpublished, study examining the distribution of postmortem damage in 81 ancient bison (*Bos bison*) individuals (Gilbert et al. 2005).

As expected, the distribution of the human hotspots across the region of interest (Fig. 4) holds many similarities to those identified in the earlier studies. Although, when comparing modern and ancient data sets, difficulties arise owing to the methods of calculation and the measurements used to identify hotspots, some striking similarities are noted between mutational and damage hotspots. Table 1 contains reduced data sets from ten studies identifying mutation hotspots through phylogenetic methods (Hasegawa et al. 1993;

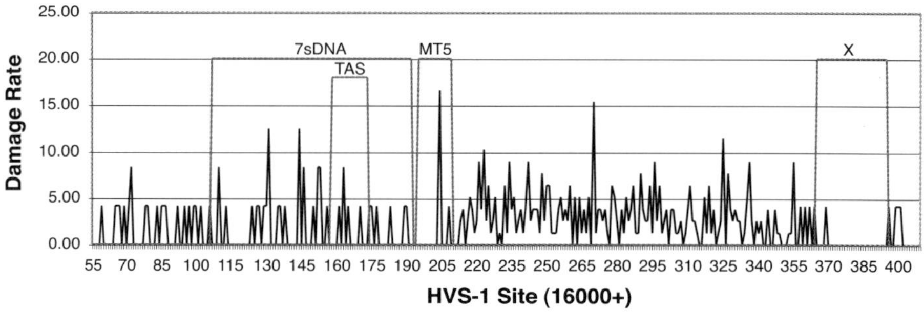

Fig. 4 Postmortem damage variation across HVS-I. Numbering is with reference to the Cambridge reference sequence (*CRS*; Anderson et al. 1981). Damage is measured as a relative rate across HVS-I, calculated as in Gilbert et al. (2003b). Four areas of structural interest are marked: *7sDNA*, the termination associated sequence (*TAS*), the putative control element *MT5*, and the LDR1 region of low variation (*cross*) postulated by Gilbert et al. (2003b). (Reprinted and modified with permission from Gilbert et al. 2003b, copyright 2003 University of Chicago Press)

Wakeley 1993; Excoffier and Yang 1999; Meyer et al. 1999; Allard et al. 2002, 2004, 2005; Malyarchuk et al. 2002; Malyarchuk and Rogozin 2004; Macaulay, www.stats.gla.ac.uk/~vincent/fingerprint/table.html; Chap. 4), one study examining postmortem damage hotspots (Gilbert et al. 2003b), and the cumulative number of sites identified as mutating in seven familial studies—that is, through the direct observation of their generation through comparisons between related individuals (Bendall et al. 1996; Mumm et al. 1997; Parsons et al. 1997; Sigurðardóttir et al. 2000; Heyer et al. 2001; Forster et al. 2002; Howell et al. 2003). (For full details, refer to the table legend.)

Of the 50 sites presented, only three are reported as mutating or receiving damage at very high rates in all data sets (16189, 16311, and 16362). In contrast, nine of 17 of the sites identified by Gilbert et al. (2003b) as damage hotspots are not found in any of the other data sets. While this figure appears high, it should also be noted that eight of 17 sites identified as mutational hotspots in familial studies are also not found in any of the other phylogenetic studies, and the majority of the sites identified as hypermutable in the phylogenetic studies are not supported by all the studies. Also of interest is the fact that two sites (16290 and 16298) are only identified as changing rapidly in the postmortem damage and familial data sets.

It is likely that these disagreements relate to various factors. These include sampling stochasticity, the generation of innate biases by the methods used to create the modern data (Bandelt et al. 2002), and the standardisation approaches employed here to harmonise the results (generated through many different processes). In addition, concerns have recently been voiced about what appears to be the frequent generation of artificial data at particular human HVS-I nucleotide positions in DNA sequenced on ABI377 sequencing

Table 1 Comparison of mutation and postmortem damage hotspots

Phylogenetic studies Site	JW93	MH93	LE99	SM99	BM02	MA02	MA04	BM04	VM04	MA05	MG03	Familials	Familial data source
16072	—	—	—	—	—	—	—	—	—	—	3	—	—
16085	—	—	—	—	—	—	—	—	—	—	3	—	—
16092	—	—	—	—	—	—	—	—	—	—	—	1	TP97
16093	—	—	4	3	3	3	—	4	4	3	4	5	LF02, NH03, SS00
16110	—	—	—	—	—	—	—	—	—	—	4	—	—
16111	—	—	—	—	—	—	—	—	—	—	—	1	SS00
16126	—	—	—	4	—	—	—	—	—	—	3	—	—
16129	—	4	4	4	3	3	—	4	—	—	—	—	—
16131	—	—	—	—	—	—	—	—	—	—	3	—	—
16144	—	—	—	—	—	—	—	—	—	—	3	—	—
16145	—	—	—	—	—	—	—	3	—	—	—	—	—
16148	—	—	—	3	—	—	—	—	—	—	—	—	—
16163	—	—	—	3	—	—	—	—	—	—	3	—	—
16172	—	—	4	3	—	3	3	3	—	—	—	—	—
16182	—	—	4	—	—	—	—	—	—	—	—	—	—
16183	—	—	4	3	—	3	—	—	—	—	—	—	—
16189	4	4	4	4	4	4	3	4	3	3	—	3	LF02
16192	—	—	4	3	—	4	—	3	—	—	—	1	KB96
16204	—	—	—	—	—	—	—	—	—	—	4	—	—
16209	—	—	4	—	—	—	—	—	—	—	—	—	—
16219	—	—	—	3	—	—	—	—	—	—	—	—	—
16222	—	—	—		—	—	—	—	—	—	—	1	KB96
16223	4	3	4	4	—	—	—	—	—	—	4	1	LF02

Table 1 (continued)

Phylogenetic studies Site	JW93	MH93	LE99	SM99	BM02	MA02	MA04	BM04	VM04	MA05	MG03	Familials	Familial data source
16230	—	—		4	—	—	—	—	—	—	—	—	—
16234	—	—	3		—	—	—	—	—	—	—	—	—
16239	—	—	—	—	—	—	—	—	—	—	—	1	KB96
16242	—	—	—	—	—	—	—	—	—	—	3	—	—
16256	—	—	—	—	—	3	—	—	—	—	—	1	KB96
16261	—	—	—	—	—	3	—	—	—	—	—	—	—
16262	—	—	—	—	—	—	—	—	—	—	3	1	TP97
16265	—	—	4	—	—	—	—	—	—	—	—	—	—
16270	—	—	4	3	—	—	—	—	—	—	4	—	—
16272	—	—	—	—	—	—	—	—	—	—	—	1	KB96
16274	—	—	—	3	—	—	—	—	—	—	—	—	—
16278	—	—	4	4	—	3	—	—	—	—	—	—	—
16290	—	—	—	—	—	—	—	—	—	—	3	1	SM97
16291	—	—	4		—	3	—	3	—	—	—	1	LF02
16293	—	—	4	4	—	—	—	—	—	—	—	1	LF02
16294	3	3	4	4	—	—	—	—	—	—	—	—	—
16298	—	—	—	—	—	—	—	—	—	—	4	1	KB96
16304	—	—	4	—	—	—	—	—	—	—	—	—	—
16309	—	—	4	4	—	—	—	—	—	—	—	—	—
16311	4	4	4	4	3	4	4	4	4	3	—	1	EH01
16319	—	—	—	3	—	—	—	—	—	—	—	—	—
16320	—	—	3	—	—	—	—	—	—	—	—	1	EH01

Table 1 (continued)

Phylogenetic studies Site	JW93	MH93	LE99	SM99	BM02	MA02	MA04	BM04	VM04	MA05	MG03	Familials	Familial data source
16325	—	—	—	—	—	—	—	—	—	—	4	—	—
16327	—	—	—	—	—	—	—	—	—	—	3	—	—
16343	—	—	3	—	—	—	—	—	—	—	—	—	—
16355	—	—	3	—	—	—	—	—	—	—	—	—	—
16362	4	4	4	4	3	4	3	4	3	3	—	—	—

Site-specific in vivo mutation rates taken from ten phylogenetic-based studies (JW93, MH93, LE99, SM99, BM02, MA02, MA04, BM04, VM04, MA05) were standardised into quartiles. Those sites falling into the upper two quartiles, *3* and *4*, are compared with similarly standardised postmortem damage rates as calculated in Gilbert et al. (2003b), as well as direct observations (measured as occurrences) taken from familial studies (i.e. mutations observed as parent–offspring differences). Owing to the variety of mutation rate estimation techniques used between the studies, the results are not directly comparable, but can be used to illustrate sites that are mutating or receiving damage at high rates. Details of the studies are as follows: BM02 Malyarchuk et al. (2002), BM04 Malyarchuk and Rogozin (2004), EH01 Heyer et al. (2001), JW93 Wakeley (1993), KB96 Bendall et al. (1996), LE99 Excoffier and Yang (1999), LF02 Forster et al. (2002), MA02 Allard et al. (2002), MA04 Allard et al. (2004), MA05 Allard et al. (2005), MG03 Gilbert et al. (2003b), MH93 Hasegawa et al. (1993), NH03 Howell et al. (2003), SM97 Mumm et al. (1997), SM99 Meyer et al. (1999), SS00 Sigurðardóttir et al. (2000), TP97 Parsons et al. (1997), VM04 Macaulay (Chap. 4) machines—particularly positions 16085, 16131, and 16270 (Brandstätter et al. 2005). As some of the aforementioned data (including a proportion of the damage data) were generated on such machines, some caution should be excised when considering the results.

4.4 Sequence Motifs with Limited DNA Damage

Within the postmortem damage data of Gilbert et al. (2003b), several other interesting findings were discovered that hint further to the importance of secondary structure of the mtDNA in base-specific in vivo mutation. Meyer et al. (1999) have remarked on the correlation of HVS-I site-specific mutation rates with the three known features of structural interest within HVS-I. The first, 7sDNA, is a short fragment of H-strand DNA which provides the mitochondrial D-loop with its characteristic triple-stranded structure, extending from a trinucleotide stop codon at 16104 to at least 110 (Doda et al. 1981; Meyer et al. 1999). The stop codon itself exhibits neither postmortem damage nor high mutation rates in this study or in the data from previous studies (Meyer et al. 1999; Excoffier and Yang 1999), and the 7sDNA itself appears to mutate and receive postmortem damage at average rates.

Two other HVS-I regions with known function exhibit quite a different story. The termination associated sequence (TAS), and the putative control element MT5 have low observed in vivo mutation rates, which have previously been suggested to result from functional constraints (Meyer et al. 1999). TAS is located upstream of the trinucleotide stop codon, and interacts with sequence-specific binding factors (Doda et al. 1981; Wallace et al. 1995), while MT5 is postulated as a protein binding site (Meyer et al. 1999). In their dataset, Gilbert et al. (2003b) observed that both TAS and MT5 appear to display lower-than-average postmortem damage rates (Fig. 4). Certain proteins have been shown to survive for much longer time scales than DNA (Bada et al. 1999), and it is possible that the low amount of postmortem damage observed in the MT5 region is related to the continued binding and protection offered by the putative control element.

However, the story does not stop there. Surprisingly, Gilbert et al. (2003b) also identified a previously unidentified fourth region of low postmortem damage, LDR1, between positions 16365 and 16395 of the HVS-I (Fig. 4). Although most comparable HVS-I mutation-rate studies do not include this entire region, three studies do (Allard et al. 2002, 2004, 2005). Encouragingly, these studies also appear to characterise relatively low rates of in vivo mutation at this 5′ segment, with regards to the rest of HVS-I. The absence of postmortem damage over such a large region (and over 600 clones) is unusual, and suggests that the DNA sequence may be protected from hydrolytic damage in some way. One possibility would be a protein–DNA association, similar to that postulated for MT5 (Meyer et al. 1999), and if further studies

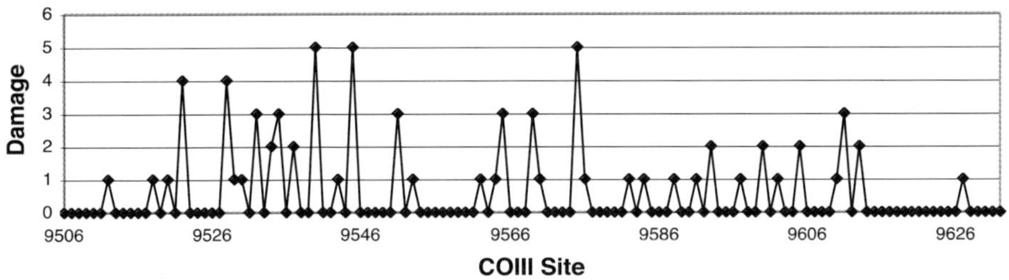

Fig. 5 Absolute damage variation across a section of mitochondrial COIII (9500–9632). Numbering is with reference to the CRS (Anderson et al. 1981). (Reprinted and modified with permission from Gilbert et al. 2003b, copyright 2003 University of Chicago Press)

verify this observation, LDR1 may be the first structural feature identified by means of aDNA studies.

A further insight into innate mitochondrial predispositions to mutational processes is enabled through the relaxation of the constraints imposed by natural selection following death. Hotspots for postmortem damage are noted by Gilbert et al. (2003b) in a coding-region segment of the human mtDNA, cytochrome oxidase subunit III (COIII), between CRS positions 9506 and 9632 (Fig. 5). Unsurprisingly, their postmortem damage results do not show the selection against first and second codon position mutations that is seen in coding-region sequences of modern populations. Thus, the ancient sequences may actually reflect the innate (i.e. prerepair and selection) mutation rates of COIII sites. Oddly, in both the humans and bear species studied, despite the fact that postmortem damage undergoes no selection, the damage rate in the HVS-I was still higher than in the coding regions examined (COIII in humans, apocytochrome *b*, Cyt *b*, in bears). This corresponds to the observations of Finnilä et al. (2001) that in vivo mutation rates of mitochondrial third codon positions are also lower than those of the HVS-I. A plausible explanation is that the secondary structural conformation of the HVS-I promotes increased rates of both in vivo mutation and postmortem damage, whereas in the coding region this is lacking, or there has been some selection to constrain mutation rates. Structural models of the human HVS-I may provide a useful test of this hypothesis.

5
Implications of Postmortem Damage Hotspots on Sequence Authenticity

The discovery that postmortem damage may be modifying human aDNA sequences at the very same HVS-I sites that are used to identify mitochondrial haplogroups has important implications. Current aDNA human phylogenetic research principally focuses on the identification of mtDNA haplogroups through either restriction digests (Merriwether et al. 1994; Kaestle and Glenn

Smith 2001) or, more commonly now, sequence analysis (Hänni et al. 1994; Krings et al. 1997, 1999; Stone and Stoneking 1998; Adcock et al. 2001). I have already mentioned how, owing to low template copy numbers in ancient extracts, sufficient levels of hydrolytic deamination may lead to the determination of an incorrect mitochondrial sequence. If one accepts the hypothesis that a disproportionate amount of this damage occurs at specific, phylogenetically important, hotspots, then it stands to reason that not only is the sequence modified, but potentially in such a manner as to artificially mislead some phylogeneticists. It is not inconceivable that a Native Amerindian haplogroup D HVS-I sequence could result from a haplogroup H Viking sample, through a few select base changes. The implications of such a misinterpretation are not hard to imagine!

These findings should not be misinterpreted as implying that all such ancient human studies are at fault. For the reasons detailed earlier (Fig. 3), well-preserved specimens are at little risk (e.g. Ötzi, the 'Ice Man' of the Tyrol; Handt et al. 1994). And if low numbers of templates are present, other techniques help prevent falling into a misidentification trap. Even though damage hotspots exist, it is very unlikely that two templates within a low-yield extraction will be damaged in exactly the same spots. Thus, simple steps such as examining multiple extractions and amplifications per sample, combined with cloning, can help identify the correct endogenous sequence.

An alternative, seemingly simple, method is to follow the lead of many of the phylogenetic studies that are based on modern mtDNA, and coamplify coding-region markers in the mtDNA along with HVS-I (Endicott et al. 2003); however, this is a luxury that is not easily afforded to the aDNA researcher. Limited by damage to small fragments of DNA, it requires the painstaking, and if correct aDNA authentication techniques are used (Cooper and Poinar 2000), expensive, amplification of a large number of the individual regions containing the single nucleotide polymorphisms of interest, in order to accurately diagnose a haplogroup. For example, in their analysis of ancient Andaman islander teeth, in order to correctly identify the mitochondrial haplogroups of the inhabitants, Endicott et al. (2003) deemed it necessary to both amplify and clone three independent fragments of aDNA template per sample.

6 Conclusion

If present at all, endogenous DNA within an ancient sample will have experienced two general damage processes, analogous to those seen in vivo. The quantity of potential PCR template and the size of the amplifiable fragment will be limited by cross-linking, radiation-induced double-strand breaks, hydrolytic and oxidative depurination, and the oxidative formation of hydantoins. In addition to this, even when amplification is successful, the DNA sequence is likely to contain miscoding lesions as a result of further hydrolytic and oxidative damage. While these processes are a hindrance to the aDNA researcher, they have also provided us with new insights into the significance of secondary structure on in vivo mitochondrial mutation, and hence mitochondrial evolution. As further studies into mitochondrial postmortem degradation are undertaken, it can be hoped that they will help resolve both the hotly contested debate on mitochondrial heteroplasmy, and the presence and significance of mitochondrial mutational hotspots.

References

Adcock G, Dennis E, Easteal S, Huttley G, Jermelin L, Peacock W, Thorne A (2001) Mitochondrial DNA sequences in ancient Australians: implications for modern human origins. Proc Natl Acad Sci USA 98:537–542

Allard MW, Miller K, Wilson M, Monson K, Budowle B (2002) Characterization of the Caucasian haplogroups present in the SWGDAM forensic mtDNA dataset for 1771 human control region sequences. Scientific Working Group on DNA Analysis Methods. J Forensic Sci 47:1215–1223

Allard MW, Wilson MR, Monson KL, Budowle B (2004) Control region sequences for East Asian individuals in the Scientific Working Group on DNA Analysis Methods forensic mtDNA data set. Legal Med (Tokyo) 6:11–24

Allard MW, Polanskey D, Miller K, Wilson MR, Monson KL, Budowle B (2005) Characterization of human control region sequences of the African American SWGDAM forensic mtDNA data set. Forensic Sci Int 148:169–179

Anderson S, Bankier A, Arrell B, de Bruijn M, Coulson A, Drouin J, Eperon I, Nierlich D, Roe B, Sanger F, Schreier P, Smith A, Staden R, Young I (1981) Sequence and organisation of the human mitochondrial genome. Nature 290:457–465

Aris-Brisou S, Excoffier L (1996) The impact of population expansion and mutation rate heterogeneity on DNA sequence polymorphism. Mol Biol Evol 13:494–504

Austin JJ, Ross AJ, Smith AB, Fortey RA, Thomas RH (1997a) Problems of reproducibility—does geologically ancient DNA survive in amber-preserved insects? Proc R Soc Lond Ser B 264:467–474

Austin JJ, Smith AB, Thomas RH (1997b) Paleontology in a molecular world: the search for authentic ancient DNA. Trends Ecol Evol 12:303–306

Bada J, Wang X, Hamilton H (1999) Preservation of key biomolecules in the fossil record: current knowledge and future challenges. Philos Trans R Soc Lond Ser B 354:77–87

Bandelt H-J, Lahermo P, Richards M, Macaulay V (2001) Detecting errors in mtDNA data by phylogenetic analysis. Int J Legal Med 115:64–69

Bandelt H-J, Quintana-Murci L, Salas A, Macaulay V (2002) The fingerprint of phantom mutations in mitochondrial DNA data. Am J Hum Genet 71:1150–1160

Bendall KE, Macaulay VA, Baker JR, Sykes BC (1996) Heteroplasmic point mutations in the human mtDNA control region. Am J Hum Genet 59:1276–1287

Bogenhagen DF (1999) Repair of mtDNA in vertebrates. Am J Hum Genet 64:1276–1281

Brandstätter A, Sänger T, Lutz-Bonengel S, Parson W, Béraud-Colomb E, Wen B, Kong Q-P, Bravi CM, Bandelt H-J (2005) Phantom mutation hotspots in human mitochondrial DNA. Electrophoresis 26:3414–3429

Burger J, Hummel S, Herrmann B, Henke W (1999) DNA preservation: a microsatellite-DNA study on ancient skeletal remains. Electrophoresis 20:1722–1728
Cooper A, Poinar H (2000) Ancient DNA: do it right or not at all. Science 289:1139
Dianov GL, Souza-Pinto N, Nyaga SG, Thybo T, Stevnsner T, Bohr VA (2001) Base excision repair in nuclear and mitochondrial DNA. Prog Nucleic Acids Res Mol Biol 68:285–297
Doda ND, Wright CT, Clayton DA (1981) Elongation of displacement-loop strands in human and mouse mitochondrial DNA is arrested near specific template sequences. Proc Natl Acad Sci USA 10:6116–6120
Douglas MP, Rogers SO (1998) DNA damage caused by common cytological fixatives. Mutat Res 401:77–88
Endicott P, Gilbert MTP, Stringer C, Lalueza-Fox C, Willerslev E, Hansen AJ, Cooper A (2003) The genetic origins of the Andaman islanders. Am J Hum Genet 72:178–184
Excoffier L, Yang Z (1999) Substitution rate variation among sites in mitochondrial hypervariable region I of humans and chimpanzees. Mol Biol Evol 16:1357–1368
Finnilä S, Lehtonen M, Majamaa K (2001) Phylogenetic network for European mtDNA. Am J Hum Genet 68:1475–1484
Forster L, Forster P, Lutz-Bonengel S, Willkomm H, Brinkmann B (2002) Natural radioactivity and human mitochondrial DNA mutations. Proc Natl Acad Sci USA 99:13950–13954
Fujikawa K, Kamiya H, Kasai H (1998) The mutations induced by oxidatively damaged nucleotides, 5-formyl-dUTP and 5-hydroxy-dCTP, in Eshcherichia coli Nucleic Acids Res 26:4582–4587
Gilbert MTP, Hansen AJ, Willerslev E, Rudbeck L, Barnes I, Lynnerup N, Cooper A (2003a) Characterisation of genetic miscoding lesions caused by postmortem damage. Am J Hum Genet 72:48–61
Gilbert MTP, Willerslev E, Hansen AJ, Rudbeck L, Barnes I, Lynnerup N, Cooper A (2003b) Distribution patterns of postmortem damage in human mitochondrial DNA. Am J Hum Genet 72:32–47
Gilbert MTP, Shapiro BA, Drummond A, Cooper A (2005) Post mortem DNA damage hotspots in Bison (Bison bison and B. bonasus) provide supporting evidence for mutational hotspots in human mitochondria. J Archaeol Sci 32:1053–1060
Greer S, Zamenhof S (1962) Studies on depurination of DNA by heat. J Mol Biol 4:123
Hagelberg E (2003) Recombination or mutation rate heterogeneity? Implications for Mitochondrial Eve. Trends Genet 19:84–90
Halliwell B (1991) DNA damage by oxygen-derived species. Its mechanisms and measurement in mammalian systems. FEBS Lett 281:9–19
Handt O, Richards M, Trommsdorff M, Kilger C, Simanainan J, Georgiev O, Bauer K, Stone A, Hedges R, Schaffner W, Utermann G, Sykes B, Pääbo S (1994) Molecular genetic analyses of the Tyrolean Ice Man. Science 264:1775–1778
Hänni C, Laudet V, Coll J, Stehelin D (1994) An unusual mitochondrial DNA sequence variant from an Egyptian mummy. Genomics 22:487–489
Hansen A, Willerslev E, Wiuf C, Mourier T, Arctander P (2001) Statistical evidence for miscoding lesions in ancient DNA templates. Mol Biol Evol 18:262–265
Hasegawa M, Horai S (1991) Time of the deepest root for polymorphism in human mitochondrial DNA. J Mol Evol 32:37–42
Hasegawa M, Di Rienzo A, Kocher T, Wilson A (1993) Towards a more accurate time scale for the human mitochondrial gene tree. J Mol Evol 37:347–354
Heyer E, Zietkiewicz E, Rochowski A, Yotova V, Puymirat J, Labuda D (2001) Phylogenetic and familial estimates of mitochondrial substitution rates: study of control region mutations in deep-rooting pedigrees. Am J Hum Genet 69:1113–1126

Higuchi R, Bowman B, Freiberger M, Ryder OA, Wilson AC (1984) DNA sequences from the quagga, and extinct member of the horse family. Nature 312:282–284

Hofreiter M, Jaenicke V, Serre D, von Haeseler A, Pääbo S (2001a) DNA sequences from multiple amplifications reveal artefacts induced by cytosine deamination in ancient DNA. Nucleic Acids Res 29:4693–4799

Hofreiter M, Serre D, Poinar HN, Kuch M, Pääbo S (2001b) Ancient DNA. Nat Rev Genet 2:353–358

Höss M, Jaruga P, Zastawny T, Dizdaroglu M, Pääbo S (1996) DNA damage and DNA sequence retrieval from ancient tissue. Nucleic Acids Res 24:1304–1307

Howell N, Smejkal CB, Mackay DA, Chinnery PF, Turnbull DM, Herrnstadt C (2003) The pedigree rate of sequence divergence in the human mitochondrial genome: there is a difference between phylogenetic and pedigree rates. Am J Hum Genet 72:659–670

Jones AS, Mian AM, Walker RT (1966) The action of alkali on some purines and their derivatives. J Chem Soc C 692–695

Kaestle F, Glenn Smith D (2001) Ancient mitochondrial DNA evidence for prehistoric population movement: the numic expansion. Am J Phys Anthropol 115:1–12

Karran P, Lindahl T (1980) Hypoxanthine in deoxyribonucleic acid: generation by heat-induced hydrolysis of adenine residues and release in free form by a deoxyribonucleic acid glycosylase from calf thymus. Biochemistry 19:6005–6011

Krings M, Stone A, Schmitz R, Krainitzki H, Stoneking M, Pääbo S (1997) Neanderthal DNA sequences and the origin of modern humans. Cell 90:19–30

Krings M, Geisert H, Schmitz R, Krainitzki H, Pääbo S (1999) DNA sequence of the mitochondrial hypervariable region II from the Neanderthal type specimen. Proc Natl Acad Sci USA 96:5581–5585

Lindahl T (1979) DNA glycosylases, endonucleases for apurinic/apyrimidinic sites and base excision-repair. Prog Nucleic Acid Res Mol Biol 22:135–192

Lindahl T (1993) Instability and decay of the primary structure of DNA. Nature 362:709–715

Lindahl T, Andersson A (1972) Rate of chain breakage at apurinic sites in double-stranded deoxyribonucleic acid. Biochemistry 11:3610–3618

Lindahl T, Karlström O (1973) Heat-induced depyrimidination of deoxyribonucleic acid in neutral solution. Biochemistry 12:5151–5154

Lindahl T, Nyberg B (1972) Rate of depurination of native deoxyribonucleic acid. Biochemistry 11:3610–3618

Lindahl T, Nyberg B (1974) Heat induced deamination of cytosine residues in deoxyribonucleic acid. Biochemistry 12:3405–3410

Lundstrom R, Tavaré S, Ward R (1992) Estimating substitution rates from molecular data using the coalescent. Proc Natl Acad Sci USA 89:5961–5965

Macaulay V, Richards M, Forster P, Bendall K, Watson E, Sykes B, Bandelt H-J (1997) mtDNA mutation rates—no need to panic. Am J Hum Genet 61:983–986

Malyarchuk BA, Rogozin IB (2004) Mutagenesis by transient misalignment in the human mitochondrial DNA control region. Ann Hum Genet 68:324–339

Malyarchuk BA, Rogozin IB, Berikov VB, Derenko MV (2002) Analysis of phylogenetically reconstructed mutational spectra in human mitochondrial DNA control region. Hum Genet 111:46–53

Marota I, Basile C, Ubaldi M, Rollo F (2002) DNA decay rate in papyri and human remains from Egyptian archaeological sites. Am J Phys Anthropol 117:310–318

Mason PA, Matheson EC, Hall AG, Lightowlers RN (2003) Mismatch repair activity in mammalian mitochondria. Nucleic Acids Res 31:1052–1058

Merriwether D, Rothhammer F, Ferrel R (1994) Genetic variation in the New World: ancient teeth, bone, and tissue as sources of DNA. Experientia 50:592–601
Meyer S, Weiss G, von Haeseler A (1999) Pattern of nucleotide substitution and rate heterogeneity in the hypervariable regions I and II of human mtDNA. Genetics 152:1103–1110
Mumm S, Whyte MP, Thakker RV, Buetow KH, Schlessinger D (1997) mtDNA analysis shows common ancestry in two kindreds with X-linked recessive hypoparathyroidism and reveals a heteroplasmic silent mutation. Am J Hum Genet 60:153–159
Nielsen-Marsh C (2000) Patterns of diagenesis in bone I: effects of site environments. J Archeol Sci 27:1139–1150
Notari RE (1967) A mechanism for the hydrolytic deamination of cytosine arabinoside in aquesous buffer. J Pharm Sci 56:804
Pääbo S (1989) Ancient DNA: extraction, characterisation, molecular cloning and enzymatic amplification. Proc Natl Acad Sci USA 86:1939–1943
Parsons TJ, Muciec DS, Sullivan K, Woodyatt N, Alliston-Greiner R, Wilson MR, Berry DL, Holland KA, Weedn VW, Gill P, Holland MM (1997) A high observed substitution rate in the human mitochondrial DNA control region. Nat Genet 15:363–368
Poinar H (2002) The genetic secrets some fossils hold. Acc Chem Res 35:676–684
Poinar H, Stankiewicz B (1999) Protein preservation and DNA retrieval from ancient tissues. Proc Natl Acad Sci USA 96:8426–8431
Poinar H, Höss M, Bada J, Pääbo S (1996) Amino acid racemization and the preservation of ancient DNA. Science 272:864–866
Rogan PK, Salvo J (1992) Study of nucleic acids isolated from ancient remains. Ybk Phys Anthropol 33:195–214
Rogers SO, Langenegger K, Holdenrieder O (2000) DNA changes in tissues entrapped in plant resins (the precursors of amber). Naturwissenschaften 87:70–75
Seo Y, Stradmann-Bellinghausen B, Rittner C, Takahama K, Schneider PM (1998) Sequence polymorphism of mitochondrial DNA control region in Japanese. Forensic Sci Int 97:155–164
Shaaper RM, Kunkel TA, Loeb LA (1983) Infidelity of DNA synthesis associated with bypass of apurinic sites. Proc Natl Acad Sci USA 80:847–891
Shapiro R (1981) Damage to DNA caused by hydrolysis. Seeberg E, Kleppe K (eds) In: Chromosome damage and repair. Plenum, New York, p 3–12
Shapiro HS, Klein RS (1966) The deamination of cytidine and cytosine by acidic buffer solutions. Mutagenic implications. Biochemistry 5:2358–2362
Schuster H (1960) Die Reaktionsweise der Desoxyribonucleinsäure mit salpetriger Säure. Z Naturforsch B 15:298–304
Sidow A, Wilson AC, Pääbo S (1991) Bacterial DNA in Clarkia fossils. Philos Trans R Soc Lond Ser B 333:429–433
Sigurðardóttir S, Helgason A, Gulcher JR, Stefansson K, Donnelly P (2000) The mutation rate in the human mtDNA control region. Am J Hum Genet 66:1599–1609
Smith CI, Chamberlain AT, Riley MS, Cooper A, Stringer CB, Collins MJ (2001) Neanderthal DNA: not just old but old and cold? Nature 410:772–773
Stankiewicz B, Poinar H, Briggs D, Evershed R, Poinar G (1998). Chemical preservation of plants and insects in natural resins. Proc R Soc Lond Ser B 265:641–647
Stenico M, Nigro L, Bertorelle G, Calafell F, Capitanio M, Corrain C, Barbujani G (1996) High mitochondrial sequence diversity in linguistic isolates of the Alps. Am J Hum Genet 59:1363–1375
Stone A, Stoneking M (1998) mtDNA analysis of a prehistoric Oneota population: implications for the peopling of the New World. Am J Hum Genet 62:1153–1170

Stoneking M (2000) Hypervariable sites in the mtDNA control region are mutational hotspots. Am J Hum Genet 67:1029–1032

Thomas RH, Schaffner W, Wilson AC, Pääbo S (1989) DNA phylogeny of the extinct marsupial wolf. Nature 340:465–467

Vachot A-M, Monnerot M (1996) Extraction, amplification and sequencing of DNA from formalin-fixed specimens. Ancient Biomol 1:3–16

Vass AA (2001) Beyond the grave—understanding human decomposition. Microbiol Today 28:109–192

Vigilant L, Stoneking M, Harpending H, Hawkes K, Wilson AC (1991) African populations and the evolution of human mitochondrial DNA. Science 253:1503–1507

Wakeley J (1993) Substitution rate variation among sites in hypervariable region I of human mitochondrial DNA. J Mol Evol 37:613–623

Wallace DC, Lott MT, Brown MD, Huopenen K, Torroni A (1995) Report on the committee on human mitochondrial DNA. Cuticchia AJ (ed) In: Human gene mapping 1995, a compendium. Johns Hopkins University Press, Baltimore, p 910–954

Yoshida M, Makino K, Morita H, Terato H, Ohyama Y, Ide H (1997) Substrate and mispairing properties of 5-formyl-2′-deoxyuridine 5′-triphosphate assessed by in vitro DNA polymerase reactions. Nucleic Acids Res 25:1570–1577

Zhang Q-M, Sugiyama H, Miyabe I, Matsuda S, Saito I, Yonei S (1997) Replication of DNA templates containing 5-formyluracil, a major oxidative lesion of thymine in DNA. Nucleic Acids Res 25:3969–3973

Zhang Q-M, Sugiyama H, Miyabe I, Matsuda S, Kino K, Saito I, Yonei S (1999) Replication in vitro and cleavage by restriction endonuclease of 5-formyluracil- and 5-hydroxymethyluracil-containing oligonucleotides. Int J Radiat Biol 75:59–65

Zischler H, Geisert H, von Haeseler A, Pääbo S (1995) A nuclear fossil of the mitochondrial D-loop and the origin of modern humans. Nature 378:489–492

Nucleic Acids and Molecular Biology, Vol. 18
Hans-Jürgen Bandelt, Vincent Macaulay, Martin Richards (Eds.)
Human Mitochondrial DNA and the Evolution of *Homo sapiens*

Lab-Specific Mutation Processes

Hans-Jürgen Bandelt (✉) · Toomas Kivisild · Jüri Parik · Richard Villems · Claudio Bravi · Yong-Gang Yao · Anita Brandstätter · Walther Parson

Dept. of Mathematics, University of Hamburg,
Bundesstr. 55, 20146 Hamburg, Germany
bandelt@math.uni-hamburg.de

You must learn from the mistakes of others.
You can't possibly live long enough to make them all yourself.

Samuel Levenson

1 Introduction

Mutations hit mitochondrial DNA (mtDNA) in the germ line and (as somatic mutations) in postmitotic tissues. After the death of the organism or extraction of an mtDNA sample, postmortem damage could set in and slightly alter the DNA (Gilbert et al. 2003; Chaps. 5, 9). But the mutational process does not stop here—it continues in a rather unexpected way with sample handling, the sequencing procedure in the laboratory, and during the subsequent documentation phase, where biases and errors can radically transform the DNA results. In contrast to the natural mutation processes in the cell that are of interest, for example, to the population geneticist or the medical geneticist, the artificial mutation processes in the laboratory have largely been ignored and have therefore been investigated much less. Yet the latter can affect mtDNA studies to such an extent that the main conclusions drawn are built entirely on the basis of artefacts in extreme cases, and may thus be completely invalid. This is often particularly dramatic with ancient mtDNA studies, which are believed to permit a more direct insight into the evolution of modern humans or past population histories. The situation may be even worse in the medical field since phylogenetic analysis is hardly ever employed there as a tool to check whether sequencing results are consistent with the worldwide mtDNA database. It is therefore necessary to recall the stages of the sequencing process and understand the causes of misrepresentation of DNA results. The numerous pitfalls are well reflected throughout the mtDNA literature.

2
The Sequencing Process and Data Handling

The first stage in the whole procedure is the sampling phase, where buccal swabs, or blood samples, and—in rare cases—hairs are taken from volunteers. Ideally, some information is attached to each of these individual samples, viz. ethnic or geographical background (for two to three generations back in the matriline). When the sample arrives in the laboratory, it runs through a multistage process for DNA typing (for a more detailed review, see specific chapters in Carracedo 2005, e.g. Köchl et al. 2005).

1. *DNA extraction.* In this step the sample is prepared for the polymerase chain reaction (PCR).
   (a) *Lysis.* Cell membranes are destroyed, so all components of the cells as well as substances in the sample external to the cells are obtained in solution.
   (b) *Purification.* The most effective method for DNA extraction in terms of sensitivity and purity of the DNA extract from forensic samples is the phenol/chloroform method coupled with a subsequent purification step: lipophile constituents (proteins, lipids, etc.) are separated from the hydrophile fraction that includes the DNA (for a detailed practical overview, see Köchl et al. 2005). In an alternative method using magnetic beads, DNA is bound to the polymorphic magnetic particles from which it is released in a purified form. There are some commercial kits available for quick DNA extraction, and it takes less than 1 h to get the genomic DNA. High-quality forensic samples such as full blood or epithelial cells retrieved from buccal scrapes may be extracted using quicker and cheaper methods, for example Chelex extraction (Walsh et al. 1991). However, these methods do not actually extract the DNA from the residual components but only separate the DNA from the cellular environment; the proteolytic component remains in solution. It is known that the quality of such DNA suffers much more from (frozen) storage in terms of DNA degradation compared with DNA extracted by real extraction methods. Furthermore, some proteins in the DNA extract are known to cause PCR inhibition if their concentration is high.
2. *PCR.* This constitutes the amplification phase for the extracted DNA, which normally runs in about 25–40 cycles, depending on the amount of starting DNA. Typically, it comprises three phases: first, the double-stranded DNA is broken up into single strands; second, a pair of primers is annealed complementary to the ends of the targeted region of DNA; third, the thermostable *Taq* DNA polymerase incorporates activated single nucleotides (dATP, dGTP, dCTP, and dTTP) complementary to the existing DNA strand. At the end of this amplification phase the amplicon is puri-

fied from the PCR reagents and is ready for the sequencing reaction. In some cases, the purified PCR products are cloned into a vector and amplified further before sequencing.

3. *Sequencing*. The nucleotide sequence of the amplicon is derived from a PCR-like process involving only one primer (for each strand separately) using commercial kits, such as the ABI PRISM BigDye Terminator Cycle Sequencing Ready Reaction Kit. A subsequent purification step eliminates the residual sequencing reagents, including the fluorescent-labelled terminators.
4. *Electrophoresis*. During this phase the sequenate is separated according to the length of the fluorescent-labelled products, which is depicted by the sequencing software as an electropherogram and automated basecalling.
5. *Interpretation and documentation*. Finally, the automated basecalling is checked visually and interpreted accordingly. Then, the final sequence is transferred to a table and subsequently to a databank.

3 Sources of Error

During laboratory stages 1–5 all kinds of problems can occur that lead to errors in the final results. The extraction phase is in general a crucial step for the analysis of biological material with unknown DNA content. The sensitivity of the method applied determines whether a sample will contain enough DNA to be amenable to PCR amplification and thus will produce sufficient amplicon in order to give a meaningful sequencing result. A potential source of error here is contamination, the admixture of external DNA to the targeted sample. The distorting effect of contamination is greatly enhanced when only minute quantities of DNA, as typically encountered in degraded and ancient DNA (aDNA) samples, are investigated. Good laboratory practice involves a number of safety precautions which keep contamination at very low level; however, avoiding contamination remains a permanent challenge because of the sensitivity of the PCR process.

Contamination at the level of DNA extraction can have different consequences. Actual mixtures of mtDNA sequences (from different individuals) can only be identified when a sufficient number of polymorphic positions are involved. A mixture of two very similar mtDNA sequences bears the risk of being falsely interpreted as heteroplasmy. Numerous ambiguous nucleotides (scored 'N') could indicate mixture between quite different mtDNA sequences. In the case where the contaminant suppresses and eventually overruns the correct sequence, the produced amplicon no longer corresponds to the donor of the sample. When this happens repeatedly in the process of generating population data, a considerably skewed variation may result. In anticipating the most frequent causes of contamination, it is important to set up an internal database of mtDNA sequences of staff members and to maintain a list of recently typed samples for monitoring and cross-checking sequencing results.

The stages of PCR, sequencing, and electrophoresis are also prone to contamination and sample mix-up (which could be viewed as an extreme form of contamination), especially when old amplification and sequencing strategies, such as the separated analyses of the two hypervariable segments, are employed. This early approach of mtDNA sequence analysis bears the disadvantage that several laboratory steps (PCR setup, amplicon control gel electrophoresis, PCR purification, cycle sequencing setup, purification of the sequencing products, setup for electrophoresis) are performed independently and in parallel for the entire sample set. The risk of introducing sample mix-up in any of these processes is high, and mix-up is practically unavoidable when samples are handled manually, unless double-safety strategies are followed rigorously. In particular, if the samples were electrophoresed by loading manually, such as on an ABI 377 DNA sequencer, and were not automatically run with a capillary electrophoresis sequencer, the chances for sample mix-up would increase when some lanes gave only a poor signal for the sequencer and the automated basecalling by the sequencing analysis software was not verified by hand.

The choice of primers in PCR is also crucial. Ideally the primers are designed to cover evolutionary conservative regions of the sequence that is being analysed. This is to guarantee the lowest probability of having a mutation in the primer region that would reduce its affinity to the template. The 3′ ends of the primer sequences are most decisive—a single mismatch here can have the cost of no amplification in certain PCR conditions. This specificity of primer annealing is actually taken advantage of in a method called allele-specific PCR. However, in routine laboratory work that is designed to resequence hundreds or thousands of amplicons, the use of one definite pair of primers can be precarious for a locus with a high mutation rate, given the risk of having some mutations in the primer region. For example, the 3′ nucleotide at position 15928 of the most widely used primer A (Vigilant et al. 1991) for amplifying the hypervariable segment I (HVS-I) region of the human mtDNA control region carries a derived nucleotide state in all mtDNAs belonging to haplogroup T, a major branch of the West Eurasian mtDNA phylogeny. Therefore employing this primer may inadvertently give a bias towards decreased frequency of this haplogroup. Moreover, the second nucleotide (at 15927) from the 3′ end of the same primer is variable as well. The variant nucleotides are, for example, shared by all members of haplogroups X2b (Reidla et al. 2003) and B5b (Kong et al. 2003).

The development of new versions of sequencing chemistry during the past 10 years and the corresponding methods for purification of the sequencing products left their footprints in mtDNA sequence data. The revolution of cycle sequencing kits improved the quality of the sequence electropherograms significantly; nevertheless, particular phantom mutations, i.e. mutations that are generated artificially, seem to be associated with some kits (Bandelt et al. 2002; Herrnstadt et al. 2003; Brandstätter et al. 2005). For example, a well-known artefact of an older version of the dye terminator sequencing chemistry was that the first guanosine in a nucleotide sequence 3′ adjacent to an adenosine was displayed with considerably reduced peak height (Parson et al. 1998). Under certain circumstances (e.g. with long sequence runs) this effect can even cause the total disappearance of the G peak, so some background noise would be read instead.

Premature stops during cycle sequencing may arise when the polymerase is released from the strand as a consequence of stable secondary DNA structures (e.g. hairpins). This is usually accompanied by non-template ddNTP addition, which manifests itself by several overlapping peaks in a dye terminator sequence electropherogram. Although these artefacts are usually described in the sequencing manuals, the positions affected are sometimes misassigned as heteroplasmic basecalls or false nucleotide insertions.

The purification of the sequencing products is important for the removal of unincorporated terminators, which otherwise would form stable dye-labelled clusters (so-called dye blobs) that migrate during electrophoresis, and mask nucleotide sequences predominantly within the first approximately 50 base pairs. Automated basecalling by the sequence analysis software does not necessarily account for dye blobs, which may still remain in the sequenate owing to ineffective purification. Such phenomena contribute to the class of phantom mutations, which could easily be avoided if raw data are inspected manually after automated basecalling by the sequencing software (see e.g. Fig. 4 of Bandelt et al. 2002).

In order to identify positions that are prone to be hit by artefacts, a collection of 5400 sequence electropherograms for both strands of the control region (produced in several laboratories) were systematically screened for sequencing artefacts (Brandstätter et al. 2005). Artefacts become manifest in two discordant readings of the separate strands. In particular, an ambiguous basecall for a position on one strand that is not confirmed in the same way as ambiguous for the corresponding position on the other strand cannot be explained by natural heteroplasmy and has therefore to be regarded as an artefact. The number of artefacts (phantom mutations) depends to some extent on the sort of automated sequencer and sequencing chemistry employed, but also on other laboratory-specific factors. The lowest numbers of artefacts per sequence were observed in sequences generated with BigDye v2.0 sequencing chemistry run on an AB3700 capillary electrophoresis instrument and in electropherograms with BigDye v1.0 chemistry on an AB3100 platform. A number of positions in the control region were repeatedly hit by phantom mutations: the top 12 (with a relative frequency of 0.1–4%) were 16030, 16085, 16095, 16239, 16358, 16368, 84, 85, 87, 253, 317, and 320. It is interesting to note that all of these positions except 16358 and 16368 are also occasionally recorded as undetermined in a number of sequences from the SWGDAM mtDNA database (Monson et al. 2002); there, positions 16030, 16085, 84, 85, 87, and 320 are never reported with a variant nucleotide and exclusively occur undetermined with a relative frequency up to 1.2%.

The final stage of analysis—data interpretation—is prone to a number of potential pitfalls. The generated sequences undergo automated analysis and basecalling by the sequencing analysis software. It seems that this output is sometimes taken at face value without critical evaluation and is thus forwarded to a data table or databank. The problem here is that the settings in the software do not completely reflect the preconditions for correct mtDNA sequence analysis. The proper action would be that all sequences generated from a sample are aligned relative to the reference sequence and the complementary sequences are scrutinized on a base-to-base level. For this task several additional software packages are provided commercially, such as the DNAstar package (DNASTAR), SeqScape (Applied Biosystems), or Sequencher (GeneCodes). Compared with the time required for the laboratory processing of the samples the thorough inspection of the raw data is much more time-consuming. Even in the phase of data interpretation, contamination can become effective when the raw sequence data are of poor quality, so unambiguous basecalling becomes difficult. It can then be observed that the data are biased towards the reference sequence with which the mtDNA sequence data are compared. Reference bias is more frequently observed in sequence regions that are difficult to read, for example long tracts of cytosines and towards the end of long sequence runs, where the heights of the sequence peaks are usually reduced.

Documentation of the sequencing results can be riddled with transcription errors: mutated positions are overlooked or misrepresented (e.g. digits are shifted or switched), nucleotides are misreported (e.g. C and G are interchanged, or by using a partially corrected or otherwise modified version of the original Cambridge reference sequence), and standard nomenclature rules are violated, which can provoke misunderstanding by the reader. In the case where some trivial software is used that would only indicate whether a nucleotide position is changed relative to the reference sequence or not, so that the variant nucleotide has to be filled in manually, the change would be realized as a transition almost by default (because transversions are generally quite rare), thus creating a transition bias. Since data tables are assembled manually, additional errors can slip in (e.g. nucleotides do not occupy their proper place in a dot table, or rows and columns are shifted in part, etc.). Finally, tables may be distorted in the printing process without being corrected at proofreading, or even after the proofreading stage by the publisher. For instance, "309+C" was enigmatically changed to "309 C" in sample LN7578 (Yao et al. 2002) and, moreover, a space separating two polymorphisms was omitted in many places in the table during the final publishing process.

Table 1 Stage–cause–phenotype error classification

Stage	Cause	Phenotype	Process
1			DNA extraction
2			PCR
3			Sequencing
4			Electrophoresis
5			Interpretation and documentation
	C		Contamination
	M		Sample mix-up
	A		Sequencing artefact
	S		Sample manipulation and bias
	L		Misalignment or incorrect reference sequence
	B		Basecall misinterpretation
	N		Nomenclature violation
	T		Transcription error
		I	Base shift
		II	Reference bias
		III	Phantom mutation
		IV	Base misreporting
		V	Artificial recombination
		VI	Skewed variation

Table 1 briefly summarizes the potential pitfalls by keywords and extends the classification schemes proposed by Bandelt et al. (2001), Parson et al. (2004), and Bandelt and Parson (2004).

Not all combinations of stage–cause–phenotype are possible. For instance, at stage 1 only contamination (C) and sample mix-up (M) are conceivable, whereas stages 2–4 can be combined with causes C, M,and A, and stage 5 with causes M, S, L, B, N, and T. Cause C can lead to phenotype III or V, cause A to phenotype III, S to phenotype VI, L to phenotype I or II, B to phenotypes I, II, or IV, N to phenotype IV, and finally, T to phenotypes I, II, IV, or V. When a certain error phenotype is inferred from a data table, the exact reconstruction of the cause and stage at which the error happened is not always possible, since poor laboratory work and poor documentation seem to go hand-in-hand in many cases.

4 Pitfalls of mtDNA Sequencing

The pitfalls of the sequencing and documentation processes are manifest in most mtDNA data sets published in forensics, molecular anthropology, and medical genetics (Röhl et al. 2001; Forster 2003; Bandelt et al. 2005a; Salas et al. 2005c). A particularly rich source of nearly all kinds of errors is constituted by the Korean mtDNA data of Lee et al. (1997, 2002); see the analyses by Bandelt et al. (2001, 2002, 2004b), Yao and Zhang (2003), and Bandelt (2005a). The laboratory conditions for sequencing in the studies of Lee et al. were quite typical of the time: as in many other studies of the late 1990s, the primer pair L15997 (or L15996) and H16410 was employed and a DNA sequencer based on gel electrophoresis was used, namely an ABI 377 or its predecessor, an ABI 373, together with some version of the dye terminator sequencing chemistry. Especially, early versions could have had an adverse effect on the readability of the initial part (about 16016–16046) of the sequences, depending on the purification method employed (Brandstätter et al. 2005). This can very well be seen in the data compiled by Lee et al. (1997), where a number of otherwise unobserved transversions are recorded at positions 16023, 16028, 16037, and 16045. In other instances, this may invoke just a single position, as was the case for the data published by Roychoudhury et al. (2001): artificial transversions at position 16039 were randomly inflicted on 13 of 115 sequences, irrespective of the phylogenetic location of the sequence in the mtDNA phylogeny. In total, this data set certainly harbours more than 40 phantom mutations (or other misreadings of nucleotides). Symptomatically, these flawed data were incorporated in a study to reveal "diverse histories of tribal populations from India" (Cordaux et al. 2003).

At the end of the sequence run, signal height and base separation of the electropherograms may fade out with gel electrophoresis sequencing. Then, if the interpretable part of the sequence is not cut down in a very conservative way, erroneous basecalls could easily occur, leading to phantom mutations. For example, such effects can be seen again in the data of Roychoudhury et al. (2001) beyond 16357 or in the table of Prasad et al. (2001) with ambiguities and transversions beyond 16394. The most dramatic effect of such erroneous basecalls, however, can be seen in the mtDNA data produced by Nasidze and Stoneking (2001) from samples of various Caucasus populations, where chunks of false mutations were distributed beyond 16357 over a large number of the 353 sequences analysed. The latter data set also shows an amazing number of phantom mutations at the otherwise extremely conservative positions 16280 and 16281, and there are further 'tandem' mutations of this kind, hitting adjacent positions (Table 2).

The intriguing feature of these data is that the numerous undetermined nucleotides (scored 'N') actually signpost the artefacts, inasmuch as the obvious phantom mutations mainly occurred at those positions that were scored 'N'; see Bandelt and Kivisild (2006) for a detailed analysis. In particular, the three sequences GE488, AZ205c, and GE242 (see "Appendix") testify to this remarkable pattern. These and other affected sequences from Nasidze and Stoneking (2001) clearly must have had enormous background noise, where the artificial signals then competed with the true signals, sometimes leading to an irresolvable ambiguity and sometimes overrunning the correct base-

Table 2 Phantom tandem mutations in the data from Nasidze and Stoneking (2001)

Sequence code	Mutations (16000+)[a]	Potential haplogroup
AR31	**067** *279G 280 281* **355**	**HV1**
AR483	**069 126 145** *280 281 367C*	**J**
AZ2	*280 281*	**H/HV*/R***
AZ342	*280 281* **298**	**pre-V**
AZ6	154 *168A 280 281* **356** *384*	**H/U4**
CH444	**111 214G 249** *280 281* **327** *388*	**U1b**
CH451	*280 281* **292**	**H/W**
DAR23	**129 223 278** *280 281*	**X**
DAR36	**258** *280 281 384*	**H/HV*/R***
KAB408	**224** *280 281* **311**	**K**
CH453	**256 352** *387 388*	**H**
CH580	**067 355** *387 388*	**HV1**
CH583	**224** 289 *387 388*	**K**
GE346	**126 163 186 243** 249 **294** 356 360 *387 388*	**T1**
AZ208	145 **223** 239 354 360 *390 391 392*	**?**
IN823	*085G* **223** 353 *390 391*	**?**
IN826	**192 223 292** 311 *363 390 391*	**W**
IN827	**069 126 193 256 335** *390 391*	**J**
IN828	**126** *390 391*	**H**

[a]Numbers refer to transitions relative to the revised Cambridge reference sequence(*rCRS*; Andrews et al. 1999); suffixes specify transversions; mutation motifs matched in the published West Eurasian database are shown in *bold face*; *italics* indicate likely phantom mutations and *underlining* highlights tandem mutations.

call. Certainly, only the light strand of mtDNA was regularly sequenced in this case. Although the early version of an automated gel electrophoresis based DNA sequencer was used, which is known for elevated background noise, the quality of the mtDNA sequencing result should have been much better. It is surprising that such junk data were and still are being used for population genetics studies (e.g., Bulayeva et al. 2003; Vernesi et al. 2004). Most recently, Nasidze et al. (2004) aimed at disguising the sequencing disaster of the earlier study with a fake error analysis, contending that there was nothing suspicious with those data (except for a single sequence sacrificed as erroneous).

Although the data set of Nasidze and Stoneking (2001) seems to be quite unique with its specific profile of artefacts clustering mainly in short stretches around 16280 and beyond 16357, there is one recently published parallel case, less extreme though. Namely, the 267 North African mtDNA sequences from Plaza et al. (2004) include several phantom mutations in the stretches 16279–16285 and 16369–16383. Some of these mutations occur repeatedly on different haplogroup backgrounds, such as the notorious 16281 transitions and C to G transversions at 16382.

Position 16085 was found to be most frequently hit by artefacts in 5400 screened electropherograms (Brandstätter et al. 2005). Interestingly, many laboratories seem to have notorious problems with this position. 16085N, for example, was reported in four out of 180 Apache mtDNAs (sample nos. 55, 56, 57, and 58 from Budowle et al. 2002). Instances of artificial transversions (to G) and transitions at 16085 were detected in a number of sequences obtained by different laboratories. A few affected sequencing results (from half a dozen different studies) were reanalysed: in no case did the changes at 16085 as reported in the publications turn out to be authentic (Brandstätter et al. 2005). It therefore seems that a 16085 change is a phantom mutation par excellence, so every mutation to T or G ever observed at 16085 would very likely constitute an artefact, just as in the case of the Caucasus data (Table 2; "Appendix"). It is then quite alarming that position 16085 was also found to show a high postmortem damage rate according to Gilbert et al. (2003). It is unclear whether this finding can be explained by real postmortem damage of DNA or by sequencing artefacts alone. With 'ancient' mtDNA data, it is very difficult in general to distinguish between different contaminant sequences and the action of postmortem damage, cloning errors, and phantom mutations; see, for example, the six human sequences retrieved from dog specimen no. 14 listed in Table 4 of Malmström et al. (2005).

The bulk of the errors in published mtDNA sequences certainly constitute clerical errors. This is supported by recent ring tests of several forensic institutes (Parson et al. 2004; Salas et al. 2005b) and is well reflected by the SWGDAM database (Bandelt et al. 2004a, b; Salas et al. 2005a). The manual transcription of the raw sequence information to a final data table is prone to error and can misrepresent the original sequences in many ways. The most straightforward error one can commit is to overlook mutations relative to the revised Cambridge reference sequence (rCRS; Andrews et al. 1999). This is a frequent phenomenon in the medical field, where the full coding region is the main target of sequencing (Bandelt et al. 2005a). Many mutations may then be missed. For instance, the haplogroup D4a sequence determined by Tawata et al. (2000) misses as many as eight out of 25 mutations along the evolutionary pathway separating the putative ancestral sequence of haplogroup D4a from the rCRS, namely, transitions at 4769, 7028, 8701, 10398, 10400, 10873, 11719, and 14766 (complete sequence kindly communicated to C.B. and T.K. by M. Tawata); see sequence Taw in Fig. 1 of Kivisild et al. (2002), where, however, only three of those missed mutations were highlighted as private backmutations.

Base shifts are introduced by casual reading of mutations; for example, Fig. 1A of Kurtz et al. (2004) testifies to a – 1 shift of the 16222 transition to position 16221. Even a systematic base shift (by – 1) of all variant positions may occur (Chap. 3). Similar misreadings within a short stretch of repeated nucleotides in the reference sequence have often occurred with coding-region sequences reported in the literature, such as the – 1 shifts of positions 4248 and 11719 in Zhao et al. (2004) or the + 1 shifts of 2706 and 14783 in Tawata et al. (2000), which all affect mutations deep in the mtDNA phylogeny and are thus easy to spot by database comparisons. In other cases, reading shifts could have been induced by outdated reading software for the sequencer output, so, especially after long tracts of cytosines, the sequences get distorted by partial reading shifts. This can be seen, for instance, in the data table of Hutter et al. (2004), where the variation beyond 16400 in HVS-I and at 396 and 410 in HVS-II was artificially created by reading a misaligned nucleotide. Moreover, comparison with the reference sequence and the database shows that the true positions 318, 321, 456, 462, 497, and 489 were all shifted by + 2. Similar effects (and other errors) can be spotted in the table in Taylor et al. (2003).

The preparation of data tables, whether in dot format or as a listing of mutation motifs relative to the rCRS, can fatally affect the represented sequence information. For instance, Lee et al. (2002) exchanged T with C in the reference sequence at position 16362 in part of the motif table, so confusion about the mutational status at this position could arise. Nucleotides in a dot table can easily slip into adjacent rows and columns, or inadvertent copy-and-paste operations leave their traces in false rows and columns. For example, some of the sequences read off from the tables in Mogentale-Profizi et al. (2001) appear to be quite bizarre, for example one sequence is reported to have mutations 16069N, 16097A, 16111, 16129, 16211T, 16231, 16319C, 146G, and 188'A', for which the restriction fragment length polymorphism (RFLP) analysis indicated haplogroup K status. Nucleotide A at position 188, however, cannot constitute a change relative to the rCRS since the rCRS does bear A at 188. Closer inspection of the dot table then gives a clue to a reconstruction of how the correct sequence might have looked: at least five nucleotide positions for this sequence have been shifted to the corresponding neighbouring columns in the data table, so the potentially correct scoring would then list mutations 16069N, 16097A, 16111, 16129, 16192, 16224, 16311, 73, and 185 (whilst still being suspicious with regard to the ambiguity at 16069 and the unusual transversion at 16097).

Contamination has always been a challenge for mtDNA sequencing, especially with hair shafts, or poorly preserved samples harbouring only degraded mtDNA, or ancient mtDNA samples (Bandelt 2005b). A classic case of cross-contamination between samples analysed in a study can be seen in the Native American mtDNA data published by Santos et al. (1996) that comprise HVS-I sequences and haplogroup-specific RFLP sites. In particular, parts of two haplogroup C sequences crept into three haplogroup B sequences, and, in turn, a sequence from haplogroup B (or some other haplogroup) erased most of the characteristic mutations of one C sequence. Moreover, some of those RFLP results constitute mosaic patterns as well; see Bandelt et al. (2000) for more details. A case of contamination incurred by laboratory personnel, in particular by the investigator himself or herself, is nicely documented in Table 1 of García-Bour et al. (2004), where the researcher's HVS-I sequence (with transitions 16294 16296, and 16304, from haplogroup T2) invaded several ancient HVS-I sequences from Native American mtDNA haplogroups (C, D), thereby partially eliminating the expected haplogroup-diagnostic mutations (Bandelt 2005b). Wherever contamination has the potential to play a role, cloning is mandatory in order to disentangle the different mtDNA contributions.

When more than one mtDNA segment is analysed there is a real risk for sample mix-up. This can then become a tremendous problem for sequencing of the total mitochondrial genome (or a considerable part of it; Yao et al. 2003b), because this involves many (approximately 30 or more) independent steps of amplifying fragments from the same sample. A similar problem arises when the same mtDNA segment is sequenced from different tissues of the same individual. Here sample mix-up will typically result in a bundle of (homoplasmic) mutations for mtDNA obtained from one tissue compared with other tissues or body fluids, which could then innocently be interpreted as somatic mutations. There are many instances of this kind of trivial error in cancer research. For instance, Kurtz et al. (2004) reported an array of homoplasmic mutations in a neurofibroma (associated with neurofibromatosis type 1) compared with normal tissue in case no. T173, viz. 64 73 152 195 16163 16186 16189 16222 (erroneously scored at 16221) 16519, which were deemed to be 'somatic'. However, except for the 64 and 16222 transitions, which seem to be private, all of these mutations plus 263, 16126, and 16294 (not reported in case T173, presumably for systematic reasons) would match typical haplotypes from haplogroup T1 (Finnilä et al. 2001); compare this, for example, with sample AUT21 from Parson et al. (1998). Therefore, we conclude that case T173 possibly had a haplogroup H mtDNA sequence, but the neurofibroma mtDNA was taken from another case by mistake, carrying a haplogroup T1 mtDNA. Alonso et al. (2005) reported a case where two different tissue sections associated with the same patient actually turned out to contain quite different mtDNAs (from haplogroups K1a and U5a, respectively), so sample mix-up during block processing or slide preparation had to be concluded. Numerous cases of perceived multiple somatic mutations can be spotted in the medical literature that were taken at face value by the authors of those papers but can in fact be explained as sample mix-up or contamination (Bandelt et al. 2005b; Salas et al. 2005c).

Different mtDNA segments, such as the two hypervariable segments of the control region, composed into one haplotype could inadvertently correspond to different individuals. This kind of error is widespread (Bandelt et al. 2000, 2001, 2004a, b; Bandelt and Parson 2004; Yao et al. 2004). Many of these artificial recombination instances can be discovered by comparing the separate segments with the total database of published sequences. Indeed, several mtDNA haplogroups bear unmistakeable mutational patterns of several mutations, for which it would be very unlikely that they would switch en bloc to another complex motif without leaving a connecting trace in the database.

It is ironic that forensic studies were so prone to sample mix-up in the past. In forensic case work, contamination is a real challenge because some stains may be difficult to analyse, such as hair shafts (compared with saliva or whole blood), because the number of available mtDNA copies for amplification is much lower. Some laboratories were then tempted to employ a nested PCR approach, which is now known to be quite susceptible to contamination invading the sequencing products (Brandstätter and Parson 2003; Grzybowski et al. 2003).

Normally, forensic case work remains unpublished, but the study of Allen et al. (1998) allows us a glimpse of the forensic practice at the time. For case no. 1 (which concerned a series of three armed bank robberies that were carried out in Stockholm in 1990 and 1991) a number of different evidence materials (stains) were investigated. Among the different sequences obtained by nested PCR, three clear instances of artificial recombinants are obvious. First, the combined HVS-I&II sequence from stain 2345/93 (hair, from robbery 1) constitutes a haplogroup H1a haplotype with respect to HVS-I (cf. Loogväli et al. 2004) and a haplogroup W haplotype for HVS-II (cf. Finnilä et al. 2001). Second, stain 2321/93 (hair, from robbery 2) represents a combination of a haplogroup L2a HVS-I sequence (cf. Salas et al. 2002) with a haplogroup X2c HVS-II sequence (cf. Reidla et al. 2003). Third, stain 1749/93 (saliva, from robbery 3) constitutes a sequence mix from haplogroups U5a and J (cf. Finnilä et al. 2001). These hybrid sequences were generated either through contamination (assisted by nested PCR) or through careless handling of tubes in the laboratory that led to sample confusion.

Long tracts of (more than seven) cytosines ('long C-stretches') regularly cause slippage in amplification, so beyond such a long C-stretch reading is inhibited by an overlay of shifted sequences. Laboratories with little experience in mtDNA sequencing would repeatedly try to generate a sequence anyway and then eventually may decide either (1) to take the base pairs of the reference sequence beyond the C-stretch as default or (2) to report nothing and omit the sequence from further analysis. Either strategy has an adverse effect on forensic case work and population genetics studies, in that either sequences are produced that constitute artificial recombinants with the reference sequence or information is lost (for case work) and a strong sampling bias (in population studies) is incurred. The more standard strategy would either generate the second half of the sequence by reverse sequencing of the other strand or, preferably, use a second internal primer pair after the long C-stretch and thus capture (most of) the second part. HVS-I harbours such a notorious C-stretch of approximately ten cytosines whenever the T at 16189 is mutated to C. The 'proficiency testing' trial of Salas et al. (2005b) documents that only one out of 14 laboratories consistently applied a professional strategy to deal with a long C-stretch, while all other laboratories variably adhered to strategies 1 or 2 in most cases.

The 16189 transition is actually very frequent and constitutes one of the top mutational hotspots in the control region; hence, there are many branches in the mtDNA phylogeny which bear a long C-stretch, so in every population study sequences with a long C-stretch will naturally show up. In western Europe, for instance, one can expect that approximately 15% of the mtDNA lineages have C at 16189. This percentage may rise to some 25% towards the Near East and the Caucasus, because some haplogroups (e.g. X and U1) for which the long C-stretch is inherited are somewhat more frequent in western Asia than in Europe. When a study does not report any polymorphism or length heteroplasmy in the region 16182–16194 (represented by AACCCC-CTCCCCA in the rCRS), this strongly indicates that the laboratory followed strategy 2 by systematically suppressing all samples with a long C-stretch (to save primer, sequencing kit, and time). For instance, Tommaseo-Ponzetta et al. (2002) aimed at studying the mtDNA variation in West New Guinea populations, but their 202 sequences show absolutely no variation in the region 16182–16194. This sharply contrasts with mtDNA data obtained by other laboratories: C is observed at 16189 in five out of 21 complete mtDNA sequences from New Guinea (Ingman et al. 2000; Ingman and Gyllensten 2003) and three out of 48 HVS-I sequences from Irian Jaya (Martin Richards, personal communication). A similarly drastic effect of the 16189 bias can be encountered in the data from Nasidze and Stoneking (2001). The part of the total data set that was most severely affected by nucleotide ambiguities and phantom mutations (as discussed earlier) consists of the Armenian, Azeri, and Georgian mtDNA sequences: none of those 140 sequences has a C at 16189. Comparison with mtDNA data published by other laboratoties sampled from the same national groups suggests that a range of 15–30% for sequences bearing C at 16189 would be expected in unbiased mtDNA samples from the southern Caucasus (Bandelt and Kivisild 2006).

The fact that the long C-stretch in HVS-I needs an extra sequencing step makes those sequences susceptible to artificial recombination. Since the reading of the H-strand can be more difficult to decipher, there is also a good opportunity for phantom mutations to enter; see Bandelt et al. (2001) for a pertinent case study. The data set of Prasad et al. (2001) serves as another excellent case in question. In these data comprising 33 HVS-I sequences taken from Nicobarese Islanders, the two most prominent haplogroups are F1a and B5a (cf. Kong et al. 2003), represented by the consensus motifs 16108 16129 16162 16172 16189 16304 and 16140 16183C 16189 16266A, respectively. These two sequences were also found in the Nicobarese by Thangaraj et al. (2003). In Prasad et al. (2001), however, two of the F1a and one of the B5a sequences additionally harbour the following complex chunk of mutations after the long C-stretch: 16242A 16275 16287A 16316T 16321 16360. None of these mutations were detected by Thangaraj et al. (2003) and each of them is either extremely rare or known to be involved in phantom mutations on other occasions (such as 16321 and 16360; Bandelt et al. 2002). This indicates that the final part of those sequences, which must have been read off from the H-strand alone, had suffered from multiple phantom mutations in one of the electrophoresis runs.

Also the C-stretch around position 310 in HVS-II can cause reading problems beyond position 315 if length heteroplasmy has occurred. In that case, the true nucleotide at certain positions then competes with the predecessor (or successor) nucleotide, which especially becomes manifest in apparent 317G, 320, 330G, 343, and 345 mutations, all affecting nucleotide C (Brandstätter et al. 2005). Sequence data displaying 311, 317G, and 320 (owing to length heteroplasmy of the C-stretch 303–309) can also be inspected in Tan et al. (2003) and Wong et al. (2004). The full mutation array 311, 317G, 320, 330G, 343, 345 can be found in two haplogroup A2 sequences in the data table in Vona et al. (2005), where also 5-mutation subarrays can be found in two haplogroups. The transitional motif 317, 320, 343, 345 was earlier reported by Vives-Bauza et al. (2002) in one case.

5
'Ancient' DNA

The mtDNA analysis of ancient material (especially bones, teeth, or hairs) has become a popular enterprise in the last 10 years. There seems to be a widespread belief in anthropology that one could get a direct grip on the past with aDNA analyses, despite well-known caveats (Kaestle and Horsburgh 2002). But, as Alan Cooper emphasized, "we know that contamination is almost impossible to avoid" (cited in Abbott 2003). Pääbo et al. (2004) made clear that "in cases where a DNA sequence identical or similar to contemporary humans is determined, it is impossible to establish its authenticity even with rigorous application of the criteria" (for authenticity, as e.g. published by Cooper and Poinar 2000). In particular, 'innate' (or 'endogenous') sources of contamination represent one of the greatest problems facing aDNA studies (Gilbert et al. 2005a–c; Malmström et al. 2005). Innate contamination arises through the direct contact of a sample with other sources of DNA, prior to its arrival in the laboratory.

A further problem with aDNA studies is that in many situations, the retrieval of accurate DNA sequences is not enough to make the results useful (Hofreiter and Vigilant 2003; Gilbert et al. 2005a). Naturally, aDNA studies are hampered by factors that limit the samples that can be obtained. Often only a few hundred nucleotides from the control region are analysed for a handful of ancient individuals. The comparison with modern mtDNA samples can be problematic (Yao et al. 2003a). Most standard population genetics methods are not designed to cope with small numbers of samples; thus, they lack the statistical basis necessary for reasonable conclusions to be drawn. For example, at the very extreme, the (rather indirect) method of principal coordinate analysis has been exercised on a single ancient individual contrasted with some modern population samples in order to persuade the reader of the ancient individual's origin (Vernesi et al. 2001). In the lack of sufficient data, meagre results are then interwoven into a narrative quite arbitrarily, just to fit an interesting story. For instance, the 'evangelist Luke' HVS-I sequence is one mutational step away from a modern Syrian sequence, and the Syrian origin of St. Luke is then considered to be the most likely (Vernesi et al. 2001), thus supporting the religious myth. On the other hand, an approximately 14 000 year old HVS-I sequence from the Alps is one mutational step away from an extant European haplotype, which was turned into support for the idea of population discontinuity according to the authors (Di Benedetto et al. 2000).

Not only the interpretation of a putative ancient mtDNA datum appears to be problematic. The compound haplotype itself ascribed to the ancient individual can be suspiciously mosaic. For example, Vernesi et al. (2001) claimed to have obtained the following "genetic characterization of the body attributed to the evangelist Luke": transitional differences at sites 16235 and 16291 in HVS-I as well as the restriction-site change + 7025 *Alu*I (thus excluding haplogroup H membership). This information is, of course, too meagre to infer the matrilineal origin of the mtDNA carrier, especially as Vernesi et al. (2001) made no attempt to search for all existing matches of the HVS-I motif in the West Eurasian database. In fact, haplogroup H status (– 7025 *Alu*I) has been confirmed for a number of modern sequences bearing the motif 16235 and 16291, so it appears questionable whether the compound haplotype (16235, 16291/ + 7025 *Alu*I) is authentic. While Vernesi et al. (2001) cloned pieces of HVS-I and obtained several different sequences (positively indicating the presence of some contaminant mtDNAs or postmortem damage or PCR errors), they did not clone the surrounding region of position 7028 but just executed *Alu*I digestion right away on amplified products. The result is therefore on very shaky grounds, as the *Alu*I band is possibly created from a minority fraction of the amplicon, thus biasing towards presence of the band. Moreover, RFLP analysis and its interpretation may even be tricky with modern mtDNA because of incomplete digestion or because bands may not be readily discernible and assignable to single mutations (Bandelt et al. 2005c). Its use in aDNA studies promotes ample scope for contaminating sequences to make their way into the results (Hofreiter and Vigilant 2003).

An abnormal mutational spectrum is often identifiable by several single extremely rare mutations or novel mutation pairs, which are distributed across an aDNA data set. The table that is supposed to display 'Etruscan' mtDNA variation (Vernesi et al. 2004) includes, for instance, the transition 16334, which has so far been observed in only one published data set—the 'Ladins' (Stenico et al. 1996), infamous for the high accumulation of sequencing artefacts (Bandelt et al. 2002). In the Etruscans the 16334 transition appears twice, but on different HVS-I backgrounds and connected with different RFLP results, viz. with + 14766 *Mse*I and – 14766 *Mse*I, respectively, so one is forced to assume two independent mutations at position 16334 in this single small data set. There are further mutations in the Etruscan data which are otherwise rare (such as mutations at 16098). In particular, transition 16229 is reported only three times in the European mtDNA pool, but in the Etruscans it appears on two different branches of the estimated mtDNA phylogeny, and in one of the sequences even jointly with the extremely rare transition 16228 (Bandelt 2004, 2005b; Malyarchuk and Rogozin 2004).

An odd pattern of recurrent mutations at haplogroup-specific sites in an ancient mtDNA study can always be 'defended' with the presumed hypervariability of the positions involved by reference to systematically flawed studies such as that of Meyer et al. (1999), which essentially count the degrees of polymorphism and therefore strongly bias the rates at all haplogroup-specific sites (Chap. 4). For instance, Barbujani et al. (2004) contended that position 16069 (specific for haplogroup J) has a mutation rate above average and that position 16219 (characteristic for haplogroup U6ab and a subhaplogroup of H6; Bandelt et al. 2004a) is a mutational hotspot since Meyer et al. (1999) claimed that the rates there are approximately 1.5 and 2.7 times the average rate. In the more carefully designed study of Excoffier and Yang (1999) both positions were estimated at about the average rate, despite the fact that this estimation was based only on HVS-I information, which on its own is prone to phylogenetic misplacement of sequences owing to recurrent mutations at mutational hotspots. With additional coding-region variation added, our estimation (based on a similar number of sequences as the Excoffier and Yang study) gives a lower count (Chap. 4).

A hidden case of mosaic structure can be uncovered in Caramelli et al. (2003), who claimed to have retrieved authentic mtDNA information for two approximately 24 000 year old European human specimens. In particular, the mtDNA of specimen Paglicci-12, with claimed mutations at positions 73, 10873, and 16223 in HVS-I but none in the stretch 10397–10400 relative to the rCRS, was regarded as a member of haplogroup N. The authors, however, confused the roles of C and T at 10873 in the mtDNA phylogeny—in fact, C at both positions 10400 and 10873, as observed in Paglicci-12, indicates that this mtDNA haplotype clearly does not belong to either of the Eurasian/Oceanian haplogroups M and N, which completely cover the non-African mtDNA pool of today (Chap. 7). Therefore, we would be led to sort this mtDNA lineage into a yet unknown African subhaplogroup of the superhaplogroup L3; but this does not sit easily with the claimed nucleotide A at 10398. The most plausible explanation then is that we are seeing here a mosaic origin of the compound mtDNA haplotype for Paglicci-12. This cannot come as a surprise because the mtDNA information other than for HVS-I was obtained in only one laboratory, so even external contamination could have easily acted upon the screening of those single mtDNA sites.

The oldest modern human specimens from which mtDNA is claimed to have been retrieved were from Australia (Adcock et al. 2001). This is now well understood to be a study of aDNA that might better have not been done at all (Cooper and Poinar 2000; Cooper et al. 2001). In such circumstances there is always a risk of getting nuclear inserts of mtDNA (numts) from contaminants to which the primers would bind. In a way, numts are also a kind of aDNA, as they constitute (nearly) frozen copies of the mtDNA at the time of incorporation into the slowly mutating nuclear DNA (Wallace et al. 1997; Thalmann et al. 2004; Chap. 3). The data of Adcock et al. show an alarmingly mosaic pattern of recurrent mutations shared with one specific numt (Zischler et al. 1995). Phantom mutations have likely acted on the results as well; for example, the extremely rare 16387 transition (seen as a frequent phantom mutation in the data of Nasidze and Stoneking 2001, however) must have been inflicted three times independently on those ten sequences (deemed to be ancient) in Table 1 of Adcock et al. (2001). In summary, the a priori expectation of sample preservation, the actual laboratory work carried out (no cloning!), and the a posteriori analysis of the sequences obtained all lead one to reject the authenticity of the aDNA.

6 Conclusion

Sequencing of mtDNA may seem to be a routine exercise nowadays—yet it is still fraught with pitfalls that are not sufficiently acknowledged by most researchers. Population and medical genetics seems to be partly immune against questioning the quality of mtDNA data since mtDNA sequences are often generated in a minimal way and are not always subjected to detailed phylogenetic analyses but rather provide the grist for coarse genetic distance measures.

Typically, mtDNA variation is inferred from reading no more than one strand of HVS-I from one amplicon each. When Pakendorf and Stoneking (2005) say that "we notice to our dismay a trend for more and more laboratories to sequence only one strand of the PCR product ..., which in our experience does not ensure adequate detection of sequencing artefacts" they seem to refer to their own experience with the 2001 sequencing disaster (see "Appendix"). But apart from that, their statement is plainly wrong: in population genetics single-strand sequencing has almost always been the rule, as everybody knows in the field (although many used to pretend that they would analyse both strands routinely). The heavy strand was sequenced and read just for those samples which had a long C run in order to obtain a full HVS-I sequence. Only in exceptional cases, both strands were entirely analysed for the purpose of quality control (e.g. Helgasson et al. 2000) in order to confirm that everything was well and fine with the L-strand sequences.

It is vital, however, for the field to present sufficient and reliable information from the mitochondrial genome. In an attempt to analyse ancient mtDNA this requirement can hardly be fulfilled. In particular, tracing early migration routes out of Africa and the pioneer settlement of Eurasia would involve specimens that very likely will not bear any DNA because of inferior preservation conditions (high temperatures, etc.). Therefore, the genetic reconstruction of the evolution of modern humans will have to rely essentially on samples of extant humans. With any mtDNA data generated, it is then obligatory to keep the aforementioned caveats in mind and to filter out the sequencing and documentation errors before interpretation of the results sets in.

Once, however, errors have happened and have found their way into a published article, a speedy correction by the authors of that article would be most helpful to prevent future misinterpretation of the incorrect mtDNA results. It is then encouraging to see corrections of inadvertent sequencing results in print (e.g. as by Da Pozzo and Federico 2005 in response to Bandelt et al. 2005a)—but the usual reaction of authors is, alas, to do absolutely nothing. Worse, the same errors may even be committed again and again by one and the same laboratory (or by the sequencing service employed), such as using a wrong reference sequence, overlooking and misdocumenting mtDNA variants—as is the case with all the studies produced by the laboratory of Min-Xin Guan (see Bandelt et al. 2005a and Yao et al. 2006 for thorough reanalyses). The worst possible reaction from the authors of studies that have been criticized in regard to low sequence quality, however, would be the plain rejection of having committed any sequencing errors at all, by arguing away the concerns about artefacts—instead of properly carrying out resequencing and providing the raw data for public inspection. Such a sad reaction is apparently now under way with the Caucasus data of Nasidze and Stoneking (2001), since Mark Stoneking (in his lecture at the symposium "Stories DNA tells" in Shanghai, 6–10 December 2005, organized by the CAS–MPG Partner Institute for Computational Biology) rejected all the findings of Bandelt and Kivisild (2006) and announced a corresponding statement by him and Ivane Nasidze to be published in the *Annals of Human Genetics*.

Appendix

In Table 3 we list the HVS-I data (scoring frame 16024–16400) from the original publication of the 353 Caucasus sequences (Nasidze and Stoneking 2001). Samples are designated as in that publication. Each mutation motif comprises the mutations relative to the rCRS; when the motif is empty (i.e. no mutation is found), the sequence is designated as rCRS. The mutations are listed minus 16000 and are transitions in the case of three digits; suffixes 1–6 encode the following transversions: 1 is A→C; 2 is A→T; 3 is G→C; 4 is G→T; 5 is C→A;

Table 3 Phantom mutations in the Caucasus data set of Nasidze and Stoneking (2001)

Code	HVS-I variation (+16000)	Code	HVS-I variation (+16000)
AB370	168 192 343	DAR01	292
AB450	168 192 343	DAR010	292
AB452	126 294 296	DAR011	256 270
AB461	042 126 145 244 261 311 ***375*** ***388***	DAR012	189
AB462	129 327 ***388***	DAR013	261 356
AB464	067 ***075*** ***076*** 157 1831 284	DAR014	069 126 145 261
AB467	093 249 365	DAR015	234 ***388***
AB468	192 256 270 311	DAR016	150 354 ***384***
AB469	254 262 263 311	DAR017	rCRS
AB471	093 129 2665	DAR018	129 223 391
AB472	129 223	DAR02	189 192 256 270
AB475	129	DAR03	126 163 186 189 294 320
AB478	126 294	DAR04	093
AB480	129	DAR05	129 223 ***281***
AB481	211	DAR06	126 213 269 294
AB484	253	DAR07	189 249 ***373*** ***381*** ***384***
AB486	078 126 292 294 296	DAR08	223
AB487	rCRS	DAR09	129 223 391
AB488	rCRS	DAR1	rCRS
AB489	304 365	DAR11	051 148 184 234 294 342 357
AB490	223 362	DAR14	114 129 223 ***280*** 391
AB519	093 223 227 234 278 362	DAR16	189
AB747	rCRS	DAR18	051 093 148 184 210 233 234 266 294
AR119	366	DAR2	rCRS
AR13	***110N*** ***118N*** ***195N*** ***279N*** ***280N*** ***281*** ***343N*** ***384***	DAR20	129 192 223 298 327 362
AR132	069 ***1516*** 193	DAR23	129 223 278 ***280*** ***281***
AR165	126 163 186 198 2021 294	DAR29	rCRS
AR170	162	DAR3	256 270
AR172	172	DAR31	093 224 278 291 311
AR187	223 240 292	DAR35	294 ***3914*** ***392***
AR188	***060N*** 069 126 193 ***195N*** 300 ***339N***	DAR36	258 ***280*** ***281*** ***384***
AR209	069 ***110N*** 126 156N ***281*** ***384***	DAR37	256 270
AR216	rCRS	DAR38	rCRS
AR230	069 126 145 172 222 261	DAR42	126 192 294 296
AR235	242 353 356 ***384***	DAR46	188 192 223 292 325
AR278	224 311	DAR49	093
AR285	093 126 294 296 304	DAR51	234
AR295	343 ***384***	GE10	256 270 390
AR297	067 092 234	GE114	217
AR3	069 126 145 222 261 274	GE119	rCRS
AR31	067 ***2796*** ***280*** ***281*** 355	GE120	093 145 1766 223 390
AR315	223 284 292	GE134	067 2201 287 355 390
AR320	224	GE157	129 223 264 311 359 ***3935***

Table 3 (continued)

Code	HVS-I variation (+16000)	Code	HVS-I variation (+16000)
AR324	129 223 264 311 319 362 391	GE17	rCRS
AR33	rCRS	GE171	***388***
AR344	***384***	GE172	rCRS
AR35	145 1766 223 390	GE218	356 362
AR370	186 356	GE220	093 ***224N*** ***311N***
AR397	093 126 289 294 296 324	GE221	126 140 153 294 296
AR4	077N 311	GE222	093 224 311
AR406	192 223N 292 325	GE223	069 126 147 242 311 ***384N***
AR407	192 223 292 325	GE224	249 354
AR41	077N 311	GE227	126 140 153 294 296
AR446	093 224 311	GE242	067 287 355 ***391N*** ***392N***
AR483	069 126 145 ***280*** ***281*** ***3671***	GE243	223 266 274 278 ***384N*** 390
AR503	rCRS	GE244	1831 184 223 266 274 278 390 ***3983***
AR504	093	GE264	126 153 294 296
AR513	145 176N 223 258 272 291 ***384*** 390	GE297	298 ***389*** ***393N***
AR54	rCRS	GE298	051 126
AR57	rCRS	GE299	298
AR6	rCRS	GE300	223 362
AR72	242 249 296 311 ***3606***	GE303	172 192
AR8	2931 325 ***373***	GE305	086 356
AR89	287	GE308	172 192
AR92	223 292 ***3755*** ***388***	GE312	129 223 264 270 311 319 362 ***371N***
AZ103	223 298 327 357	GE329	224 311 327
AZ105	192 261	GE330	148
AZ106	192 261	GE338	140
AZ10c	***060N*** 067 ***275N*** ***306N*** ***313N*** ***337N*** ***339N*** 355 360 ***388N*** ***389N***	GE339	067 355
AZ11	067 183 311 3275	GE346	126 163 186 243 249 294 356 360 ***387*** ***388***
AZ111c	207 ***265N*** ***269N*** ***281*** ***3065***	GE363	rCRS
AZ12	325 360	GE379	126 294 296
AZ16	311 360	GE394	126 186 249 287 360
AZ17	126 209 ***281*** 309 3182 ***3755*** 390	GE403	129 223 264 270 311 319 362 391
AZ18	298	GE409	129 223 264 270 311 319 362 391
AZ19	069 126 193	GE414	223 292
AZ197	129 223 294 391	GE426	256 270 325 399
AZ2	***280*** ***281***	GE430	086 249 287 360
AZ204	298	GE44	148 223 288 298 319 327
AZ205c	067 183 311 3275 ***387N*** ***388N***	GE456	rCRS
AZ208	145 223 239 354 360 ***390*** ***391*** ***392***	GE46	126 2705 294 296
AZ20c	***059N*** ***064N*** 069 ***122N*** ***157N*** ***162N*** 278	GE478	137N 176
AZ21	067 355	GE479	256 270 293 399

Table 3 (continued)

Code	HVS-I variation (+16000)	Code	HVS-I variation (+16000)
AZ22	093 224 311	GE484	rCRS
AZ23	***281***	GE488	***0604*** ***1958*** ***279N*** ***280N*** ***281N*** ***384***
AZ24	234	GE490	rCRS
AZ26	209 309 3182	GE528	rCRS
AZ274	117 126 294 296	GE59	224 2462 311 ***3711***
AZ28	224 311	GE63	067 355
AZ29	rCRS	GE64	067 355
AZ3	354	GE84	256 352 399
AZ31	086 356	GE88	067 129
AZ32	126 163 186 294 ***382***	GE98	rCRS
AZ342	***280*** ***281*** 298	GEVN	126 163 186 2056 294
AZ365	218	IN303	069 129 223 311
AZ396	069 126 193 274	IN816	066 192 256 270 304
AZ4	311	IN820	366
AZ408	169 290	IN821	304 335
AZ40c	183 ***195N*** 242 ***263N*** ***280N*** ***281N*** ***282N*** ***288N*** ***3034*** ***3221*** ***375N***	IN822	179 356
AZ41	126 163N 186 294	IN823	***0856*** 223 353 ***390*** ***391***
AZ514	***1685*** ***281***	IN826	192 223 292 311 ***363*** ***390*** ***391***
AZ58c	***042N*** ***060N*** 207 230N 356 ***384***	IN827	069 126 193 256 335 ***390*** ***391***
AZ5c	rCRS	IN828	126 ***390*** ***391***
AZ6	154 ***1685*** ***280*** ***281*** 356 ***384***	IN862	192 256 270 311
AZ8	223 227 278 362	IN864	066 192 256 270 304
AZ9	***110N*** ***1184*** ***1958*** 239 ***275*** 3182 ***3395***	IN865	rCRS
CH361	256 270 293 399	IN866	rCRS
CH438	067 355	IN867	066 192 256 270 304
CH439	rCRS	IN873	069 126 145 261 290
CH440	354	IN878	129
CH441	126 294 296 304	IN879	rCRS
CH442	051	IN881	168 192 343
CH443	rCRS	IN883	168 192 343
CH444	111 2146 249 ***280*** ***281*** 327 ***388***	IN884	129 223 391
CH445	223 256 275 292 325 ***372*** ***388***	IN885	223 278 290 292 391
CH446	224 296 311	IN886	rCRS
CH447	093 129 212 223	IN887	223 249 311 359
CH448	129 223	IN889	093 129 249 365
CH450	078 126 294 296	IN890	***3872*** ***388***
CH451	***280*** ***281*** 292	IN891	218 297
CH453	256 352 ***387*** ***388***	IN892	066 192 256 270 304
CH454	092	IN895	256 270
CH455	rCRS	IN897	192 223 292 311
CH457	067 216 355	IN898	356
CH465	212 224 311	IN899	069 126
CH473	167 224 311 359	IN900	192 223

Table 3 (continued)

Code	HVS-I variation (+16000)	Code	HVS-I variation (+16000)
CH509	129 223 391	IN901	rCRS
CH510	067 355 391	IN902	261 271 343 368
CH511	292	IN989	356
CH515	rCRS	KAB121	069 126 193 256 335
CH516	178 216 304	KAB124	161 249 288
CH518	069 126 193 256	KAB160	223 362 ***382***
CH520	rCRS	KAB161	069 126 ***382***
CH521	093 129 223 298 327	KAB165	235 291 399
CH557	067 069 355	KAB170	278 311 399
CH564	051 129 1831 362	KAB171	1476 172 223 248 295 297 355
CH565	129	KAB185	078 126 292 294 296
CH569	207 223	KAB186	192 256 270 311
CH570	111 2145 249 290 327	KAB187	093 129 189 213 249 ***281***
CH574	067	KAB230	192 256 270 311
CH578	093 129 186 249 365 ***3936***	KAB236	343
CH579*	093 129 220 298 318 327 359	KAB308	162
CH580	067 355 ***387*** ***388***	KAB315	291 294 295 296
CH583	224 289 ***387*** ***388***	KAB330	271 292
CH597	129	KAB331	129 223 298 327
CH600	***085*** 129 223 298 327	KAB332	223 311
CH604	126 163 186 189 ***281*** 294	KAB333	354
CH605	175	KAB336	rCRS
CH606	126	KAB338	278 311
CH612	223 297 362	KAB340	356
CHE446	037 069 126 145 270 290	KAB341	278 311
CHE471	069 126	KAB342	129
CHE730	192	KAB343	***122*** 138 ***1516*** 224 311 360 ***382*** ***3956***
CHE731	288 362	KAB344	rCRS
CHE732	256 270	KAB345	126 294 296 304
CHE733	129 145 223 269 355	KAB346	129
CHE735	rCRS	KAB347	223 325 362
CHE736	rCRS	KAB348	223 325 355 362 ***367*** ***371*** ***372***
CHE737	126 294 296	KAB349	093 224 311
CHE738	245 294 356	KAB361	343
CHE739	179 1831	KAB363	168 192 343 ***3755***
CHE740	069 126	KAB364	192 256 270 311
CHE742	129 145 223 2876 309	KAB365	304
CHE744	rCRS	KAB366	***025*** ***026*** 111
CHE745	356 362	KAB371	129
CHE749	093 209 224 311	KAB372	rCRS
CHE751	126 294 296	KAB373	129
CHE754	129 1942	KAB380	rCRS
CHE755	223 278 316	KAB381	278 311
CHE763	245 294	KAB385	129 356

Table 3 (continued)

Code	HVS-I variation (+16000)	Code	HVS-I variation (+16000)
CHE765	309 3182	KAB388	362
CHE767	126 294 296	KAB398	111 2145 249 290 327
CHE768	179 356	KAB399	129 223 391
		KAB405	2201 292 ***3805***
		KAB408	224 ***280*** ***281*** 311
		KAB409	rCRS
		KAB410	093 129 186 189 249 365
		KAB419	354
		KAB423	rCRS
		KAB522	223

6 is C→G; 7 is T→A; 8 is T→G. Suffix N means undetermined or ambiguous nucleotide. An asterisk marks the single sequence from this data that has been asserted to be incorrect according to Nasidze et al. (2004).

In underlined bold italics we highlight all mutations that are very likely phantom mutations, as inferred from a comparison with the worldwide mtDNA database and an appreciation of repetitive patterns involving ambiguous nucleotides. There are probably quite a few more (e.g. at position 16360, or some phantom mutations that have erased mutations from the true motifs) as well as other artificial mutations (including documentation errors). The near absence in this data set of the otherwise very frequent 16189 transition signposts yet another kind of problem with these data. A conservative estimate would then posit that these samples were hit by at least approximately 0.5 phantom mutation per sequence—just about the same level of phantom mutations that were inflicted on the Indian HVS-I data set of Roychoudhury et al. (2001).

References

Abbott A (2003) Anthropologists cast doubt on human DNA evidence. Nature 423:468

Adcock G, Dennis E, Easteal S, Huttley G, Jermelin L, Peacock W, Thorne A (2001) Mitochondrial DNA sequences in ancient Australians: implications for modern human origins. Proc Natl Acad Sci USA 98:537–542

Allen M, Engström A-S, Meyers S, Handt O, Saldeen T, von Haeseler A, Pääbo S, Gyllensten U (1998) Mitochondrial DNA sequencing of shed hairs and saliva on robbery caps: sensitivity and matching probabilities. J Forensic Sci 43:453–464

Alonso A, Alves C, Suárez-Mier MP, Albarrán C, Pereira L, Fernández de Simón L, Martín P, García O, Gusmão L, Sancho M, Amorim A (2005) Mitochondrial DNA haplotyping revealed the presence of mixed up benign and neoplastic tissue sections from two individuals on the same prostatic biopsy slide. J Clin Pathol 58:83–86

Andrews RM, Kubacka I, Chinnery PF, Lightowlers RN, Turnbull DM, Howell N (1999) Reanalysis and revision of the Cambridge reference sequence for human mitochondrial DNA. Nat Genet 23:147

Bandelt H-J (2004) Etruscan artifacts. Am J Hum Genet 75:919–920

Bandelt H-J (2005a) Exploring reticulate patterns in DNA sequence data. In: Bakker FT, Chatrou LW, Gravendeel B, Pelser PB (eds) Plant species-level systematics: new perspectives on pattern and process. Regnum Vegetabile 142. Koeltz, Königstein, pp 245–270

Bandelt H-J (2005b) Mosaics of ancient mitochondrial DNA: positive indicators of nonauthenticity. Eur J Hum Genet 13:1106–1112

Bandelt H-J, Kivisild T (2006) Quality assessment of DNA sequence data: autopsy of a mis-sequenced mtDNA population sample. Ann Hum Genet. DOI 10.1111/j.1529-8817.2005.00234.x

Bandelt H-J, Parson W (2004) Fehlerquellen mitochondrialer DNS-Datensätze und Evalution der mtDNS-Datenbank D-Loop-BASE. Rechtsmedizin 14:251–257

Bandelt H-J, Macaulay V, Richards M (2000) Median networks: speedy construction and greedy reduction, one simulation, and two case studies from human mtDNA. Mol Phylogenet Evol 16:8–28

Bandelt H-J, Lahermo P, Richards M, Macaulay V (2001) Detecting errors in mtDNA data by phylogenetic analysis. Int J Legal Med 115:64–69

Bandelt H-J, Quintana-Murci L, Salas A, Macaulay V (2002) The fingerprint of phantom mutations in mitochondrial DNA data. Am J Hum Genet 71:1150–1160

Bandelt H-J, Salas A, Bravi C (2004a) Problems in FBI mtDNA database. Science 305:1402–1404

Bandelt H-J, Salas A, Lutz-Bonengel S (2004b) Artificial recombination in forensic mtDNA population databases. Int J Legal Med 118:267–273

Bandelt H-J, Achilli A, Kong Q-P, Salas A, Lutz-Bonengel S, Sun C, Zhang Y-P, Torroni A, Yao Y-G (2005a) Low "penetrance" of phylogenetic knowledge in mitochondrial disease studies. Biochem Biophys Res Commun 333:122–130

Bandelt H-J, Kong Q-P, Parson W, Salas A (2005b) More evidence for non-maternal inheritance of mitochondrial DNA? J Med Genet 42:957–960

Bandelt H-J, Yao Y-G, Kivisild T (2005c) Mitochondrial genes and schizophrenia. Schizophr Res 72:267–269

Barbujani G, Vernesi C, Caramelli D, Castrì L, Lalueza-Fox C, Bertorelle G (2004) Etruscan artifacts: much ado about nothing. Am J Hum Genet 75:923–927

Brandstätter A, Parson W (2003) Mitochondrial DNA heteroplasmy or artefacts—a matter of the amplification strategy? Int J Legal Med 117:180–184

Brandstätter A, Sänger T, Lutz-Bonengel S, Parson W, Béraud-Colomb E, Wen B, Kong Q-P, Bravi CM, Bandelt H-J (2005) Phantom mutation hotspots in human mitochondrial DNA. Electrophoresis 26:3414–3429

Budowle B, Allard MW, Fisher CL, Isenberg AR, Monson KL, Stewart JE, Wilson MR, Miller KW (2002) HVI and HVII mitochondrial DNA data in Apaches and Navajos. Int J Legal Med 116:212–215

Bulayeva K, Jorde LB, Ostler C, Watkins S, Bulayev O, Harpending H (2003) Genetics and population history of Caucasus populations. Hum Biol 75:837–853

Caramelli D, Lalueza-Fox C, Vernesi C, Lari M, Casoli A, Mallegni F, Chiarelli B, Dupanloup I, Bertranpetit J, Barbujani G, Bertorelle G (2003) Evidence for a genetic discontinuity between Neandertals and 24 000-year-old anatomically modern Europeans. Proc Natl Acad Sci USA 100:6593–6597

Carracedo A (ed) (2005) Forensic DNA typing protocols. Methods in molecular biology, vol 297. Humana, Totowa

Cooper A, Rambaut A, Macaulay V, Willerslev E, Hansen AJ, Stringer C (2001) Human origins and ancient human DNA. Science 292:1655–1656

Cooper AR, Poinar H (2000) Ancient DNA: do it right or not at all. Science 289:1139

Cordaux R, Saha N, Bentley GR, Aunger R, Sirajuddin SM, Stoneking M (2003) Mitochondrial DNA analysis reveals diverse histories of tribal populations from India. Eur J Hum Genet 11:253–264

Da Pozzo P, Federico A (2005) Commentary to mitDNA research for the pathogenesis of mitochondrial disorders. Biochem Biophys Res Commun 336:1003–1004

Di Benedetto G, Nasidze IS, Stenico M, Nigro L, Krings M, Lanzinger M, Vigilant L, Stoneking M, Pääbo S, Barbujani G (2000) Mitochondrial DNA sequences in prehistoric human remains from the Alps. Eur J Hum Genet 8:669–677

Excoffier L, Yang Z (1999) Substitution rate variation among sites in mitochondrial hypervariable region I of humans and chimpanzees. Mol Biol Evol 16:1357–1368

Finnilä S, Lehtonen MS, Majamaa K (2001) Phylogenetic network for European mtDNA. Am J Hum Genet 68:1475–1484

Forster P (2003) To err is human. Ann Hum Genet 67:2–4

García-Bour J, Pérez-Pérez A, Álvarez S, Fernández E, López-Parra AM, Arroyo-Pardo E, Turbón D (2004) Early population differentiation in extinct aborigines from Tierra del Fuego-Patagonia: ancient mtDNA sequences and Y-chromosome STR characterization. Am J Phys Anthrop 123:361–370

Gilbert MTP, Willerslev E, Hansen AJ, Barnes I, Rudbeck L, Lynnerup N, Cooper A (2003) Distribution patterns of postmortem damage in human mitochondrial DNA. Am J Hum Genet 72:32–47 (erratum 72:779)

Gilbert MTP, Bandelt H-J, Hofreiter M, Barnes I (2005a) Assessing ancient DNA studies. Trends Ecol Evol 20:541–544

Gilbert MTP, Hansen AJ, Willerslev E, Turner-Walker G, Collins M (2005b) Insights into the processes behind the contamination of degraded human teeth and bone samples with exogenous sources of DNA. Int J Osteoarchaeol 15:1–9

Gilbert MTP, Rudbeck L, Willerslev E, Hansen AJ, Smith C, Penkman KEH, Prangenberg K, Nielsen-Marsh CM, Jans ME, Arthur P, Lynnerup N, Turner-Walker G, Biddle M, Kjølbye-Biddle B, Collins MJ (2005c) Biochemical and physical correlates of DNA contamination in archaeological human bones and teeth excavated at Matera, Italy. J Archaeol Sci 32:785–793

Grzybowski T, Malyarchuk BA, Czarny J, Miśicka-Śliwka D, Kotzbach R (2003) High levels of mitochondrial DNA heteroplasmy in single hair roots: reanalysis and revision. Electrophoresis 24:1159–1165

Helgason A, Sigurðardóttir S, Gulcher JR, Ward R, Stefánsson K (2000) mtDNA and the origin of the Icelanders: deciphering signals of recent population history. Am J Hum Genet 66:999–1016

Herrnstadt C, Preston G, Howell N (2003) Errors, phantom and otherwise, in human mtDNA sequences, Am J Hum Genet 72:1585–1586

Hofreiter M, Vigilant L (2003) Ancient human DNA: phylogenetic applications. In: Encyclopedia of the human genome, Macmillan, London.

Hutter G, Nickenig C, Garritsen H, Hellenkamp F, Hoerning A, Hiddemann W, Dreyling M (2004) Use of polymorphisms in the noncoding region of the human mitochondrial genome to identify potential contamination of human leukemia-lymphoma cell lines. Hematol J 5:61–68

Ingman M, Gyllensten U (2003) Mitochondrial genome variation and evolutionary history of Australian and New Guinean aborigines. Genome Res 13:1600–1606

Ingman M, Kaessmann H, Pääbo S, Gyllensten U (2000) Mitochondrial genome variation and the origin of modern humans. Nature 408:708–713

Kaestle FA, Horsburgh KA (2002) Ancient DNA in anthropology: methods, applications, and ethics. Yearbk Phys Anthrop 45:92–130

Kivisild T, Tolk H-V, Parik J, Wang Y, Papiha SS, Bandelt H-J, Villems R (2002) The emerging limbs and twigs of the East Asian mtDNA tree. Mol Biol Evol 19:1737–1751 (erratum 20:162)

Köchl S, Niederstätter H, Parson W (2005) Deoxyribonucleic acid extraction and quantitation of forensic samples using the phenol-chloroform method and real-time polymerase reaction. In: Carracedo A (ed) Forensic DNA typing protocols. Methods in molecular biology, vol 297. Humana, Totowa, pp 13–30

Kong Q-P, Yao Y-G, Sun C, Bandelt H-J, Zhu C-L, Zhang Y-P (2003) Phylogeny of East Asian mitochondrial DNA lineages inferred from complete sequences. Am J Hum Genet 73:671–676 (erratum 75:157)

Kurtz A, Lueth M, Kluwe L, Zhang T, Foster R, Mautner V-F, Hartmann M, Tan D-J, Martuza RL, Friedrich RE, Driever PH, Wong L-JC (2004) Somatic mitochondrial DNA mutations in neurofibromatosis type 1-associated tumors. Mol Cancer Res 2:433–441

Lee SD, Shin CH, Kim KB, Lee YS, Lee JB (1997) Sequence variation of mitochondrial DNA control region in Koreans. Forensic Sci Int 87:99–116

Lee SD, Lee YS, Lee JB (2002) Polymorphism in the mitochondrial cytochrome B gene in Koreans. An additional marker for individual identification. Int J Legal Med 116:74–78

Loogväli E-L, Roostalu U, Malyarchuk BA, Derenko MV, Kivisild T, Metspalu E, Tambets K, Reidla M, Tolk H-V, Parik J, Pennarun E, Laos S, Lunkina A, Golubenko M, Barać L, Peričić M, Balanovsky OP, Gusar V, Khusnutdinova EK, Stepanov V, Puzyrev V, Rudan P, Balanovska EV, Grechanina E, Richard C, Moisan J-P, Chaventré A, Anagnou NP, Pappa KI, Michalodimitrakis EN, Claustres M, Gölge M, Mikerezi I, Usanga E, Villems R (2004) Disuniting uniformity: a pied cladistic canvas of mtDNA haplogroup H in Eurasia. Mol Biol Evol 21:2012–2021

Malmström H, Storå J, Dalén L, Holmlund G, Götherström A (2005) Extensive human DNA contamination in extracts from ancient dog bones and teeth. Mol Biol Evol 22:2040–2047

Malyarchuk BA, Rogozin IB (2004) On the Etruscan mtDNA contribution to modern humans. Am J Hum Genet 75:920–923

Meyer S, Weiss G, von Haeseler A (1999) Pattern of nucleotide substitution and rate heterogeneity in the hypervariable regions I and II of human mtDNA. Genetics 152:1103–1110

Mogentale-Profizi N, Chollet L, Stévanovitch A, Dubut V, Poggi C, Pradié MP, Spadoni JL, Gilles A, Béraud-Colomb E (2001) Mitochondrial DNA sequence diversity in two groups of Italian Veneto speakers from Veneto. Ann Hum Genet 65:153–166

Monson KL, Miller KWP, Wilson MR, DiZinno JA, Budowle B (2002) The mtDNA Population Database: an integrated software and database resource for forensic comparison. Forensic Sci Commun 4(2). http://www.fbi.gov/hq/lab/fsc/backissu/april2002/miller1.htm

Nasidze I, Stoneking M (2001) Mitochondrial DNA variation and language replacements in the Caucasus. Proc R Soc Lond Ser B 268:1197–1206

Nasidze I, Ling EYS, Quinque D, Dupanloup I, Cordaux R, Rychkov S, Naumova O, Zhukova O, Sarraf-Zadegan N, Naderi GA, Asgary S, Sardas S, Farhud DD, Sarkisian T, Asadov C, Kerimov A, Stoneking M (2004) Mitochondrial DNA and Y-chromosome variation in the Caucasus. Ann Hum Genet 68:205–221

Pääbo S, Poinar H, Serre D, Jaenicke-Després V, Hebler J, Rohland N, Kuch M, Krause J, Vigilant L, Hofreiter M (2004) Genetic analyses from ancient DNA. Annu Rev Genet 38:645–679

Pakendorf B, Stoneking M (2005) Mitochondrial DNA and human evolution. Annu Rev Genomics Hum Genet 6:165–183

Parson W, Brandstätter A, Alonso A, Brandt N, Brinkmann B, Carracedo A, Corach D, Froment O, Furac I, Grzybowski T, Hedberg K, Keyser-Tracqui C, Kupiec T, Lutz-Bonengel S, Mevag B, Ploski R, Schmitter H, Schneider P, Syndercombe-Court D, Sørensen E, Thew H, Tully G, Scheithauer R (2004) The EDNAP mitochondrial DNA population database (EMPOP) collaborative exercises: organisation, results and perspectives. Forensic Sci Int 139:215–226

Parson W, Parsons TJ, Scheithauer R, Holland MM (1998) Population data for 101 Austrian Caucasian mitochondrial DNA d-loop sequences: application of mtDNA sequence analysis to a forensic case. Int J Legal Med 111:124–132

Plaza S, Calafell F, Helal A, Bouzerna N, Lefranc G, Bertranpetit J, Comas D (2003) Joining the pillars of Hercules: mtDNA sequences show multidirectional gene flow in the western Mediterranean. Ann Hum Genet 67:312–328

Prasad BVR, Ricker CE, Watkins WS, Dixon ME, Rao BB, Naidu JM, Jorde LB, Bamshad M (2001) Mitochondrial DNA variation in Nicobarese Islanders. Hum Biol 73:715–725

Reidla M, Kivisild T, Metspalu E, Kaldma K, Tambets K, Tolk H-V, Parik J, Loogväli EL, Derenko M, Malyarchuk B, Bermisheva M, Zhadanov S, Pennarun E, Gubina M, Golubenko M, Damba L, Fedorova S, Gusar V, Grechanina E, Mikerezi I, Moisan JP, Chaventre A, Khusnutdinova E, Osipova L, Stepanov V, Voevoda M, Achilli A, Rengo C, Rickards O, De Stefano GF, Papiha S, Beckman L, Janicijevic B, Rudan P, Anagnou N, Michalodimitrakis E, Koziel S, Usanga E, Geberhiwot T, Herrnstadt C, Howell N, Torroni A, Villems R (2003) Origin and diffusion of mtDNA haplogroup X. Am J Hum Genet 73:1178–1190

Röhl A, Brinkmann B, Forster L, Forster P (2001) An annotated mtDNA database. Int J Legal Med 115:29–39

Roychoudhury S, Roy S, Basu A, Banerjee R, Vishwanathan H, Usha Rani MV, Sil SK, Mitra M, Majumder PP (2001) Genomic structures and population histories of linguistically distinct tribal groups of India. Hum Genet 109:339–350

Salas A, Richards M, De la Fe T, Lareu M-V, Sobrino B, Sánchez-Diz P, Macaulay V, Carracedo A (2002) The making of the African mtDNA landscape. Am J Hum Genet 71:1082–1111

Salas A, Carracedo Á, Macaulay V, Richards M, Bandelt H-J (2005a) A practical guide to mitochondrial DNA error prevention in clinical, forensic, and population genetics. Biochem Biophys Res Commun 335:891–899

Salas A, Prieto L, Montesino M, Albarrán C, Arroyo E, Paredes-Herrera MR, Di Lonardo AM, Doutremepuich C, Fernández-Fernández I, de la Vega AG, Alves C, López CM, López-Soto M, Lorente JA, Picornell A, Espinheira RM, Hernández A, Palacio AM, Espinoza M, Yunis JJ, Pérez-Lezaun A, Pestano JJ, Carril JC, Corach D, Vide MC, Álvarez-Iglesias V, Pinheiro MF, Whittle MR, Brehm A, Gómez J (2005b) Mitochondrial DNA error prophylaxis: assessing the causes of errors in the GEP'02-03 proficiency testing trial. Forensic Sci Int 148:191–198

Salas A, Yao Y-G, Macaulay V, Vega A, Carracedo Á, Bandelt H-J (2005c) A critical reassessment of the role of mitochondria in tumorigenesis. PloS Med 2:e296

Santos SEB, Ribeiro-dos-Santos AKC, Meyer D, Zago MA (1996) Multiple founder haplotypes of mitochondrial DNA in Amerindians revealed by RFLP and sequencing. Ann Hum Genet 60:305–319

Stenico M, Nigro L, Bertorelle G, Calafell F, Capitanio M, Corrain C, Barbujani G (1996) High mitochondrial sequence diversity in linguistic isolates of the Alps. Am J Hum Genet 59:1363–1375

Tan D-J, Chang J, Chen W-L, Agress LJ, Yeh K-T, Wang B, Wong L-J (2003) Novel heteroplasmic frameshift and missense somatic mitochondrial DNA mutations in oral cancer of betel quid chewers. Genes Chromosomes Cancer 37:186–194

Tawata M, Hayashi JI, Isobe K, Ohkubo E, Ohtaka M, Chen J, Aida K, Onaya T (2000) A new mitochondrial DNA mutation at 14577 T/C is probably a major pathogenic mutation for maternally inherited type 2 diabetes. Diabetes 49:1269–1272

Taylor RW, McDonnell MT, Blakely EL, Chinnery PF, Taylor GA, Howell N, Zeviani M, Briem E, Carrara F, Turnbull DM (2003) Genotypes from patients indicate no paternal mitochondrial DNA contribution, Ann Neurol 54:521–524

Thalmann O, Hebler J, Poinar HN, Pääbo S, Vigilant L (2004) Unreliable mtDNA data due to nuclear insertions: a cautionary tale from analysis of humans and other great apes. Mol Ecol 13:321–335

Thangaraj K, Singh L, Reddy AG, Rao VR, Sehgal SC, Underhill PA, Pierson M, Frame IG, Hagelberg E (2003) Genetic affinities of the Andaman Islanders, a vanishing human population. Curr Biol 13:86-93

Tommaseo-Ponzetta M, Attimonelli M, De Robertis M, Tanzariello F, Saccone C (2002) Mitochondrial DNA variability of West New Guinea populations. Am J Phys Anthrop 117:49–67

Vernesi C, Di Benedetto G, Caramelli D, Secchieri E, Katti E, Malaspina P, Novelletto A, Terribile Wiel Marin A, Barbujani G (2001) Genetic characterization of the body attributed to the evangelist Luke. Proc Natl Acad Sci USA 98:13460–13463

Vernesi C, Caramelli D, Dupanloup I, Bertorelle G, Lari M, Cappellini E, Moggi-Cecchi J, Chiarelli B, Castrì L, Casoli A, Mallegni F, Lalueza-Fox C, Barbujani G (2004) The Etruscans: a population-genetic study. Am J Hum Genet 74:694–704

Vigilant L, Wilson AC, Harpending H (1991) African populations and the evolution of human mitochondrial DNA. Science 253:1503–1507

Vives-Bauza C, Andreu AL, Manfredi G, Beal MF, Janetzky B, Gruenewald TH, Lin MT (2002) Sequence analysis of the entire mitochondrial genome in Parkinson's disease. Biochem Biophys Res Commun 290:1593–1601

Vona G, Falchi A, Moral P, Calò CM, Varesi L (2005) Mitochondrial sequence variation in the Guahibo Amerindian population from Venezuela. Am J Phys Anthropol 127:361–369

Wallace DC, Stugard C, Murdock D, Schurr T, Brown MD (1997) Ancient mtDNA sequences in the human nuclear genome: a potential source of errors in identifying pathogenic mutations. Proc Natl Acad Sci USA 94:14900–14905

Walsh PS, Metzger DA, Higuchi R (1991) Chelex 100 as a medium for simple extraction of DNA for PCR-based typing from forensic material. Biotechniques 10:506–513

Wong L-J, Tan D-J, Bai R-K, Yeh K-T, Chang J (2004) Molecular alterations in mitochondrial DNA of hepatocellular carcinomas: is there a correlation with clinicopathological profile? J Med Genet 41:e65

Yao Y-G, Zhang Y-P (2003) Pitfalls in the analysis of ancient human mtDNA. Chin Sci Bull 48:826–830

Yao Y-G, Kong Q-P, Bandelt H-J, Kivisild T, Zhang Y-P (2002) Phylogeographic differentiation of mitochondrial DNA in Han Chinese. Am J Hum Genet 70:635–651

Yao Y-G, Kong Q-P, Man X-Y, Bandelt H-J, Zhang Y-P (2003a) Reconstructing the evolutionary history of China: a caveat about inferences drawn from ancient DNA. Mol Biol Evol 20:214–219

Yao Y-G, Macaulay V, Kivisild T, Zhang Y-P, Bandelt H-J (2003b) To trust or not to trust an idiosyncratic mitochondrial data set. Am J Hum Genet 72:1341–1346

Yao Y-G, Bravi CM, Bandelt H-J (2004) A call for mtDNA data quality control in forensic science. Forensic Sci Int 141:1–6

Yao Y-G, Salas A, Bravi CM, Bandelt H-J (2006) A reappraisal of complete mtDNA variation in East Asian families with hearing impairment. Hum Genet (in press). DOI 10.1007/s00439-006-0154-9

Zhao H, Li R, Wang Q, Yan Q, Deng J-H, Han D, Bai Y, Young W-Y, Guan M-X (2004) Maternally inherited aminoglycoside-induced and nonsyndromic deafness is associated with the novel C1494T mutation in the mitochondrial 12S rRNA gene in a large Chinese family. Am J Hum Genet 74:139–152

Zischler H, Geisert H, von Haeseler A, Pääbo S (1995) A nuclear fossil of the mitochondrial D-loop and the origin of modern humans. Nature 378:489–492

Part II
Evolution of Human mtDNA

Nucleic Acids and Molecular Biology, Vol. 18
Hans-Jürgen Bandelt, Vincent Macaulay, Martin Richards (Eds.)
Human Mitochondrial DNA and the Evolution of *Homo sapiens*

The World mtDNA Phylogeny

Toomas Kivisild (✉) · Mait Metspalu · Hans-Jürgen Bandelt ·
Martin Richards · Richard Villems

Institute of Molecular and Cell Biology, Tartu University and Estonian Biocentre,
Ria 23, 51010 Tartu, Estonia
tkivisild@ebc.ee

1
Haplotypes and Trees

What makes DNA attractive for those interested in questions about human evolutionary history is its inherent nature to reproduce itself imperfectly. After all, it was Darwin who pointed out that if there were no variation evolution would have been impossible. After the maternal inheritance of mitochondrial DNA (mtDNA) was first shown in humans (Giles et al. 1980) it took another half a decade, when another copy machine—the PCR method–was developed, and an avalanche of genetics studies started using the molecule as a tool to investigate the origins of human populations.

Maternal inheritance governs the haploid nature of this genome. Because all nuclear loci, except for the Y chromosome, are diploid, any newly arising mutation in one of the sister chromosomes generates a diplotype, a combined pair of sequence types that comprise their carrier's genotype. It is usually very difficult, if not impossible, to deduce whether the newly arisen mutation occurs on the background of the haplotype inherited from the individual's mother or father. If there are informative positions relatively close to the new mutation that distinguish the paternal and maternal alleles then it is, in principle, possible to establish directly the allelic associations, which are required for inferring phylogenetic trees of individual sequences. However, for the routine work of population genetics this approach is, at least for the time being, far too costly and time-consuming.

And, after all, who cares? Population trees, which organize the populations under study in a hierarchical fashion, can be drawn from variable sources of genetic data without the need to infer haplotypes (Cavalli-Sforza et al. 1994). Yet the interpretation of such trees can often be quite knotty, as it relies on hypotheses about population histories that are hard to justify in practice (Hey 1998). Population trees are based on the general assumption that for each branching event the parent population splits instantaneously into two daughter subpopulations (again of the same size as the parent population), with no subsequent gene flow. These daughter populations are then supposed to diverge genetically (i.e. their gene frequency distributions become gradually more different) at an equal rate as a result of the action of random genetic drift. In reality, the effective population sizes, and therefore the impact of drift, inevitably vary over time—consider, for example, the dramatic effects of the Quaternary ice ages on the global biosphere (Hewitt 2000), or the genetic-outlier status of the European Saami, as compared with other Finno-Ugric populations of larger effective size (Tambets et al. 2004). Prehistoric gene flows from other populations are even harder to take into account.

Another issue not always adequately taken into account is that, while in cladistic analysis *synapomorphic* characters—those that are in a derived state in at least two taxa—underlie the structure of the hierarchy of the tree, in the case of population trees the branching structure is determined by the averaged differences between the populations, irrespective of the nature of the characters that define them. Consider, for example, trees obtained from a selection of 28 individual complete mtDNA sequences (Ingman et al. 2000). These can be grouped into five continental 'populations' where the effect of gene flow and difference in total effective population sizes can be considered minimal, in comparison with cases of small groups living in close proximity (Fig. 1). The clustering pattern in a population tree constructed from these data (Fig. 1a) suggests that Europeans and Oceanian populations (from New Guinea and the remote Pacific) are more closely related to each other than either is to Asian populations, which might lead to the inference that the former two share a more recent common ancestral population. The same is true for Siberians and Native Americans. Whilst the second cluster is consistent with the widely accepted view on Asian origins of Native American populations, the clustering of Europeans together with Oceanians might appear to warrant somewhat drastic revisions to accounts of Eurasian prehistory.

The tree of individual mtDNA sequences (Fig. 1b), on the other hand, whilst rather more complex at first glance, helps to explain this unexpected grouping at the population level. It shows that Oceanians, Native Americans, and Siberians all share a common ancestry within several clades of mtDNA sequences, haplogroup B and haplogroup M. Besides these, some New Guineans fall into a further clade that is not shared with any other populations. It is important to note here that the branching pattern emerging from this tree is not an 'accident' resulting from small sample sizes for each individual region. This is clear, because the phylogeography of these common region-specific clades was explored extensively in earlier (lower-resolution) studies that consistently show this distribution of these clades amongst these populations. The unexpected pattern can, instead, be explained by the fact that the Oceanian- and European-specific clades are nested, as one can see when drawing out the phylogeny of individual sequences (Fig. 1b), within the same parental haplogroup R. But haplogroup R has an essentially worldwide (non-African) distribution. As already noted, it is also found in Siberian and Native American lineages (in the derived form of haplogroup B). This

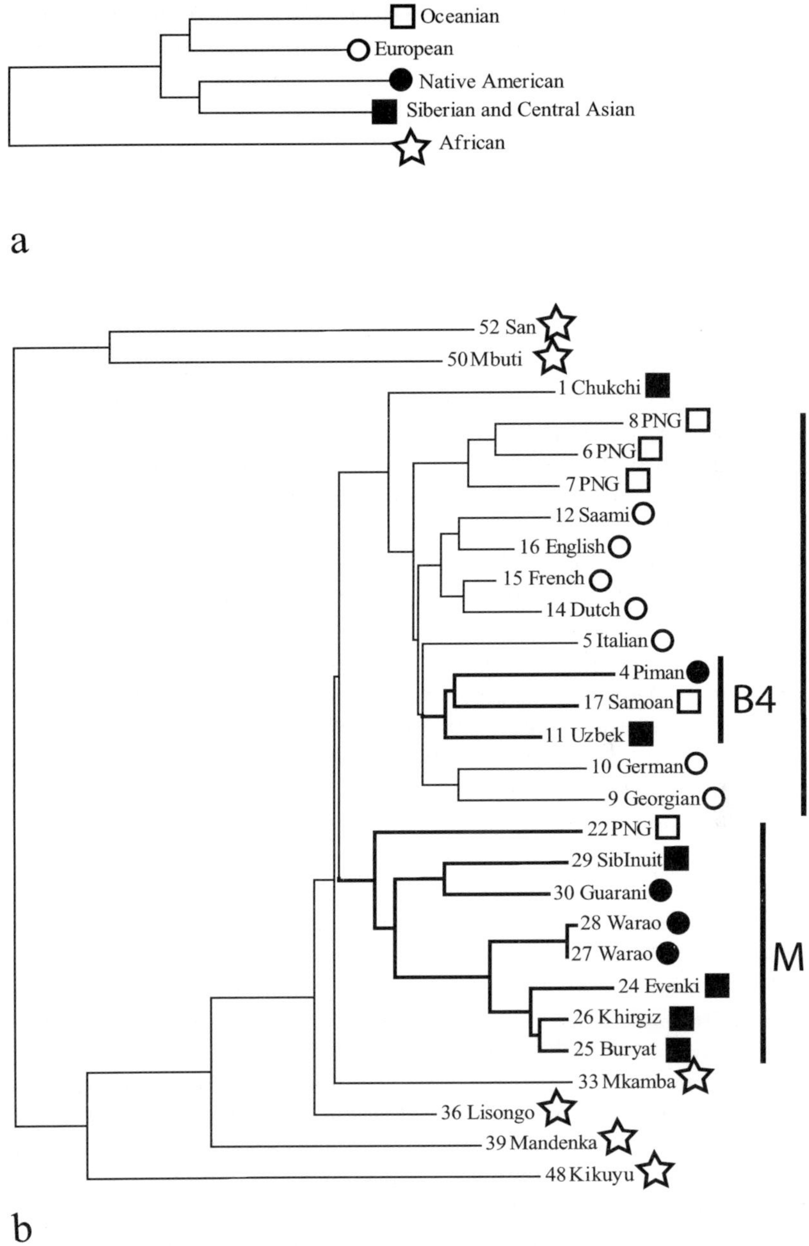

Fig. 1 A tree of populations and a tree of individuals. **a** Neighbour-joining (*NJ*) population tree, constructed from net nucleotide distances between populations, and **b** NJ tree of individuals. Both trees are unrooted (but drawn midpoint rooted) and were constructed using MEGA (Kumar et al. 2001) from the same subset of 28 complete sequences, representing five continental populations. *PNG* Papua New Guinea. (Taken from Ingman et al. 2000)

makes it uninformative in a cladistic sense as a *symplesiomorphic* (i.e. ancestrally shared) group. Unfortunately, most population genetics approaches overlook the phylogenetic signals in the data when drawing relationships between populations based on (averaged) genetic distances, no matter how and under which model these distances are computed.

The phylogeographic analysis of mtDNA variation, departing as it does from a reconstructed gene tree or network such as that in Fig. 1b, resembles more the analysis of human immunodeficiency virus (HIV) lineages from a group of carriers (Myers et al. 1995) than it does the reconstruction of evolutionary history of species or genera. Like mtDNA, HIV also follows a clonal pattern when transmitted from individual to individual. The major difference in inheritance is that mtDNA evolves by the accumulation of point mutations, whereas HIV also undergoes recombination. Although mtDNA might exceptionally undergo somatic recombination (as claimed by Kraytsberg et al. 2004; but see Chap. 2), a number of events are needed in tandem for a recombinant to enter the gene pool: paternal penetrance, the opportunity for recombination to take place, the passing on of the recombinant into the next generation, and its spread in the population. There has been no convincing evidence to date of recombinants having been inherited, or affecting the tree topology (Elson et al. 2001; Kivisild and Villems 2000).

As mitochondria are transmitted from one generation to the next, mutations occur in their DNA molecules that accumulate over time, provided that they are successfully passed through bottlenecks of egg-cell replications (Chap. 2). The mutation rate is not uniform across the molecule: the noncoding control region, or D-loop, shows on average more than 5 times higher sequence variation than the coding region (Ingman et al. 2000; Chap. 4). This suggests that the information needed to control mtDNA replication and gene expression may depend less on precise conservation of the primary structure of the DNA sequence behind these functions than do the regions encoding ribosomal proteins and RNAs. Within the hypervariable segments (HVS) of the control region (Chap. 1), several sites can be considered as mutational hotspots as they are highly recurrent in the phylogenetic reconstructions, mutating many times on the tree (Hasegawa et al. 1993; Malyarchuk and Rogozin 2004; Chap. 4). The same set of mutations has also been shown to be the most recurrent ones in pedigree studies (Heyer et al. 2001). At the same time, even in the hypervariable parts of the control region, there are sites that rarely if ever undergo mutation, although some have been known to undergo mutation in vitro or in silico as a result of errors generated in the laboratory, in the computer or on the printed page (Chap. 6). Because the mutations at hypervariable sites tend to become saturated (i.e. mutated more than once back and forth) in a relatively shorter evolutionary time span than those occurring on conservative sites, molecular divergence is not linear over time. Molecular clock estimates based on control-region divergence probably saturate well within the last 100 000 years.

2
Haplogroup Structure and Definition

Phylogenetic trees are graphs generated in an attempt to group species—or indeed any identifiable taxonomic units—by the hierarchic order of their descent. Trees constructed from two phylogenetic studies of a single locus can, in principle, be identical in their overall branching order (or topology), even if the underlying sequences that were used to draw them are quite different. In turn, seemingly dissimilar branching patterns between two trees do not necessarily tell us that the corresponding sets of sequences are related only distantly.

A way to overcome the puzzle of comparing trees obtained from different studies, but using the same locus, is to label the branches they are composed of. The currently accepted nomenclature of mtDNA haplogroups (groups of haplotypes) was initiated by Torroni et al. (1993) by distillation of the mtDNA restriction fragment length polymorphism (RFLP) data set of Native Americans to a restricted number of mutations that defined the four basal (or primary) branches in the tree, named alphabetically as A, B, C, and D. Later, the haplogroup structures of other continental populations were characterized (Torroni et al. 1994a, b) and the cladistic rules for the hierarchical ordering of haplogroups and subhaplogroups were overtly established by Richards et al. (1998). The same principles were later used as the basis of the (long-overdue) reformed nomenclature of the Y chromosomal tree (YCC 2002).

This simple hierarchical haplogroup labelling is illustrated in Fig. 2a. We use the example of two haplogroups, M and N, that are the two most prominent haplogroups in the global pool of mtDNA variation. The subhaplogroup nomenclature moves from the root to the tips, with the gain of extra identifying symbol (either a letter or a number, in alternating order) attached to the label at each branching point: viz. M > M7 > M7a > M7a1. Each branch is defined by one or (preferably) more mutations, combination of which tags it for direct genotyping within a population. Most commonly the mutations are reported as variants or deviations from the Cambridge reference sequence (CRS), the first sequenced mtDNA (Anderson et al. 1981), or its correction (Andrews et al. 1999), the revised CRS (rCRS), which, using the current haplogroup nomenclature, falls into an H2b subclade (Loogväli et al. 2004) of the most frequent haplogroup in Europe, haplogroup H. The star symbol is used to denote members of a haplogroup that do not belong to any named subhaplogroup (relative to the current classification). For example, the single M* haplotype in Fig. 2a carries the star symbol because it belongs neither to M1 nor to M7. The star is not attached to the two haplogroup N lineages on this figure because no downstream haplogroup of N was defined or genotyped by the researcher who provided us with the data to draw the plot. Similarly, all M1, M7, and M* lineages can be pooled under a common signifier M, without the star.

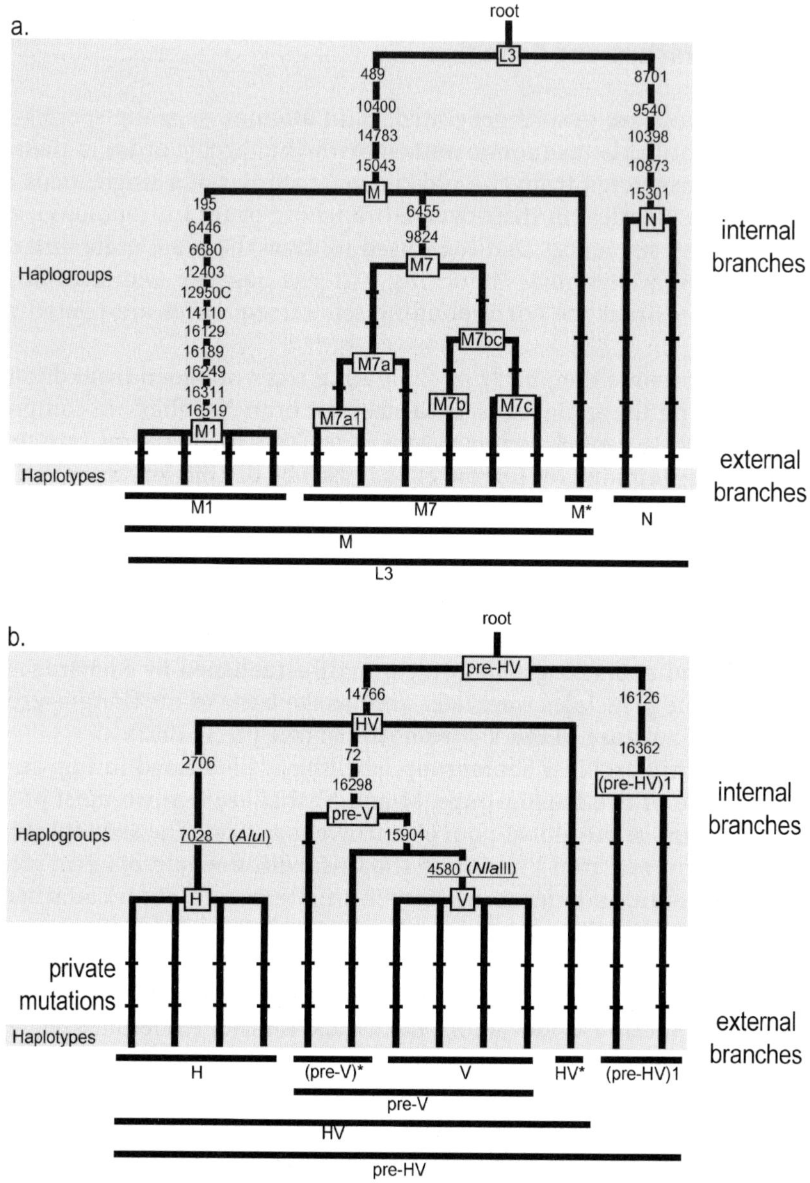

Fig. 2 Two examples of the application of the rules of haplogroup nomenclature. **a** Root-to-tips labelling of subclades of a sample of clades within haplogroups M and N. Haplogroup-defining mutations, at nucleotide positions numbered relative to the revised Cambridge reference sequence (Andrews et al. 1999) are shown on each inner branch of the tree, with character change specified only for transversions. Definitions of M1 are due to A. Torroni (unpublished data) and of M7 are according to Kong et al. (2003). **b** Tips-to-root labelling of ancestral haplogroups nesting haplogroups H and V. The *star* denotes haplotypes remaining unclassified under current nomenclature

Mitochondrial types that share only a single polymorphism, however, should not necessarily warrant the proper definition of a mitochondrial haplogroup, despite the fact that, in the early days of mtDNA analysis, a single site was often all there was to go on. When, however, the limit of resolution is reached with complete sequences, one may decide to erect a haplogroup that is based on a single site—provided that this is not a mutational hotspot. Mutational recurrence in mtDNA, even in its coding regions, is frequent enough (Herrnstadt et al. 2002; Kivisild et al. 2006) to often dazzle the phylogenetic signal when using only a limited number of polymorphic sites. Mutation at position 10398, for example, which is one of the defining markers of haplogroup N (Fig. 2a), frequently leads to reversion to the ancestral state (common throughout Africa). This takes place, for example, on the defining branches of haplogroup B5 in Asia and two daughter clades of haplogroup N, J and K, in Europe, which led to the suggestion that haplogroups U and K occupy disjoint branches and which caused a misplacement of haplogroup J lineages that did not allow for the identification of the JT clade in the early RFLP-based trees (e.g. Torroni et al. 1996). Now, however, haplogroup K is well identified as a part of haplogroup U (sharing three coding-region mutations with all other U clades), although K and U are still treated in the medical literature as if they were disjoint phylogenetic units (e.g. http://www.mitomap.org/WorldMigrations.pdf; Giacchetti et al. 2004; Huerta et al. 2005) and the designation UK (Shen et al. 2004) or Uk (Coskun et al. 2004; Wallace 2005), or KU (Ruiz-Pesini et al. 2000) for haplogroup U still occasionally occurs as an archaism. Similarly, a variant from the CRS at position 15607 is one of the defining markers for both haplogroup T in Europe and haplogroup P in Oceania. As another example, haplogroup L5 shares the ancestral state at a number of defining sites with haplogroups L0, L1, and L2 relative to haplogroups L3 and L4. Yet, one ancestor of a subclade of L5 has experienced a parallel mutation at position 3594, which is commonly taken as the key site by which to distinguish haplogroups L3 and L4 from the remaining sub-Saharan specific mtDNA haplogroups (Kivisild et al. 2004). It is therefore the combination of multiple defining markers, embracing the information from the whole set of branches of the tree, rather than the status at any single point mutation, that defines a haplogroup.

mtDNA data continued to accumulate in the late 1990s, at an improved molecular resolution. It gradually became apparent that, besides the most common branches that had already been assigned an alphabetic haplogroup label, there existed multiple intermediate phylogenetic branching points that could not be easily adapted to the simple pre-existing root-to-tips oriented hierarchy of the finite Latin alphabet, without distorting the existing scheme of mtDNA haplogroup nomenclature. As an example (Fig. 2b), consider the two common European haplogroups, H and V, initially characterized by their defining RFLP motifs (Torroni et al. 1994a, 1998). It turned out that these two haplogroups share a common ancestor, or to be precise, a single common variant at position 14766 of the mtDNA molecule, and that there are several other related lineages derived from this ancestor that are spread at moderate or low frequencies in the Near East, Europe, and North Africa. To avoid drastically reshuffling the existing nomenclature, it was therefore necessary to create combined haplogroup names, following the tips-to-root order, to cope with the existing labels (Richards et al. 1998; Saillard et al. 2000; Torroni et al. 2001a). As a result, haplogroups H and V can be seen now as subgroups of the superhaplogroup HV, which itself is derived from the parental group pre-HV. The first sister group of HV in pre-HV is called (pre-HV)1 (and not, by the way, pre-HV1, which would signify, potentially, the first sister group of HV1 nested over a common ancestor within the HV clade, if such be found in the future). Anyway, prefixing with 'pre' for naming haplogroups is now rather avoided (with pre-HV and pre-V as the only surviving instances) since otherwise cumbersome circumfixes would eventually arise. In contrast, names such as UKJT, occasionally found in the medical literature, that are supposed to designate haplogroup clusters (Pyle et al. 2005) constitute polyphyletic entities and are therefore meaningless.

The current nomenclature thus developed, mainly in the context of the rather complex European mtDNA variation, as the synthesis (Torroni et al. 1996; Richards et al. 1998; Macaulay et al. 1999) of both control-region (Richards et al. 1996) and high-resolution RFLP-based (Torroni et al. 1994a) classifications. Alone, although the RFLP-based system provided better resolution than control-region sequencing, both systems had weaknesses: for example, control-region sequences were unable to distinguish some members of haplogroup M from members of haplogroup N in East Asia, and many members of haplogroup H from members of haplogroups HV or U in West Eurasia; whereas high-resolution RFLP analysis failed to resolve haplogroup R. The two systems have been broadly mutually supporting when brought together in combination, whilst in detail being continuously updated by accumulating data from the complete mtDNA sequences (Finnilä et al. 2001; Herrnstadt et al. 2002; Achilli et al. 2004; Coble et al. 2004; Loogväli et al. 2004; Palanichamy et al. 2004; Quintana-Murci et al. 2004; Quintáns et al. 2004; Tambets et al. 2004; Friedlaender et al. 2005; Merriwether et al. 2005; Trejaut et al. 2005; Kivisild et al. 2006).

In summary, the notational system for mtDNA haplogroups (as well as Y-chromosome haplogroups) is quite lucid—but, of course, users would need to appreciate its long genesis and understand its basic rules. The system has not "progressed in a rather haphazard manner, resulting in nonmonophyletic groupings receiving the same designation" as recently proclaimed by Pakendorf and Stoneking (2005), who, in misinterpreting Fig. 2 of Tanaka et al. (2004), pointed to "the clade comprising branches N9, Y1b, and Y2". As for the correct reading of this figure, haplogroup N9 splits into two sister branches, N9a and N9b′Y, whence the old RFLP-derived haplogroup Y (which itself is composed of two subhaplogroups, Y1 and Y2) plays the role of a haplogroup that could have been named N9c, so that N9b′Y could have become N9bc (thus perfectly paralleling the situation with haplogroups M7a and M7bc in Fig. 2; see also Chap. 8). But it is certainly preferable to retain the traditional name of haplogroup Y. The system is sufficiently flexible to allow for different naming that, for instance, could instead posit N9b as the legitimate name for the sister branch of N9a, so that the currently used name N9b would become N9b1 and, consequently, Y would serve as a shorthand for N9b2. The latter naming strategy, however, would not be prudent in this particular case, because the potential relationship of N9b and Y is only weakly supported by the quite variable position 5147 (Kivisild et al. 2006) and the notorious hypervariable position 16519 (which is quite often disregarded in phylogenetic analyses of human mtDNA). Since Y1 does not bear the 5147 transition, it would be equally parsimonious (ignoring 16519) to postulate that N9a, N9b, and Y constitute three independent branches of haplogroup N9, where N9b and Y2 independently gained the 5147 change. The current nomenclature, over which Pakendorf and Stoneking (2005) stumbled, is then optimal in that it permits both alternative views on the phylogeny without changing nomenclature: the tree favoured by Tanaka et al. (2004) would claim the existence of haplogroup N9b′Y, whereas the alternative view would not accept the latter as a legitimate haplogroup. What could possibly appear to be hazardous to the casual reader is the unfortunate reintroduction of novel names incurred by neglecting the already existing haplogroup nomenclature (e.g. Starikovskaya et al. 2005).

3 Phylogeographic Inferences

The geographic spread of a species in the light of the phylogeny is the subject of an approach known as phylogeography (Avise 2000). The process by which the most recent common ancestor (MRCA) of the whole tree or of a given haplogroup is defined consists essentially of three steps. Firstly, an unrooted tree or a network is sought that is the shortest, the most parsimonious, or the likeliest graph to describe the genetic relationship between observed haplotypes. Secondly, the closest available outgroup is chosen to define the root of the tree. Finally, the character states at every variable position are determined for the root haplotype to define the sequence characteristic for the MRCA. It is the geographic spread of the first principal branches stemming out of the MRCA node that is used in phylogeographic inferences to determine the probable source region of the variation. Even if there is no geographic structure among the sampled species, the MRCA would be defined for a genetic locus and the phylogeographic approach applied blindly would attempt to determine the region of its origin (without much success though). However, in the case of relatively long term isolation of different geographic areas, or ge-

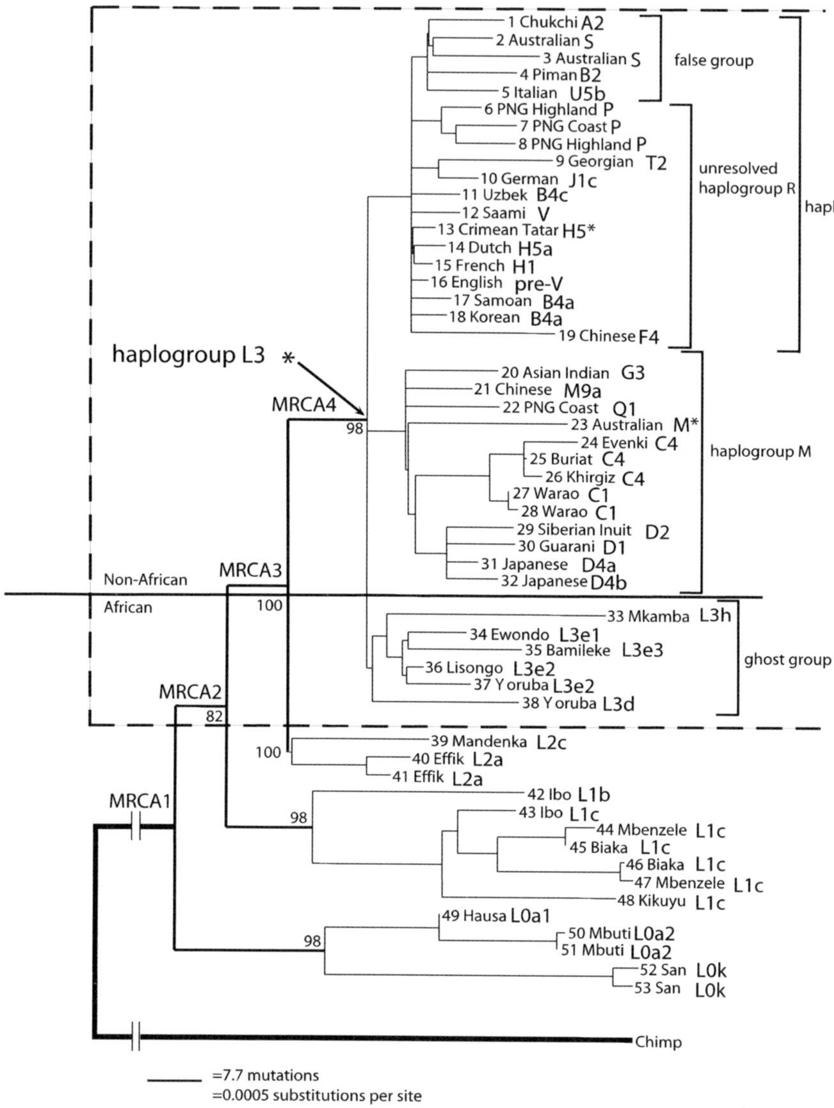

1 Chukchi A2
2 Australian S
3 Australian S
4 Piman B2
5 Italian U5b
false group
6 PNG Highland P
7 PNG Coast P
8 PNG Highland P
9 Georgian T2
10 German J1c
11 Uzbek B4c
12 Saami V
13 Crimean Tatar H5*
14 Dutch H5a
15 French H1
16 English pre-V
17 Samoan B4a
18 Korean B4a
19 Chinese F4
unresolved haplogroup R
haplogroup N
haplogroup L3 *
MRCA4
98
20 Asian Indian G3
21 Chinese M9a
22 PNG Coast Q1
23 Australian M*
24 Evenki C4
25 Buriat C4
26 Khirgiz C4
27 Warao C1
28 Warao C1
29 Siberian Inuit D2
30 Guarani D1
31 Japanese D4a
32 Japanese D4b
haplogroup M
Non-African
African
MRCA3
100
33 Mkamba L3h
34 Ewondo L3e1
35 Bamileke L3e3
36 Lisongo L3e2
37 Yoruba L3e2
38 Yoruba L3d
ghost group
MRCA2
82
100
39 Mandenka L2c
40 Effik L2a
41 Effik L2a
MRCA1
98
42 Ibo L1b
43 Ibo L1c
44 Mbenzele L1c
45 Biaka L1c
46 Biaka L1c
47 Mbenzele L1c
48 Kikuyu L1c
98
49 Hausa L0a1
50 Mbuti L0a2
51 Mbuti L0a2
52 San L0k
53 San L0k
Chimp
=7.7 mutations
=0.0005 substitutions per site

a.

Fig. 3 NJ trees of 53 human coding-region/complete mitochondrial DNA (*mtDNA*) sequences (displayed midpoint rooted). **a** Tree based on the coding region. The first four deepest splits support the African origin of human mtDNA variation. MRCA4, defining haplogroup L3 (encircled with a *dashed line*), which is the most recent common ancestor (*MRCA*) for Asian, Oceanian, and Native American populations, also gives rise to several African lineages. Haplogroup affiliations have been added to individual sequences. **b** NJ tree for the same data as in **a** but newly constructed using MEGA with the default settings (insertions and deletions ignored, Kimura 2-parameter method, uniform rates, 1000 bootstraps). **c** NJ constructed as in **b** but now including the control region. (**a** Redrawn from Ingman et al. 2000)

netic drift affecting small populations distributed over a large area, or just because of insufficient sampling, or other reasons than that, one of the principal branches may turn out to be region-specific and the phylogeographic inference would define it as the source from where the initial differentiation of the locus occurred.

Estimation of inner branches deep in the tree is prone to error and the variance of the expected inner branch lengths of the tree rises towards the root, whence it is more reasonable to consider not only the very first split of the tree but rather several of the deepest branching points of the tree in order to gain some knowledge about the robustness of the phylogeographic inference being made. In the case of a tree drawn from the coding region of human mtDNA complete sequences (Ingman et al. 2000), it is actually not only the first branching point, MRCA1, but also the next three of them that support the African origin of the global mtDNA sequence variation (Fig. 3). Similar branching order of the tree has been by now recapitulated independently using further coding-region sequences sampled worldwide, in particular, of African ancestry (Herrnstadt et al. 2002; Torroni et al. 2001b; Mishmar et al. 2003; Kivisild et al. 2006). Notice that the structure of African haplogroups in this tree is very close to that inferred by many previous studies based on RFLPs and control-region sequences (Fig. 4). Most haplogroups spread in sub-Saharan Africa have distinctive HVS-I and RFLP motifs. However, the advantage of full coding-region sequence data lies in the pulling out of the significant robustness of the principal branching order of the haplogroup tree in Africa. Yet, it is still possible that a branching pattern supporting one particular region as a source can arise also through a secondary loss of diversity because of a population bottleneck or long-term lower effective population size in a geographic region that initially hosted the root haplotype and its own region-specific descendants. Caveats in drawing phylogeographic inferences from deeply rooting branches of Y-chromosome phylogeny, where improvement of molecular resolution is still much needed, are discussed elsewhere (Macaulay 2002; Weale et al. 2003).

There are notorious problems concerning the technical aspects of tree estimation, too, as performed in some studies of mtDNA, especially with the neighbour-joining (NJ) method. The tree of Fig. 3, the long branches of which are well backed up by other data and studies, bears some idiosyncrasies that could have an adverse effect on phylogeographic interpretation. Firstly, the control region as a whole has been disregarded. Although it includes three HVS that could in principle quickly saturate, real saturation along the tree is only observed at the highly variable sites within these segments (and outside, at position 16519). Indeed, the positional mutation spectrum in HVS-I and HVS-II is known to be highly skewed. Eliminating the whole control region rather than a portion of fast mutations for phylogeny estimation is therefore not reasonable because some coding-region sites may be faster than the vast majority of control-region sites. Secondly, a distance-based method such as

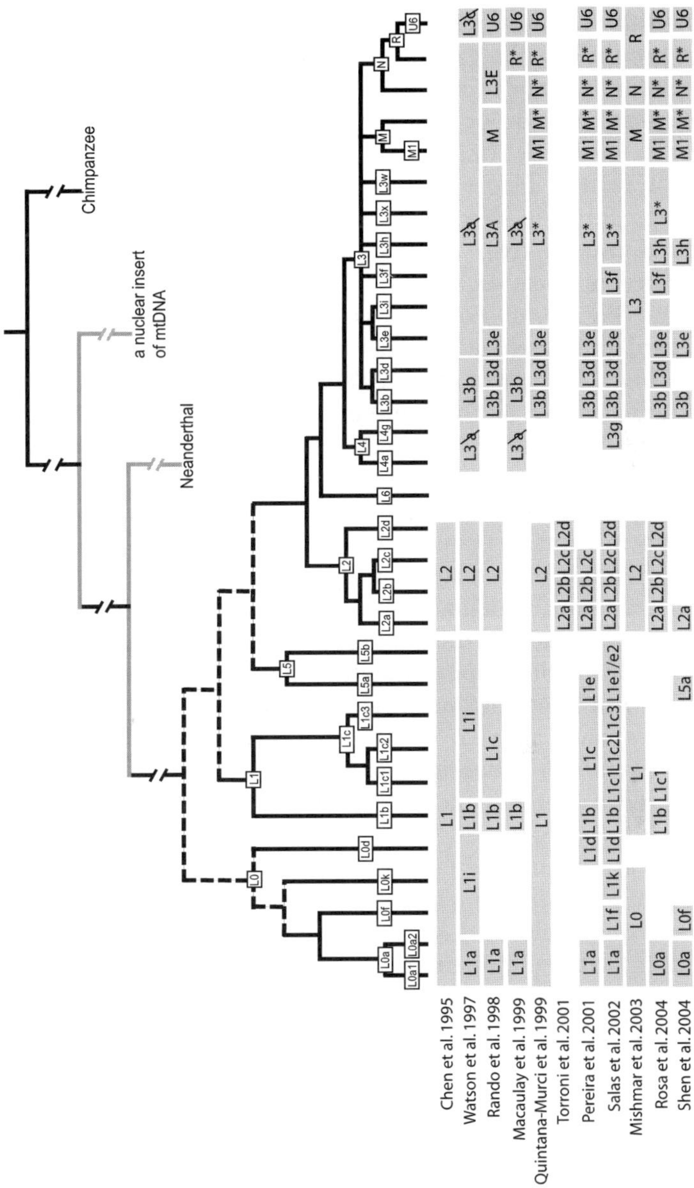

Fig. 4 The evolution of mtDNA haplogroup structure in African populations (Kivisild et al. 2006). Haplogroup definitions are as in Fig. 2 of Kivisild et al. (2004), and are based on a synthesis of Chen et al. (1995); Zischler et al. (1995); Bandelt and Forster (1997); Krings et al. (1997); Watson et al. (1997); Rando et al. (1998); Ingman et al. (2000); Pereira et al. (2001b); Torroni et al. (2001b); Herrnstadt et al. (2002); Salas et al. (2002); Mishmar et al. (2003); Rosa et al. (2004); Shen et al. (2004); Kivisild et al. (2004). Ambiguity of the placement of the root within the inner branches using all three outgroups is shown with a *dashed line*

NJ is known to produce artificial resolution of multifurcations (thus creating 'ghost groups') at the expense of erasing some branches supported by only one mutation (Appendix in Richards et al. 1996; Yao et al. 2002). Thirdly, some input manipulation or program bugs, or extravagant choices of model parameters or decisions (unintended or not) for the graphic presentation of a NJ tree, can further distort the final outcome.

To distinguish these effects, which must have been inflicted on Fig. 3a, we generated two NJ trees de novo, one from the coding region and the other from the complete sequences of Ingman et al. (2000), with rather standard settings (Fig. 3b, c). For both trees, however, deletions and insertions were disregarded beforehand in order to conform to the original approach as reflected by Fig. 4 of Ingman et al. (2000). Our NJ trees both harbour one pronounced clade, formed by the Samoan/Korean haplotypes 17 and 18, which is not found in the original NJ tree (Fig. 3a). Since this clade is supported by three coding-region sites (5465, 9123, and 10238), there is virtually no way (given the fairly low level of homoplasy here) to lose this strong signal other than through manipulation or by error. In all three trees numerous ghost groups are manifest, i.e. clusters of unrelated haplotypes in the vicinity of multifurcation points, which are generated by spurious recurrent mutations. Bootstrap values do not permit clear-cut decisions of which clusters are supported by some synapomorphic mutation and which are not. It seems that all clusters here with bootstrap values below 35% are definite ghost groups and with bootstrap values above 75% are well supported by at least two mutations. But for the clusters in between these percentages one could not tell the status without reconstructing ancestral haplotypes by maximum parsimony (or likelihood).

One ghost group and one false group are highlighted, by way of example, in Fig. 3a. Namely, there is no synapomorphic mutation to support the ghost group comprising haplotypes 33–38, which would rather form three independent branches (within haplogroup L3). The false group consisting of the first five sequences includes an 'emigrant' (the haplogroup U5b sequence) from the larger haplogroup R. Here the NJ grouping was assisted by spuriously shared mutations between different pairs, namely a mutation at 14182 is reported in the Italian and one of the two Australian sequences (haplotypes 5 and 2) and a mutation at 11177 in the Italian and (North American) Piman sequences (haplotypes 5 and 4); compare Fig. 4 of Ingman et al. (2000). This eventually led to the joining of the Italian U5b sequence with the Piman and Australian sequences at a certain stage of the NJ procedure.

The total omission of the control region from the computation of the NJ tree had no effect on the deepest (African) branchings but led to a loss of expected clades in haplogroup N. Most importantly, haplogroup R is no longer discernible through this kind of approach (see also Fig. S2 of Ruiz-Pesini et al. 2004). Further, consider, for instance, haplotypes 4, 11, 17, and 18 that are dispersed as independent branches within the haplogroup N part of the tree of Fig. 3a (in contrast to Fig. 1). These belong to haplogroup B4 (in which the Native American clade B2 is nested), as indicated by sharing transitions at three (highly variable) positions in the control region (16189, 16217, and 16519) as well as the 9-bp deletion (of 8281–8289, which belongs to the contiguous coding region but is non-encoding, i.e. not covered by any mtDNA gene; Chap. 1). Since these four mutations did not enter the NJ input, there was no chance to retrieve this clade in the trees of Fig. 3a and b. A more reasonable approach would thus have been to dismiss only hypervariable positions of the control region but retain deletions and insertions (except for unstable long C-stretches).

The popularity of NJ trees in the early days of mtDNA analysis mainly stemmed from the fact that the NJ algorithm accepts fairly large data sets and delivers one (and only one) tree for a given choice of parameters (for distance calculations) and fixed input order of sequences. NJ as a tool for analysing human mtDNA variation has still survived in a niche of human genetics up to the present day, where large HVS-I data sets are digested by NJ to produce ad hoc clusters without reference to the existing world mtDNA phylogeny (e.g. Cordaux et al. 2003; Tajima et al. 2004).

4
African mtDNA Variation and Haplogroup Structure

The work of Cann et al. (1987) used only a small sample of African Americans to represent its 'African' population, and the follow-up work from the Wilson laboratory focused on control-region variation. It was then left to Chen et al. (1995) to elucidate mtDNA variation in Africans using high-resolution restriction analysis. This first attempt to classify African mtDNAs within the common haplogroup nomenclature, as outlined already, was unfortunately based on a tree rooted with an "Asian outgroup" lineage (viz. from haplogroup F1, by following Denaro et al. 1981). Even though the study concluded that the African mtDNA variation was ancestral to that in Eurasia, the single label L for the ancestral African haplogroup was maintained in the follow-up studies on African populations until 2003. In the strict phylogenetic sense, the 'African-specific haplogroup L' would now be a misconception, as different subclades of haplogroup L embrace all mtDNA haplogroups, both in Africa and outside. In the existing literature on mtDNA variation in humans, a haplogroup is commonly perceived cladistically as any monophyletic group in the global tree, whilst 'L', if used according to its initial definition, would be a paraphyletic roofing for any group that stems from the African part of human mtDNA phylogeny and shares the non-CRS allele at position 3594.

For historical reasons, outlined already, and the desire of geneticists working with African populations to elaborate on a cladistic framework, several amendments have been made to the labelling of the major African haplogroups. For example, haplogroups L1 and L2 were described initially as the two major branches in a tree on the basis of African mtDNA variation (Chen et al. 1995). While L2, now dissected into its four extant daughter clades (Torroni et al. 2001b), has been confirmed to be a monophyletic clade, re-rooting of the African mtDNA tree has exposed L1 in its original definition as a paraphyletic group. This has resulted in a reclassification of its components into the (supposedly) monophyletic haplogroups L0, L1, and L5 (Mishmar et al. 2003; Salas et al. 2004; Shen et al. 2004; Kivisild et al. 2004, 2006). Unfortunately, this update was not noticed in the survey by Pakendorf and Stoneking (2005) when citing Mishmar et al. (2003). The different clades in L0 were previously part of the old paraphyletic group L1, but in the new nomenclature at least the suffixes were retained, so that, for example, L0a refers to the earlier L1a, etc. Figure 4 shows the revised skeleton of the mtDNA haplogroup tree in sub-Saharan Africans. At its root, the human mtDNA tree splits into two branches, one defined as haplogroup L0 and the other holding all of the rest of extant African and non-African mtDNA haplogroups.

The geographic spread of L0 lineages is largely restricted to eastern and southern parts of sub-Saharan Africa (Soodyall and Jenkins 1992; Bandelt and Forster 1997; Watson et al. 1997; Chen et al. 2000; Pereira et al. 2001b; Salas et al. 2002; Kivisild et al. 2004). Haplogroups L0a and L0f are more divergent and frequent in and around Ethiopia, while haplogroups L0k and L0d are characteristic of Khoisan populations of the south. Haplogroup L1 and several subgroups of haplogroup L2, on the other hand, are both frequent and diverse in West Africa (Graven et al. 1995; Watson et al. 1997; Rando et al. 1998; Rosa et al. 2004). Disjoint location of distinctive subclades specific to the Mbuti (in haplogroup L2) and the Biaka (in haplogroup L1c) has been taken as evidence that the so-called pygmy phenotype has originated more than once (Bandelt et al. 1995; Chen et al. 1995; Wallace et al. 1999). This stands in contrast to the similar Y-chromosome haplogroup structures observed in these two populations, with both Biaka and Mbuti being characterized predominantly by Y-chromosome haplogroups B2a, B2b, and E3a (Underhill et al. 2000; Knight et al. 2003).

Haplogroup L3 (Watson et al. 1997) can be singled out from the spectrum of African mtDNA haplogroups as the sole carrier of branches covering mtDNA variation into the rest of the world in prehistoric times. Haplogroups M and N, which are found in all non-Africans, represent just two of the eight known daughter clades of L3, the other six of which are African-specific (Fig. 4). Haplogroups L3bd and L3ei are common throughout West Africa and among Bantu speakers of Southeast Africa, but are rare or absent in Ethiopia or Egypt (Graven et al. 1995; Watson et al. 1997; Rando et al. 1998; Pereira et al. 2001b; Salas et al. 2002; Rosa et al. 2004; Stevanovitch et al. 2004; Kivisild et al. 2004). Other subclades of L3 are either restricted to or are predominantly found in populations of East and Northeast Africa.

There are sound reasons to believe that several recent episodes of gene flow have significantly reorganized the geographic substructure of African populations. One of the most influential within the last couple of thousand years, with a widespread impact on the gene pool of sub-Saharan Africa, was probably the Bantu expansion. It is likely that the Bantu expansion was the main mechanism explaining the spread of haplogroups L0a2 and L3b and fragments of haplogroups L2, L3e, and L5 from West, Central, and East Africa towards the south (Bandelt et al. 1995, 2001; Chen et al. 1995; Soodyall et al. 1996; Alves-Silva et al. 2000; Pereira et al. 2001a; Salas et al. 2002). Another important influence from outside, affecting mainly populations of northern and eastern Africa, seems to have been an influx of lineages from the Near East, deriving from haplogroup N (Passarino et al. 1998; Richards et al. 2003b; Kivisild et al. 2004).

Because this, likely more recent, gene flow has significantly influenced the haplogroup structure of many present-day northern and eastern African populations, these cannot be assumed to represent primordial population substructure in Africa, say, at the end of the Pleistocene. A population tree, for example, drawn from present-day populations of Africa, would therefore be without a clear meaning, because it would combine information about the levels of ancient population differentiation with that reflecting more recent gene flow episodes. It would be tempting to interpret the node on the tree in Fig. 5 that gives rise to the cluster containing Yemenis, Egyptians, Northwest African, and Near Eastern populations as indicating that Yemenis represent the ancient stock from which North African and Near Eastern populations diverged, but recent gene flow is more likely to be behind the branching pattern in this tree. Phylogeographic analysis of mtDNA lineages found in Yemen points to a nearly equal presence of haplogroups specific to the Near East and sub-Saharan Africa (Richards et al. 2003b; Kivisild et al. 2004). A high proportion of lineages belonging to African haplogroups found in Yemen have exact matches in Mozambique, yet neither a distance-based population tree nor a haplogroup frequency based admixture analysis is able to show the link between Yemenis and the Mozambique population from a number of shared lineages detected by phylogeographic analysis (Kivisild et al. 2004).

The complexity of historic interactions between populations emphasized in the tree combining African and Near Eastern populations can be generalized for any other region where populations have not been isolated from each other over long time periods. A phylogeographic approach tries to overcome the problem of gene flow by identifying its potential components in existing populations and estimating their probable source and time of arrival in the target population. This decomposition approach has been criticized from the standpoint that it is not the history of anonymous lineages but of populations that is of general interest (Barbujani and Bertorelle 2001; Chikhi et al. 2002): "A problem with the lineage-based approach of studying haplogroups is that it only elucidates the history of the haplogroups themselves,

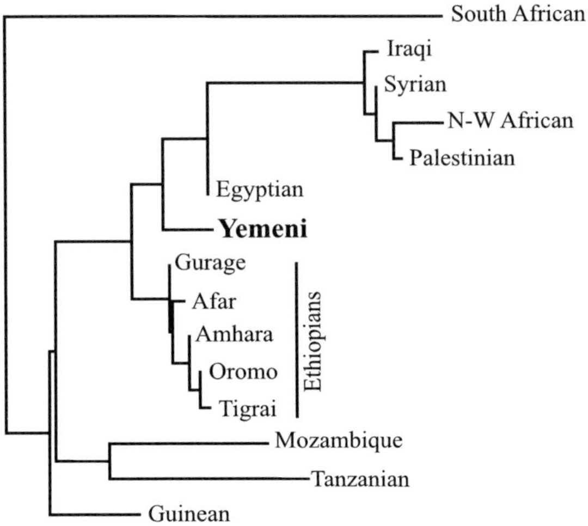

Fig. 5 Population tree of 15 African and Near Eastern populations. The NJ tree is based on F_{st} distances calculated from mtDNA haplogroup frequencies and was constructed with MEGA, using the data of Kivisild et al. (2004)

and does not provide direct insights into the history of the individual populations in which they are present" (Pakendorf and Stoneking 2005). Yet it can be argued that populations, which are often quite abstractly and synchronically defined, are always composed of bundles of individual lineages that are transmitted from generation to generation. It is true that there is no living example for the equation 'one population = one haplogroup' but this is exactly what phylogeographic analysis tries to circumvent, by attempting to reconstruct the dynamic pools of lineages in prehistoric populations diachronically. This task is certainly not easy, especially where continuous gene flow between populations has been the rule rather than the exception—in the case of the Eurasian continent, for example. In such instances, detailed population histories in the classic sense are hardly possible to reconstruct, because such populations have never existed. What an exhaustive analysis of genetic variation *can* achieve is a broad characterization of regional variation within and between the continents.

Compared with genetic distance analyses of population genetics favoured and advertised by some molecular anthropologists (e.g. Pakendorf and Stoneking 2005), coalescence methods focussing on the interplay between demography and genealogy make fuller use of the data. It has therefore been tempting to view the coalescing of lineages from any given mtDNA population sample under the convenient theoretical roof of coalescent theory, in which the pairwise coalescences are modelled as random events in a stochastic process moving backward in time; for a painless introduction, see Chap. 11 in Klein and Takahata (2001). The theoretical model is not complete without explicit assumptions about demographic parameters such as effective population size. Although inferences about demography of past populations are well within the scope of phylogeography, most direct applications of coalescence methods are hampered by naïve modelling that stipulates a closed population (admitting no introgression since the MRCA), panmixia (with no geographic substructure), and either constant population size or exponentially increasing or decreasing sizes (Weiss and von Haeseler 1998). Such premises can hardly be reconciled with what we know from prehistoric reconstruction. In consequence, fitting such a model to worldwide or regional mtDNA population samples would not contribute much to our understanding of how the spread of modern humans might have proceeded (Bandelt et al. 2003b).

5 mtDNA Variation Outside Africa

Haplogroups A, B, C, and D were the first mtDNA haplogroups to be recognized and labelled, as part of a high-resolution study of Native American populations (Torroni et al. 1992, 1993). Subsequently, each of these haplogroups has also been found in East Asia, and it has been shown that haplogroups C and D share a common ancestor within haplogroup M, while haplogroups A and B coalesce within haplogroup N. Besides these four subclades of haplogroups M and N defined in Native Americans, many other sister clades have subsequently been defined in modern Asian populations (see Fig. 1 in Chap. 8).

Haplogroups CZ (within M8), D4 and G (within M), and A and N9 (within N) broadly characterize most of the mtDNA lineages found in Northeast Asian populations; while haplogroups E and M7 (within M) and B4a and R9 (within N) are largely concentrated in populations of continental and island Southeast Asia (Forster et al. 2001; Kivisild et al. 2002 and references therein; Merriwether et al. 2005; Trejaut et al. 2005)—although there is substantial overlap. Similarly, region-specific subclades of haplogroups M and N characterize populations of Pakistan and India (Kivisild et al. 1999; Metspalu et al. 2004; Palanichamy et al. 2004; Quintana-Murci et al. 2004) and Papua New Guinea and Australia (Forster et al. 2001; Ingman and Gyllensten 2003; Friedlaender et al. 2005). For a more detailed coverage of Asian and Oceanian mtDNA variation, see Chaps. 8 and 10.

Virtually all European mtDNA lineages can be classified within the framework of six haplogroups—N1, N2, X, pre-HV, JT, and U (Palanichamy et al. 2004; see Fig 1 in Chap. 8)—whose distribution is restricted largely to West Eurasia and North Africa (except for U, which is also found throughout India). The first three of these haplogroups, with a minor frequency throughout Europe (Richards et al. 2000), derive directly from the root of haplogroup N.

The three haplogroups, pre-HV, JT, and U, that derive from a MRCA in haplogroup R cover altogether approximately 80–90% of the total variation in most European populations. Both N and R thus appear as founder haplogroups shared by Europeans with the whole range of populations outside Africa. This means that the MRCA for all European maternal lineages is at the same time the ancestor for a significant proportion of Asian, Oceanian, and Native American lineages. The reverse does not hold, however: the MRCA for haplogroups M and N that are co-spread in Asia, Oceania, and the Americas stands one step higher in the hierarchy of mtDNA haplogroups than the MRCA of Europeans. It is the root haplotype in haplogroup L3 that is the MRCA for all non-Africans, and as already mentioned, this type gives rise, at the same time, to at least six branches whose spread is restricted to Africa (Fig. 4). The different MRCAs for European and Asian mtDNA lineages also indicate that the coalescence time of Asian mtDNA lineages is older than that for Europeans.

The five basic mtDNA haplogroups (A–D and X) covering all extant Native American mtDNAs form a subset derived from East Asian variation, involving haplogroups descending both from M and N haplogroups, thereby carrying the African MRCA of Asian populations with them to the New World. Thus, even though the archaeologically accepted time frame for peopling of Americas is significantly younger than that of Europe, Native Americans taken as a mtDNA gene pool have an estimated coalescence time to their MRCA, the root of haplogroup L3, by 20 000–30 000 years older than the similar estimate for Europeans, the root of haplogroup N (Tang et al. 2002). In particular, a simplistic coalescence approach exercised on a Native American sample (such as the Nuu-Chah-Nulth sample by Weiss and von Haeseler 1998) would therefore hardly capture realistic features of Native American beginnings (Bandelt et al. 2003b).

The discrepancy between coalescence times and settlement times can, at least theoretically, be solved if the founding lineages can be reasonably well identified in a phylogeographic analysis. A study comparing Native American and Asian mtDNA haplotypes based on HVS-I sequencing (Forster et al. 1996) concluded that the molecular variation accumulated on a limited number of Native American founder lineages that have an Asian origin can be dated to 21 000–23 000 years ago. These estimates are not in themselves unproblematic, and there is a long tradition debating these: 'Clovis-first' archaeologists would opt for a settlement time closer to around 15 000 years ago (Mandryk et al. 2001). This might be accounted for by diversity accumulated in the immediate source population (perhaps in Beringia). Or it may be that a somewhat older but archaeologically undetected (because it is submerged) 'coastal route' into the Americas was followed by the first Americans before or around the Late Glacial Maximum (LGM; Dalton 2003). Archaeological evidence for the pre-LGM presence of a human population in the Arctic has started to accumulate recently (Pitulko et al. 2004). Imported mtDNA variation within the founder haplogroups in the Americas would, however, be partly detectable by comparing branches shared between extreme geographical locations (i.e northernmost America versus southernmost America in this case). Of course, postulating such explanations assumes that the molecular clock has been reasonably well calibrated and that the founder analysis used is adequate to the task of reconstruction at hand. Clearly, comparing HVS-I sequences alone or just haplogroup frequencies in the Americas, as is still a popular enterprise, cannot resolve the issue finally. It seems, though, that the few complete sequences attributed to Native Americans would indeed point to the simplest scenario: one mtDNA founder per haplogroup (Bandelt et al. 2003a). In any case, the differing coalescence times for European and Asian populations clearly do not, of themselves, indicate two different migrations out of Africa, the early one to Asia, and the second, more recent, to Europe, because the three founding lineages, the roots of haplogroups M, N, and R, display almost identical coalescence times (Kong et al. 2003; Mishmar et al. 2003; Kivisild et al. 2006), irrespective of the calibration method being used.

Although the same haplotypes at the base of haplogroups M, N, and R root the European and East Asian mtDNA haplogroup trees, the branching within the trees, i.e. the haplogroup composition, is completely different in the two regions (Kivisild et al. 2002; Kong et al. 2003). Within both continents, however, fewer differences can be observed between regional populations. The distribution of the six basic subclades of haplogroups N and R over all Europe is surprisingly uniform (Richards et al. 2002), especially when compared with the striking differences in Y-chromosome haplogroup pools between the eastern and western parts of the continent (Rosser et al. 2000; Semino et al. 2000). Using improved molecular resolution and sampling, some differentiation within Europe can be observed (Torroni et al. 2001a; Richards et al. 2002; Achilli et al. 2004; Loogväli et al. 2004; McEvoy et al. 2004; Tambets et al. 2004), but it remains much less pronounced than that of the male line of descent (Cruciani et al. 2004; Rootsi et al. 2004; Semino et al. 2004). No mtDNA haplogroup in Europe, for example, parallels the transcontinental distributions of Y-chromosome haplogroups R1a, N, or E3b (Zerjal et al. 1997; Rosser et al. 2000; Semino et al. 2000, 2004; Kivisild et al. 2003; Cruciani et al. 2004).

Several migrations originating from the Near East—from Early Upper Palaeolithic to Neolithic and more recent times—have likely brought to Europe a large number of independent founder lineages, nested within haplogroups pre-HV, JT, U, N1, N2, and X, that all have their origin in or around the Near East in the late Pleistocene (Richards et al. 2000). Current reconstructions, we might note, are rather different to the one popularized by Sykes (2001), who describes the peopling of Europe in terms of independent origins and dispersals of some of the major haplogroups, i.e. the ones amenable to romantically labelling as the 'Seven Daughters of Eve'. Such narratives should be viewed in the context of the business plan for establishing a genetic ancestry testing company, rather than in that of the mitochondrial research community.

In practice, drawing phylogeographic inferences about the peopling of Europe and the gene flow within the continent is a far from trivial task, because of multiple gene flows following similar courses, and potentially including derived clades of the same haplogroups, involvement of back migration, sampling, and other problems. A critical evaluation of the founder assumptions is outlined in Richards et al. (2000, 2003a).

6 The Role of Selection on mtDNA Variability

Howell et al. (2003) discussed a possible role for purifying selection, acting both in the coding and in the control regions, which would explain at least in part the disparity between the pedigree and phylogenetic rates of mtDNA sequence divergence (that is to say, the rates estimated empirically from family studies and gene trees, respectively—the former leading to higher estimates than the latter in several cases; Chap. 4). Different genes in the coding region may well, in addition, evolve at a different pace because of different functional constraints affecting them (Mishmar et al. 2003; Elson et al. 2004). There is a significant excess of synonymous mutations over non-synonymous ones in all of the mitochondrial protein-coding genes, highlighting their functional activity.

What concerns evolutionary geneticists is whether, and how, selection might affect the tree topology that is reconstructed, and whether a molecular clock can be applied to loci that may be under selection. It is quite a common misconception to assume that the molecular clock can only be used in cases of complete neutrality; and even that selection, in any of its forms, can distort the branching pattern of phylogenetic trees. Many, perhaps most, loci are likely to be affected by selection to some extent, whether directly or indirectly, for example by hitch-hiking with selected loci to which they are closely linked (as in the case of the 'neutral' non-coding regions of the mtDNA, all fully linked to the functional protein-coding and ribosomal genes). There is no all-embracing solution to this issue: whether selection has affected any part of the tree must be investigated empirically, case by case.

Negative or purifying selection eliminates deleterious mutations during the course of evolution. Accordingly, external branches, the twigs of a phylogenetic tree, which are defined by mutations of recent origin, should be expected to carry relatively more slightly deleterious mutations than the internal ones. Each mildly deleterious mutation, taken alone, would be characterized by marginally reduced frequency in the population as the fixation probability of its carrier haplotype is lower than the average. However, when such a substitution survives a population bottleneck or occurs in a small quickly expanding population, it can not only be maintained but may ultimately reach an elevated frequency in the population (Avise 2000). Several mutations specific for haplogroup J, for example, which probably expanded in Europe alongside the Neolithic spread of farmers (Richards et al. 1996), have been suggested to play a background role in LHON expression (Torroni et al. 1997). The possibility of cumulative chance fixation of mildly deleterious mutations (because of clonal inheritance and the lack of recombination) has led to the suggestion that mitochondria may be destined to Muller's ratchet because there is no mechanism, like recombination for most nuclear genes, to get rid of the accumulating deleterious mutations (Lynch 1996).

Assuming that human mtDNA variation is largely neutral, the most common interpretation of frequency differences seen between populations is migration and drift. An alternative hypothesis explored and discussed recently suggests rather that climatic selection may have played a significant role in shaping the haplogroup structures in different environments (Mishmar et al. 2003; Ruiz-Pesini et al. 2004). According to this hypothesis, mildly deleterious substitutions, in genes encoding mitochondrial proteins and ribosomal DNA, may decrease the efficiency of the mitochondrial oxidative phosphorylation system to generate ATP, so that more energy goes directly into heat production. The extra heat produced might be an advantage then for populations living in a cold climate. Mishmar et al. (2003) and Ruiz-Pesini et al. (2004) therefore explored whether there is an excess of potentially slightly deleterious mutations in populations living in cold climates, and found a positive correlation between the ratio of non-synonymous versus synonymous mutations (Ka/Ks) and geographic latitude. Specifically, haplogroups A, C, D, and X, arguably adapted to the cold climate of Northeast Asia before their spread to the Americas, and several haplogroups in Europe carried significantly more non-synonymous changes than haplogroup L lineages in Africa.

However, as shown by other studies (Moilanen et al. 2003; Moilanen and Majamaa 2003; Elson et al. 2004; Kivisild et al. 2006), there is a general and significant excess of non-synonymous mutations, irrespective of geographic latitude of the population, in the external versus the internal branches of the global mtDNA tree. The internal, haplogroup-defining branches of the tree are defined by mutations that have survived purifying selection in populations over a relatively longer time period than the ones emerging newly in the tips. The conclusions of Mishmar et al. (2003) and Ruiz-Pesini et al. (2004) have been criticized because (1) they compared 'old' African haplogroups versus 'young' European and North Asian haplogroups, (2) they used inappropriate statistics that artificially increase significance values, and (3) because significant differences in Ka/Ks observed between European haplogroups, which, for example, have evolved under conditions of marked climate change, are unlikely to be explained by a simplistic model of adaptation (Elson et al. 2004). Different patterns of amino acid replacement, involving mostly threonine and valine codons, that can be observed amongst human populations and between humans and other mammalian species could potentially be subject of natural selection (Kivisild et al. 2006), requiring further enquiry. Therefore, aspects of potential correlation between mtDNA phylogeography and adaptation, fascinating as they certainly might be, still await further examination.

References

Achilli A, Rengo C, Magri C et al. (2004) The molecular dissection of mtDNA haplogroup H confirms that the Franco-Cantabrian glacial refuge was a major source for the European gene pool. Am J Hum Genet 75:910–918

Alves-Silva J, da Silva Santos M, Guimarães PE, Ferreira AC, Bandelt H-J, Pena SD, Prado VF (2000) The ancestry of Brazilian mtDNA lineages. Am J Hum Genet 67:444–461 (erratum 67:775)

Anderson S, Bankier AT, Barrell BG, de Bruijn MH, Coulson AR, Drouin J, Eperon IC, Nierlich DP, Roe BA, Sanger F, Schreier PH, Smith AJ, Staden R, Young IG (1981) Sequence and organization of the human mitochondrial genome. Nature 290:457–465

Andrews RM, Kubacka I, Chinnery PF, Lightowlers RN, Turnbull DM, Howell N (1999) Reanalysis and revision of the Cambridge reference sequence for human mitochondrial DNA. Nat Genet 23:147

Avise JC (2000) Phylogeography: The history and formation of species. Harvard University Press, Cambridge

Bandelt H-J, Forster P, Sykes BC, Richards MB (1995) Mitochondrial portraits of human populations using median networks. Genetics 141:743–753

Bandelt H-J, Forster P (1997) The myth of bumpy hunter-gatherer mismatch distributions. Am J Hum Genet 61:980–983

Bandelt H-J, Alves-Silva J, Guimarães PE, Santos MS, Brehm A, Pereira L, Coppa A, Larruga JM, Rengo C, Scozzari R, Torroni A, Prata MJ, Amorim A, Prado VF, Pena SD (2001) Phylogeography of the human mitochondrial haplogroup L3e: a snapshot of African prehistory and Atlantic slave trade. Ann Hum Genet 65:549–563

Bandelt H-J, Herrnstadt C, Yao Y-G, Kong Q-P, Kivisild T, Rengo C, Scozzari R, Richards M, Villems R, Macaulay V, Howell N, Torroni A, Zhang Y-P (2003a) Identification of Native American founder mtDNAs through the analysis of complete mtDNA sequences: some caveats. Ann Hum Genet 67:512–524

Bandelt H-J, Macaulay V, Richards M (2003b) What molecules can't tell us about the spread of languages and the Neolithic. In: Bellwood P, Renfrew C (eds) Examining the farming/language dispersal hypothesis. McDonald Institute for Archaeological Research, Cambridge, pp 99–111

Barbujani G, Bertorelle G (2001) Genetics and the population history of Europe. Proc Natl Acad Sci USA 98:22–25

Cann RL, Stoneking M, Wilson AC (1987) Mitochondrial DNA and human evolution. Nature 325:31–36

Cavalli-Sforza LL, Menozzi P, Piazza A (1994) The history and geography of human genes. Princeton University Press, Princeton

Chen YS, Torroni A, Excoffier L, Santachiara-Benerecetti AS, Wallace DC (1995) Analysis of mtDNA variation in African populations reveals the most ancient of all human continent-specific haplogroups. Am J Hum Genet 57:133–149

Chen YS, Olckers A, Schurr TG, Kogelnik AM, Huoponen K, Wallace DC (2000) mtDNA variation in the South African Kung and Khwe—and their genetic relationships to other African populations. Am J Hum Genet 66:1362–1383

Chikhi L, Nichols RA, Barbujani G, Beaumont MA (2002) Y genetic data support the Neolithic demic diffusion model. Proc Natl Acad Sci USA 99:11008–11013

Coble MD, Just RS, O'Callaghan JE, Letmanyi IH, Peterson CT, Irwin JA, Parsons TJ (2004) Single nucleotide polymorphisms over the entire mtDNA genome that increase the power of forensic testing in Caucasians. Int J Legal 118:137–146

Cordaux R, Saha N, Bentley GR, Aunger R, Sirajuddin SM, Stoneking M (2003) Mitochondrial DNA analysis reveals diverse histories of tribal populations from India. Eur J Hum Genet 11:253–264

Coskun PE, Beal MF, Wallace DC (2004) Alzheimer's brains harbor somatic mtDNA control-region mutations that suppress mitochondrial transcription and replication. Proc Natl Acad Sci USA 101:10726–10731

Cruciani F, La Fratta R, Santolamazza P, Sellitto D, Pascone R, Moral P, Watson E, Guida V, Colomb EB, Zaharova B, Lavinha J, Vona G, Aman R, Cali F, Akar N, Richards M, Torroni A, Novelletto A, Scozzari R (2004) Phylogeographic analysis of haplogroup E3b (E-M215) Y chromosomes reveals multiple migratory events within and out of Africa. Am J Hum Genet 74:1014–1022

Dalton R (2003) The coast road. Nature 422:10–12

Denaro M, Blanc H, Johnson MJ, Chen KH, Wilmsen E, Cavalli-Sforza LL, Wallace DC (1981) Ethnic variation in Hpa 1 endonuclease cleavage patterns of human mitochondrial DNA. Proc Natl Acad Sci USA 78:5768–5772

Elson JL, Andrews RM, Chinnery PF, Lightowlers RN, Turnbull DM, Howell N (2001) Analysis of European mtDNAs for recombination. Am J Hum Genet 68:145–153

Elson JL, Turnbull DM, Howell N (2004) Comparative genomics and the evolution of human mitochondrial DNA: assessing the effects of selection. Am J Hum Genet 74:229–238

Finnilä S, Lehtonen MS, Majamaa K (2001) Phylogenetic network for European mtDNA. Am J Hum Genet 68:1475–1484

Forster P, Harding R, Torroni A, Bandelt H-J (1996) Origin and evolution of Native American mtDNA variation: a reappraisal. Am J Hum Genet 59:935–945

Forster P, Torroni A, Renfrew C, Röhl A (2001) Phylogenetic star contraction applied to Asian and Papuan mtDNA evolution. Mol Biol Evol 18:1864–1881

Friedlaender J, Schurr T, Gentz F, Koki G, Friedlaender F, Horvat G, Babb P, Cerchio S, Kaestle F, Schanfield M, Deka R, Yanagihara R, Merriwether DA (2005) Expanding southwest Pacific mitochondrial haplogroups P and Q. Mol Biol Evol 22:1506–1517

Giacchetti M, Monticelli A, De Biase I, Pianese L, Turano M, Filla A, De Michele G, Cocozza S (2004) Mitochondrial DNA haplogroups influence the Friedreich's ataxia phenotype. J Med Genet 41:293–295

Giles RE, Blanc H, Cann HM, Wallace DC (1980) Maternal inheritance of human mitochondrial DNA. Proc Natl Acad Sci USA 77:6715–6719

Graven L, Passarino G, Semino O, Boursot P, Santachiara-Benerecetti S, Langaney A, Excoffier L (1995) Evolutionary correlation between control region sequence and restriction polymorphisms in the mitochondrial genome of a large Senegalese Mandenka sample. Mol Biol Evol 12:334–345

Hasegawa M, Di Rienzo A, Kocher TD, Wilson AC (1993) Toward a more accurate time scale for the human mitochondrial DNA tree. J Mol Evol 37:347–354

Herrnstadt C, Elson JL, Fahy E, Preston G, Turnbull DM, Anderson C, Ghosh SS, Olefsky JM, Beal MF, Davis RE, Howell N (2002) Reduced-median-network analysis of

complete mitochondrial DNA coding-region sequences for the major African, Asian, and European haplogroups. Am J Hum Genet 70:1152–1171 (erratum 71:448–449)
Hewitt G (2000) The genetic legacy of the Quaternary ice ages. Nature 405:907–913
Hey J (1998) Population genetics and human origins—haplotypes are key! Trends Genet 14:303–305
Heyer E, Zietkiewicz E, Rochowski A, Yotova V, Puymirat J, Labuda D (2001) Phylogenetic and familial estimates of mitochondrial substitution rates: study of control region mutations in deep-rooting pedigrees. Am J Hum Genet 69:1113–1126
Howell N, Smejkal CB, Mackey DA, Chinnery PF, Turnbull DM, Herrnstadt C (2003) The pedigree rate of sequence divergence in the human mitochondrial genome: there is a difference between phylogenetic and pedigree rates. Am J Hum Genet 72:659–670
Huerta C, Castro MG, Coto E, Blázquez M, Ribacoba R, Guisasola LM, Salvador C, Martínez C, Lahoz CH, Alvarez V (2005) Mitochondrial DNA polymorphisms and risk of Parkinson's disease in Spanish population. J Neurol Sci 236:49–54
Ingman M, Kaessmann H, Pääbo S, Gyllensten U (2000) Mitochondrial genome variation and the origin of modern humans. Nature 408:708–713
Ingman M, Gyllensten U (2003) Mitochondrial genome variation and evolutionary history of Australian and New Guinean aborigines. Genome Res 13:1600–1606
Kivisild T, Villems R (2000) Questioning evidence for recombination in human mitochondrial DNA. Science 288:1931a
Kivisild T, Bamshad MJ, Kaldma K, Metspalu M, Metspalu E, Reidla M, Laos S, Parik J, Watkins WS, Dixon ME, Papiha SS, Mastana SS, Mir MR, Ferak V, Villems R (1999) Deep common ancestry of Indian and western-Eurasian mitochondrial DNA lineages. Curr Biol 9:1331–1334
Kivisild T, Tolk H-V, Parik J, Wang Y, Papiha SS, Bandelt H-J, Villems R (2002) The emerging limbs and twigs of the East Asian mtDNA tree. Mol Biol Evol 19:1737–1751 (erratum 20:162)
Kivisild T, Rootsi S, Metspalu M, Mastana S, Kaldma K, Parik J, Metspalu E, Adojaan M, Tolk H-V, Stepanov V, Gölge M, Usanga E, Papiha SS, Cinnioğlu C, King R, Cavalli-Sforza L, Underhill PA, Villems R (2003) The genetic heritage of the earliest settlers persists both in Indian tribal and caste populations. Am J Hum Genet 72:313–332
Kivisild T, Reidla M, Metspalu E, Rosa A, Brehm A, Pennarun E, Parik J, Geberhiwot T, Usanga E, Villems R (2004) Ethiopian mitochondrial DNA heritage: tracking gene flows across and around the Strait of Tears. Am J Hum Genet 75:752–770
Kivisild T, Shen P, Wall DP, Do B, Sung R, Davis KK, Passarino G, Underhill PA, Scharfe C, Torroni A, Scozzari R, Modiano D, Coppa A, de Knjiff P, Feldman MW, Cavalli-Sforza LL, Oefner PJ (2006) The role of selection in the evolution of human mitochondrial genomes. Genetics 172:373–387
Klein J, Takahata N (2001) Where do we come from? The molecular evidence for human descent. Springer, Berlin Heidelberg New York
Knight A, Underhill PA, Mortensen HM, Zhivotovsky LA, Lin AA, Henn BM, Louis D, Ruhlen M, Mountain JL (2003) African Y chromosome and mtDNA divergence provides insight into the history of click languages. Curr Biol 13:464–473
Kong Q-P, Yao Y-G, Sun C, Bandelt H-J, Zhu C-L, Zhang Y-P (2003) Phylogeny of East Asian mitochondrial DNA lineages inferred from complete sequences. Am J Hum Genet 73:671–676 (erratum 75:157)
Kraytsberg Y, Schwartz M, Brown TA, Ebralidse K, Kunz WS, Clayton DA, Vissing J, Khrapko K (2004) Recombination of human mitochondrial DNA. Science 304:981
Krings M, Stone A, Schmitz RW, Krainitzki H, Stoneking M, Paabo S (1997) Neandertal DNA sequences and the origin of modern humans. Cell 90:19–30

Kumar S, Tamura K, Jakobsen IB, Nei M (2001) MEGA2: molecular evolutionary genetics analysis software. Arizona State University, Tempe

Loogväli EL, Roostalu U, Malyarchuk BA, Derenko MV, Kivisild T, Metspalu E, Tambets K, et al. (2004) Disuniting uniformity: a pied cladistic canvas of mtDNA haplogroup H in Eurasia. Mol Biol Evol 21:2012–2021

Lynch M (1996) Mutation accumulation in transfer RNAs: molecular evidence for Muller's ratchet in mitochondrial genomes. Mol Biol Evol 13:209–220

Macaulay VA, Richards MB, Hickey E, Vega E, Cruciani F, Guida V, Scozzari R, Bonné-Tamir B, Sykes B, Torroni A (1999) The emerging tree of West Eurasian mtDNAs: a synthesis of control-region sequences and RFLPs. Am J Hum Genet 64:232–249

Macaulay V (2002) Book review of: Relethford JH (2001) Genetics and the search for modern human origins. Wiley, New York, Heredity 89:160

Malyarchuk BA, Rogozin IB (2004) Mutagenesis by transient misalignment in the human mitochondrial DNA control region. Ann Hum Genet 68:324–339

Mandryk CAS, Josenhans H, Fedje DW, Mathewes RW (2001) Late Quaternary paleoenvironments of northwestern North America: implications for inland versus coastal migration routes. Quat Sci Rev 20:301–314

McEvoy B, Richards M, Forster P, Bradley DG (2004) The longue durée of genetic ancestry: multiple genetic marker systems and Celtic origins on the Atlantic facade of Europe. Am J Hum Genet 75:693–702

Merriwether DA, Hodgson JA, Friedlaender FR, Allaby R, Cerchio S, Koki G, Friedlaender JS (2005) Ancient mitochondrial M haplogroups identified in the Southwest Pacific. Proc Natl Acad Sci USA 102:13034–13039

Metspalu M, Kivisild T, Metspalu E, Parik J, Hudjashov G, Kaldma K, Serk P, Karmin M, Behar DM, Gilbert MTP, Endicott P, Mastana S, Papiha SS, Skorecki K, Torroni A, Villems R (2004) Most of the extant mtDNA boundaries in South and Southwest Asia were likely shaped during the initial settlement of Eurasia by anatomically modern humans. BMC Genet 5:26 (erratum 6:41)

Mishmar D, Ruiz-Pesini E, Golik P, Macaulay V, Clark AG, Hosseini S, Brandon M, Easley K, Chen E, Brown MD, Sukernik RI, Olckers A, Wallace DC (2003) Natural selection shaped regional mtDNA variation in humans. Proc Natl Acad Sci USA 100:171–176

Moilanen JS, Majamaa K (2003) Phylogenetic network and physicochemical properties of nonsynonymous mutations in the protein-coding genes of human mitochondrial DNA. Mol Biol Evol 20:1195–1210

Moilanen JS, Finnilä S, Majamaa K (2003) Lineage-specific selection in human mtDNA: lack of polymorphisms in a segment of MTND5 gene in haplogroup. J Mol Biol Evol 20:2132–2142

Myers G, Korber B, Hahn B, Jeang KT, Mellors J, McCutchan FE, Henderson L, Pavlakis G (1995) Human retroviruses and AIDS: a compilation and analysis of nucleic acid and amino acid sequences. Los Alamos National Laboratory, Los Alamos

Pakendorf B, Stoneking M (2005) Mitochondrial DNA and human evolution. Annu Rev Genomics Hum Genet 6:165–183

Palanichamy Mg, Sun C, Agrawal S et al. (2004) Phylogeny of mtDNA macrohaplogroup N in India based on complete sequencing: implications for the peopling of South Asia. Am J Hum Genet 75:966–978

Passarino G, Semino O, Quintana-Murci L, Excoffier L, Hammer M, Santachiara-Benerecetti AS (1998) Different genetic components in the Ethiopian population, identified by mtDNA and Y-chromosome polymorphisms. Am J Hum Genet 62:420–434

Pereira L, Dupanloup I, Rosser ZH, Jobling MA, Barbujani G (2001a) Y-chromosome mismatch distributions in Europe. Mol Biol Evol 18:1259–1271

Pereira L, Macaulay V, Torroni A, Scozzari R, Prata MJ, Amorim A (2001b) Prehistoric and historic traces in the mtDNA of Mozambique: insights into the Bantu expansions and the slave trade. Ann Hum Genet 65:439–458

Pitulko VV, Nikolsky PA, Girya EY, Basilyan AE, Tumskoy VE, Koulakov SA, Astakhov SN, Pavlova EY, Anisimov MA (2004) The Yana RHS site: humans in the Arctic before the last glacial maximum. Science 303:52–56

Pyle A, Foltynie T, Tiangyou W, Lambert C, Keers SM, Allcock LM, Davison J et al. (2005) Mitochondrial DNA haplogroup cluster UKJT reduces the risk of PD. Ann Neurol 57:564–567

Quintana-Murci L, Chaix R, Wells S, Behar D, Sayar H, Scozzari R, Rengo C, Al-Zahery N, Semino O, Santachiara-Benerecetti A, Coppa A, Ayub Q, Mohyuddin A, Tyler-Smith C, Mehdi Q, Torroni A, McElreavey K (2004) Where West meets East: the complex mtDNA landscape of the Southwest and Central Asian corridor. Am J Hum Genet 74:827–845

Quintáns B, Álvarez-Iglesias V, Salas A, Phillips C, Lareu MV, Carracedo A (2004) Typing of mitochondrial DNA coding region SNPs of forensic and anthropological interest using SNaPshot minisequencing. Forensic Sci Int 140:251–257

Rando JC, Pinto F, González AM, Hernández M, Larruga JM, Cabrera VM, Bandelt H-J (1998) Mitochondrial DNA analysis of northwest African populations reveals genetic exchanges with European, near-eastern, and sub-Saharan populations. Ann Hum Genet 62:531–550

Richards M, Côrte-Real H, Forster P, Macaulay V, Wilkinson-Herbots H, Demaine A, Papiha S, Hedges R, Bandelt H-J, Sykes B (1996) Paleolithic and neolithic lineages in the European mitochondrial gene pool. Am J Hum Genet 59:185–203

Richards MB, Macaulay VA, Bandelt H-J, Sykes BC (1998) Phylogeography of mitochondrial DNA in western Europe. Ann Hum Genet 62:241–260

Richards M, Macaulay V, Hickey E, Vega E, Sykes B, Guida V, Rengo C et al (2000) Tracing European founder lineages in the Near Eastern mtDNA pool. Am J Hum Genet 67:1251–1276

Richards M, Macaulay V, Torroni A, Bandelt H-J (2002) In search of geographical patterns in European mitochondrial DNA. Am J Hum Genet 71:1168–1174

Richards M, Macaulay V, Bandelt H-J (2003a) Analyzing genetic data in a model-based framework: inferences about European prehistory. In: Bellwood P, Renfrew C (eds) Examining the farming/language dispersal hypothesis. McDonald Institute for Archaeological Research, Cambridge, pp 459–466

Richards M, Rengo C, Cruciani F, Gratrix F, Wilson J, Scozzari R, Macaulay V, Torroni A (2003b) Extensive female-mediated gene flow from sub-Saharan Africa into near eastern Arab populations. Am J Hum Genet 72:1058–1064

Rootsi S, Magri C, Kivisild T, Benuzzi G, Help H, Bermisheva M, Kutuev I et al. (2004) Phylogeography of Y-chromosome haplogroup I reveals distinct domains of prehistoric gene flow in Europe. Am J Hum Genet 75:128–137

Rosa A, Brehm A, Kivisild T, Metspalu E, Villems R (2004) MtDNA profile of West Africa Guineans: towards a better understanding of the Senegambia region. Ann Hum Genet 68:340–352

Rosser ZH, Zerjal T, Hurles ME, Adojaan M, Alavantic D, Amorim A, Amos W et al. (2000) Y-chromosomal diversity in Europe is clinal and influenced primarily by geography, rather than by language. Am J Hum Genet 67:1526–1543

Ruiz-Pesini E, Lapeña A-C, Díez-Sánchez C, Pérez-Martos A, Montoya J, Alvarez E, Díaz M, Urriés A, Montoro L, López-Pérez MJ, Enríquez JA (2000) Human mtDNA haplogroups associated with high or reduced spermatozoa motility. Am J Hum Genet 67:682–696

Ruiz-Pesini E, Mishmar D, Brandon M, Procaccio V, Wallace DC (2004) Effects of purifying and adaptive selection on regional variation in human mtDNA. Science 303:223–226

Saillard J, Magalhaes P, Schwartz M, Rosenberg T, Nørby S (2000) Mitochondrial DNA variant 11719G is a marker for the mtDNA haplogroup cluster HV. Hum Biol 72:1065–1068

Salas A, Richards M, De la Fe T, Lareu MV, Sobrino B, Sánchez-Diz P, Macaulay V, Carracedo Á (2002) The making of the African mtDNA landscape. Am J Hum Genet 71:1082–1111

Salas A, Richards M, Lareu MV, Scozzari R, Coppa A, Torroni A, Macaulay V, Carracedo Á (2004) The African diaspora: mitochondrial DNA and the Atlantic slave trade. Am J Hum Genet 74:454–465

Semino O, Passarino G, Quintana-Murci L, Liu A, Beres J, Czeizel A, Santachiara-Benerecetti AS (2000) MtDNA and Y chromosome polymorphisms in Hungary: inferences from the palaeolithic, neolithic and Uralic influences on the modern Hungarian gene pool. Eur J Hum Genet 8:339–346

Semino O, Magri C, Benuzzi G, Lin AA, Al-Zahery N, Battaglia V, Maccioni L, Triantaphyllidis C, Shen P, Oefner PJ, Zhivotovsky LA, King R, Torroni A, Cavalli-Sforza LL, Underhill PA, Santachiara-Benerecetti AS (2004) Origin, diffusion, and differentiation of Y-chromosome haplogroups E and J: inferences on the neolithization of Europe and later migratory events in the Mediterranean area. Am J Hum Genet 74:1023–1034

Shen P, Lavi T, Kivisild T, Chou V, Sengun D, Gefel D, Shpirer I, Woolf E, Hillel J, Oefner P, Feldman M (2004) Reconstruction of patrilineages and matrilineages of Samaritans and other Israeli populations from Y-chromosome and mitochondrial DNA sequence variation. Hum Mut 24:248–260

Soodyall H, Jenkins T (1992) Mitochondrial DNA polymorphisms in Khoisan populations from southern Africa. Ann Hum Genet 56:315–324

Soodyall H, Vigilant L, Hill AV, Stoneking M, Jenkins T (1996) mtDNA control-region sequence variation suggests multiple independent origins of an Asian-specific 9-bp deletion in sub-Saharan Africans. Am J Hum Genet 58:595–608

Starikovskaya EB, Sukernik RI, Derbeneva OA, Volodko NV, Ruiz-Pesini E, Torroni A, Brown MD, Lott MT, Hosseini SH, Huoponen K, Wallace DC (2005) Mitochondrial DNA diversity in indigenous populations of the southern extent of Siberia, and the origins of Native American haplogroups. Ann Hum Genet 69:67–89

Stevanovitch A, Gilles A, Bouzaid E, Kefi R, Paris F, Gayraud RP, Spadoni JL, El-Chenawi F, Beraud-Colomb E (2004) Mitochondrial DNA sequence diversity in a sedentary population from Egypt. Ann Hum Genet 68:23–39

Sykes B (2001) Seven daughters of Eve. Norton, New York

Tajima A, Hamaguchi K, Terao H et al. (2004) Genetic background of people in the Dominican Republic with or without obese type 2 diabetes revealed by mitochondrial DNA polymorphism. J Hum Genet 49:495–499

Tambets K, Rootsi S, Kivisild T, Help H, Serk P, Loogväli EL, Tolk H-V et al (2004) The western and eastern roots of the Saami—the story of genetic outliers told by mitochondrial DNA and Y chromosomes. Am J Hum Genet 74:661–682

Tanaka M, Cabrera VM, González AM, Larruga JM, Takeyasu T, Fuku N, Guo LJ et al. (2004) Mitochondrial genome variation in eastern Asia and the peopling of Japan. Genome Res 14:1832–1850

Tang H, Siegmund DO, Shen P, Oefner PJ, Feldman MW (2002) Frequentist estimation of coalescence times from nucleotide sequence data using a tree-based partition. Genetics 161:447–459

Torroni A, Schurr TG, Yang CC, Szathmary EJ, Williams RC, Schanfield MS, Troup GA, Knowler WC, Lawrence DN, Weiss KM et al (1992) Native American mitochondrial DNA analysis indicates that the Amerind and the Nadene populations were founded by two independent migrations. Genetics 130:153–162

Torroni A, Schurr TG, Cabell MF, Brown MD, Neel JV, Larsen M, Smith DG, Vullo CM, Wallace DC (1993) Asian affinities and continental radiation of the four founding Native American mtDNAs. Am J Hum Genet 53:563–590

Torroni A, Lott MT, Cabell MF, Chen YS, Lavergne L, Wallace DC (1994a) mtDNA and the origin of Caucasians: identification of ancient Caucasianspecific haplogroups, one of which is prone to a recurrent somatic duplication in the D-loop region. Am J Hum Genet 55:760–776

Torroni A, Miller JA, Moore LG, Zamudio S, Zhuang J, Droma T, Wallace DC (1994b) Mitochondrial DNA analysis in Tibet: implications for the origin of the Tibetan population and its adaptation to high altitude. Am J Phys Anthropol 93:189–199

Torroni A, Huoponen K, Francalacci P, Petrozzi M, Morelli L, Scozzari R, Obinu D, Savontaus ML, Wallace DC (1996) Classification of European mtDNAs from an analysis of three European populations. Genetics 144:1835–1850

Torroni A, Petrozzi M, D'Urbano L, Sellitto D, Zeviani M, Carrara F, Carducci C, Leuzzi V, Carelli V, Barboni P, De Negri A, Scozzari R (1997) Haplotype and phylogenetic analyses suggest that one European-specific mtDNA background plays a role in the expression of Leber hereditary optic neuropathy by increasing the penetrance of the primary mutations 11778 and 14484. Am J Hum Genet 60:1107–1121

Torroni A, Bandelt H-J, D'Urbano L, Lahermo P, Moral P, Sellitto D, Rengo C, Forster P, Savontaus M-L, Bonné-Tamir B, Scozzari R (1998) MtDNA analysis reveals a major late Palaeolithic population expansion from southwestern to northeastern Europe. Am J Hum Genet 62:1137–1152

Torroni A, Bandelt H-J, Macaulay V, Richards M, Cruciani F, Rengo C, Martinez-Cabrera V et al (2001a) A signal, from human mtDNA, of postglacial recolonization in Europe. Am J Hum Genet 69:844–852

Torroni A, Rengo C, Guida V, Cruciani F, Sellitto D, Coppa A, Calderon FL, Simionati B, Valle G, Richards M, Macaulay V, Scozzari R (2001b) Do the four clades of the mtDNA haplogroup L2 evolve at different rates? Am J Hum Genet 69:1348–1356

Trejaut JA, Kivisild T, Loo JH, Lee CL, He CL, Xi JR, Li ZY, Lin M (2005) Traces of archaic mitochondrial lineages persist in Austronesian-speaking Formosan populations. PLoS Biol 3(8):e247

Underhill PA, Shen P, Lin AA, Jin L, Passarino G, Yang WH, Kauffman E, Bonne-tamir B, Bertranpetit J, Francalacci P, Ibrahim M, Jenkins T, Kidd JR, Mehdi SQ, Seielstad MT, Wells RS, Piazza A, Davis RW, Feldman MW, Cavalli-Sforza LL, Oefner PJ (2000) Y chromosome sequence variation and the history of human populations. Nat Genet 26:358–361

Wallace DC, Brown MD, Lott MT (1999) Mitochondrial DNA variation in human evolution and disease. Gene 238:211–230

Wallace DC (2005) A mitochondrial paradigm of metabolic and degenerative diseases, aging, and cancer: a dawn for evolutionary medicine. Annu Rev Genet 39:359–407

Watson E, Forster P, Richards M, Bandelt H-J (1997) Mitochondrial footprints of human expansions in Africa. Am J Hum Genet 61:691–704

Weale ME, Shah T, Jones AL, Greenhalgh J, Wilson JF, Nymadawa P, Zeitlin D, Connell BA, Bradman N, Thomas MG (2003) Rare deep-rooting Y chromosome lineages in humans: lessons for phylogeography. Genetics 165:229–234

Weiss G, von Haeseler A (1998) Inference of population history using a likelihood approach. Genetics 149:1539–1546

Yao Y-G, Kong Q-P, Bandelt H-J, Kivisild T, Zhang Y-P (2002) Phylogeographic differentiation of mitochondrial DNA in Han Chinese. Am J Hum Genet 70:635–651

YCC (2002) A nomenclature system for the tree of human Y-chromosomal binary haplogroups. Genome Res 12:339–348

Zerjal T, Dashnyam B, Pandya A, Kayser M, Roewer L, Santos FR, Schiefenhovel W, Fretwell N, Jobling MA, Harihara S, Shimizu K, Semjidmaa D, Sajantila A, Salo P, Crawford MH, Ginter EK, Evgrafov OV, Tyler-Smith C (1997) Genetic relationships of Asians and northern Europeans, revealed by Y-chromosomal DNA analysis. Am J Hum Genet 60:1174–1183

Zischler H, Geisert H, von Haeseler A, Pääbo S (1995) A nuclear fossil of the mitochondrial D-loop and the origin of modern humans. Nature 378:489–492

Nucleic Acids and Molecular Biology, Vol. 18
Hans-Jürgen Bandelt, Vincent Macaulay, Martin Richards (Eds.)
Human Mitochondrial DNA and the Evolution of *Homo sapiens*

The Pioneer Settlement of Modern Humans in Asia

Mait Metspalu (✉) · Toomas Kivisild · Hans-Jürgen Bandelt · Martin Richards · Richard Villems

Institute of Molecular and Cell Biology, Tartu University and Estonian Biocentre, Riia 23, Tartu, Estonia
mait@ebc.ee

1 Introduction

Different hypotheses, routes, and the timing of the out-of-Africa migration are the focus of another chapter of this book (Chap. 10). However, in order to dig more deeply into discussions about pioneer settlement of Asia, it is necessary to emphasize here that many recent genetic, archaeological, and anthropological studies have started to favour the Southern Coastal Route (SCR) concept as the main mechanism of the primary settlement of Asia (Lahr and Foley 1994; Quintana-Murci et al. 1999; Stringer 2000; Kivisild et al. 2003, 2004); see also Oppenheimer (2003).

The coastal habitat as the medium for humans to penetrate from East Africa to Asia and Australasia was perhaps first envisaged by the evolutionary geographer Carl Sauer, who considered the populations taking this route as adapted to the ecological niche of the seashore (Sauer 1962). After reaching Southwest Asia, modern humans had a choice of two potential routes by which to colonize the rest of Asia. These two were separated by the world's mightiest mountain system—the Himalayas. The pioneer settlers could continue taking the SCR or they could change their habitat and turn instead to the north, passing through Central Asia and southern Siberia (or via the route that later became known as the Silk Road). Here, one has to avoid confusion with the 'Northern Route' of the out-of-Africa exodus and use the term 'Northern Asian Route'.

In principle, the first pioneer population of modern humans could have spread in several directions simultaneously: for example following the coast towards the east and, at the same time, cutting into the Asian inland along the river valleys (Wells et al. 2001; Oppenheimer 2003). As we argue later, a single coastal route towards the east appears to be sufficient to explain most, if not all, existing mtDNA genetic variation, not only in North and Southeast Asia, but also in Oceania. Before discussing extant mitochondrial (mtDNA) diversity in Asia, we shall briefly review the palaeoclimatological background and archaeological evidence for these events.

2 Palaeoclimatological Context

The patterns of colonization of Asia by anatomically modern human (AMH) population(s) were undoubtedly highly dependent on the surrounding environment. Our knowledge of past climates is frozen in ice sheets: polar ice cores and (ocean) sediments have been intensively studied in order to reconstruct the climate of the past. These global changes are reflected in the more detailed regional palaeovegetation surveys based, for example, on ancient pollen analysis (reviewed in Adams and Faure 1997; Adams et al. 1999; Ray and Adams 2001). As temperature change is a robust characteristic of the environment, we shall concentrate on that and will not go into details of, say, palaeovegetation.

After the Eemian interglacial, some 110 000–130 000 years ago, the global climate cooled until the period of the lowest temperatures, during the Last Glacial Maximum (LGM) 15 000–25 000 years ago. This process was not a steady one; instead, there were multiple oscillations of warmer and chillier periods. Intense cold and arid, but short-lived, Heinrich-type events characterized the otherwise gradually cooling phase between 110 000 and 70 000 years ago. This was followed by the stage 4 Glacial Maximum (also known as the Early Wisconsin Glacial) extending to about 50 000 years ago with conditions rather similar to the LGM (Adams et al. 1999). Warmer but highly variable temperatures were characteristic of the period thereafter, extending until the onset of the LGM.

As much as the changing temperatures, the peopling of Asia by AMHs was affected by the accompanying fluctuations in humidity. Lower temperatures generally mean less evaporation. The resulting global decrease in rainfall contributed to the extension of desert areas, for example, in Central and southwestern Asia. Even around 50 000 years ago, when a warmer and moister stage opened the green passage between the Arabian Sea and the Levant (the Zagros corridor), the deserts in Central Asia and northern Africa remained difficult habitats for most of the creatures—including humans (reviewed in Oppenheimer 2003).

During the colder phases of climate, which were also much more arid, the global sea level was much lower than it is today. That was mainly because enormous quantities of water were trapped in the extended polar ice caps, with a corresponding decrease in the volume of the oceans as the water cooled. The fluctuation of the sea level made crossing water obstacles easier at some times and harder at others. For example, the distance between Australia and Timor (the widest strait that had to be crossed en route to Sahul, or the Greater Australian landmass comprising both Australia and New Guinea during the Ice Age) was shortest (at 170 km) between 65 000 and 70 000 years ago, when the global sea level was about 80 m below its present level. During the subsequent warmer phases the strait lengthened, but did not exceed 220 km even during the maximum high stand (around 50 000 years ago) when the sea stood only 40 m below its current level (Chappell 2002). However, given the development of some form of water craft, there seems no strong reason to regard the time of shortest distance as the sole window of opportunity for crossing (cf. Oppenheimer 2003).

3 Archaeological and Palaeontological Evidence of the Peopling of Asia by AMH

Any fossil record is, inevitably, incomplete. Fossilization of skeletal remains is a rare event, depending on climate and probably many other factors, whereas their recovery depends on the intensity of archaeological investigation of a region. There is a particular issue when one considers the course of the likely coastal route out of Africa. If the beachcombing modern humans, being dependent on a seashore environment, indeed began colonizing Eurasia via the SCR, then many of the potential archaeological sites are at present submerged under the sea. An 80-m rise of the sea level (the difference between the sea level 70 000 years ago and today) altered the coastline considerably, shifting it hundreds of kilometres inland and probably inundating the range of beachcombing AMHs. Another important factor that needs to be considered is the tectonics of the continental shelves. Furthermore, the accuracy of fossil dating techniques is in constant dispute (Chen and Zhang 1991; Klein 1999).

Despite these problems, fossils are and will probably continue to be the best evidence of the spread of AMHs around the globe. Widely accepted datings for the earliest AMH skeletal remains outside Africa (excluding the brief 'extension' of the African range into the Near East; Chap. 10) reach approximately 45 000 years ago (Foley 1998 and references therein). Claims for considerably older dates in Asia, for example 67 000 years ago for Liujiang, China, have been heavily criticized (Etler 1996), although they continue to be made (Shen et al. 2002).

In Europe, the earliest AMH remains fall between 37 000 and 45 000 years ago (van Andel et al. 2003). Interestingly, AMH remains of similar antiquity have been found in Borneo and Australia, where the most ancient remains at Niah Cave and Lake Mungo, respectively, have recently been redated to more than 40 000 years ago (Barker et al. 2002; Bowler et al. 2003). The beginning of the human settlement at the Australian site was estimated to go back even to around 50 000 years ago (Bowler et al. 2003) or even 62 000 years ago (Thorne et al. 1999), while similar dates have been proposed for other Australian sites, such as Devil's Lair, southwestern Australia (Turney et al. 2001), and Deaf Adder Gorge, northern Australia (Roberts et al. 1990, 1994). These dates are of great importance because they also set the time boundary for the peopling of Asia.

The fact that younger remains from Inner Mongolia, with probable antiquity of not much more than 30 000 years ago, constitute the earliest widely accepted fossil findings of AMHs in mainland East Asia (Chen and Zhang 1991; Etler 1996) highlights the weakness of negative arguments raised by the lack of fossil evidence. Rather similar is the situation in South Asia, where the oldest fossils of modern humans uncovered so far, from southern Sri Lanka, are dated to 28 000 and 33 000 years ago (Kennedy et al. 1987; Kennedy and Deraniyagala 1989). The early dates of fossils in Australia imply that AMHs had to be in South Asia at least 10 000 years earlier than this.

Archaeological evidence has sometimes been interpreted as supporting modern human presence in South Asia over 60 000 years ago (references in Kumar and Reddy 2003). However, because the putative Middle Palaeolithic sites under consideration lack any human fossil evidence, it is not clear whether they can unambiguously be associated with modern humans (Joshi 1996). The spread of modern humans in the Middle East and Europe is generally coupled with the radiation of Upper Palaeolithic technology—which, however, did not reach Australia together with AMHs. The same may well be true for South Asia, where Upper Palaeolithic technology does not show up before 30 000 years ago (Chakrabarti 1999). Again, the introduction of Upper Palaeolithic technology to India postdates significantly the time frame when the carriers of the Middle Palaeolithic tools would have been walking on its shores to reach Australia. And, more importantly, when the Upper Palaeolithic technology reached India, it was contemporaneous with the pre-existing Middle Palaeolithic technology there for at least 10 000 years.

The transition from Middle to Upper Palaeolithic technology in the southern Near East occurred approximately 50 000 years ago (Gilead 1991). In Europe, it emerges almost simultaneously in both central Europe and in northern Spain before expanding through the continent (about 47 000 years ago: van Andel et al. 2003). The earliest Upper Palaeolithic technology is of nearly similar antiquity to that in the east, in the Zagros mountains (Olszewski and Dibble 1994), but slightly younger than that in the Caucasus region (30 000–32 000 years ago: Bar-Yosef 2001), where Neanderthals survived until 32 000 years ago. A similarly early transition (39 000–43 000 years ago) has been suggested for the Altai Mountain and Lake Baikal regions of southern Siberia (Dolukhanov et al. 2002; Vasil'ev et al. 2002), although these Early Upper Palaeolithic cultures share many features in common with the preceding Middle Palaeolithic Mousterian culture (Kuzmin and Keates 2004). Moreover, the archaeological record alone, with a lack of human skeletal remains, is inconclusive regarding whether or not the initial Middle to Upper Palaeolithic transition in Siberia was coupled with the influx of an AMH population from the west. Upper Palaeolithic artefacts from 18 000 years ago have been found in association with skeletal remains that bear similar morphology to contemporary AMH teeth from Europe (Scott and Turner 1997).

The Middle Palaeolithic settlement of AMH in Sahul together with the radiation of Upper Palaeolithic technology from the Middle East and its early arrival in southern Siberia have often been interpreted as supporting the existence of two different migration routes from Africa towards East Asia: an earlier one following the southern route along the coast of Asia towards Australia, carrying Middle Palaeolithic technology, followed by a migration associated with the Upper Palaeolithic via the northern route through the Levant and further along the Northern Asian Route through Central Asia and southern Siberia (Lahr and Foley 1994; Jobling and Tyler-Smith 2003). This twin-dispersal model makes different predictions about genetic patterns that are testable. If the two-route scenario (or 'pincer model') indeed explained the source of modern humans in Asia, then one should be able to find unique Northern and Central Asian specific lineages that cannot be derived from South and Southeast Asian variation, and vice versa. On the other hand, if the single southern or northern route scenario holds, then it should be possible to derive all northern variants from the gene pool of the south, and vice versa.

4 How to Infer 'Pioneer Settlement' from Extant mtDNA Variation?

An obvious starting point for deducing the patterns of the pioneer human settlement from the extant mtDNA diversity is to identify regionally autochthonous haplogroups and calculate their coalescence ages (Chap. 4). The average over the oldest of these would indicate the lower bounds for the start of the colonization. A founder type is identified as an ancestral node which is present (or may have been lost but is then phylogenetically reconstructed) both in the source and in the destination area (Richards et al. 2000). Ideally, the coalescence time of the founder type in the destination area would suggest the time of its arrival (Stoneking et al. 1990; Torroni et al. 1993a, b; Sykes et al. 1995; Forster et al. 1996; Richards et al. 2000). However, let us look at two nonideal cases. Firstly, if the founder population is small and does not disperse upon arrival, the coalescence times of the founder types may underestimate their entrance time. This is because the most recent common ancestors (MRCAs) for the future generations may successfully be replaced by younger ones, which is evidenced in phylogenetic reconstructions of branches defined by multiple mutations. Most likely these mutations occurred one by one in the evolutionary sequence but in the extant populations it is not clear which mutation occurred before and which after the founding event of that particular region of interest. For evaluation and comparison, one can also draw the tree and calculate the age of the descending haplogroups in the supposed source area. Secondly, in more massive migrations, a considerable amount of variation (within a haplogroup) may already be present among migrants and in that case their extant diversity (per haplogroup) is a sum of different periods of their demographic history. Back-migration(s) to the source area and the ability to detect founder types through adequate sampling are the other two major challenges for the founder analysis approach (Richards et al. 2000).

The strategy outlined requires good data coverage of the regions under investigation. Good in this context means both the 'depth'—phylogenetic resolution—and 'width'—geographic scope—of the data sets. Phylogenetic depth can be improved by searching for more markers until the bounds are met for the specific locus—a step that is already achieved for human mtDNA, but still lies ahead for the Y chromosome, for example. In practice, however, when surveying mtDNA diversity in different regions, one is quite often limited by the depth of the available data sets, since these often only consist of HVS-I sequences, which are still extremely popular in molecular anthropology as "the high mutation rate of this segment ensures a sufficient number of polymorphic sites for population genetic analyses" (Pakendorf and Stoneking 2005). It is exactly the recurrent nature of most mutations in this segment that inevitably destroys the more ancient signals one is usually interested in. Such data sets are thus of a limited value in the absence of coding-region information, and one should avoid making inferences based on insufficient information.

5
The Peopling of Asia as Seen Through the Lens of mtDNA Diversity

Complete and partial mtDNA coding-region sequences have been used to map the backbone and determine the fine structure of the mtDNA lineages present in Asia (Kivisild et al. 2002; Yao et al. 2002; Kong et al. 2003; Metspalu et al. 2004; Palanichamy et al. 2004; Quintana-Murci et al. 2004). The recent analysis of complete mtDNA sequences from 672 Japanese individuals has provided a significant refinement of the East Asian mtDNA phylogeny (Tanaka et al. 2004). Combining these and other published data, we summarize in Fig. 1 the Asian mtDNA tree topology. With many papers that refine the phylogeny being published almost simultaneously, it is hardly surprising that relevant literature may sometimes be missed and some confusion regarding the naming of the haplogroups and their branching order arises. The ongoing flow of complete mtDNA sequences from newly emerging basal branches within the Asian-Oceanian mtDNA phylogeny could not find place in this single diagram. A more detailed version of Fig. 1, along with commentaries that aim to overcome some of these difficulties, can be obtained from http://www.evolutsioon.ut.ee/. The macrohaplogroups M and N effectively cover the whole mtDNA pool in Asia (see also Chap. 7). The start of their dispersal has been dated to approximately 60 000–65 000 years ago (Maca-Meyer et al. 2001; Mishmar et al. 2003; Palanichamy et al. 2004). Macrohaplogroup M is slightly more frequent than N in Siberia, northern China, Japan, and South

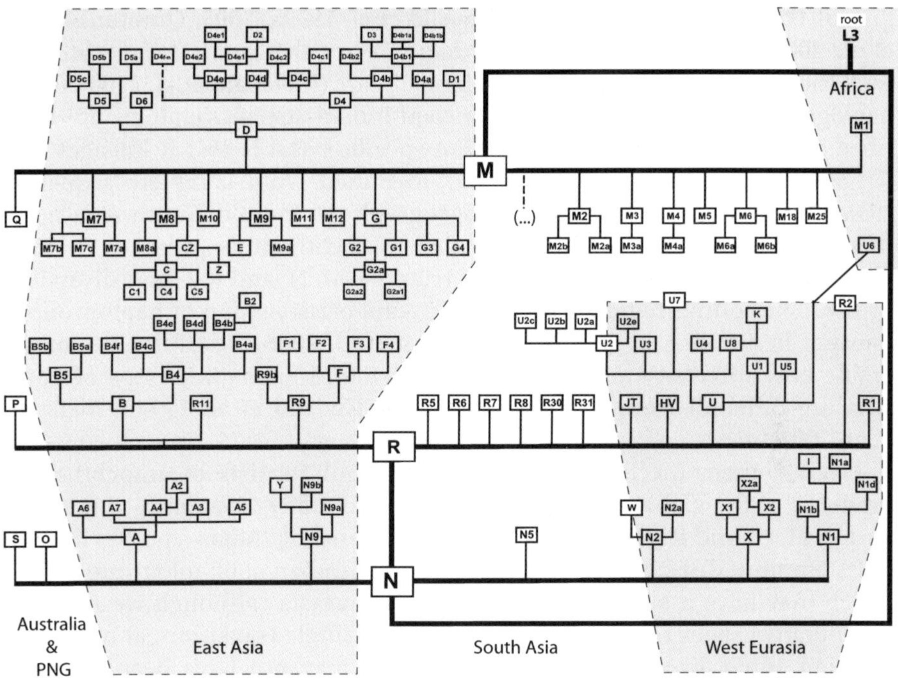

Fig. 1 Phylogenetic tree relating the Asian mitochondrial DNA (*mtDNA*) haplogroups. See http://www.evolutsioon.ut.ee for further information, particularly regarding the nomenclature of the haplogroups. *PNG* Papua New Guinea

Asia, while in Southeast Asia it is the other way around. M is nearly absent from Southwest Asia, where subhaplogroups branching from the N (including R) trunk dominate the mtDNA landscape. The N and R subbranches in West and East Eurasia do not overlap, and they form two distinct mtDNA 'domains'. With approximately similar shares, these two make up most of the mtDNA pool of Central Asia (Comas et al. 2004).

While the mtDNA makeup of the Americas represents an offshoot of the East Asian domain (Torroni et al. 1993a, b, 1994; Forster et al. 1996), Sahul (Australia/New Guinea) largely constitutes yet another autochthonous one. Stemming from the trunks R and M, haplogroups P and Q, respectively, cover more than half of the extant mtDNA pool sampled in Papua New Guinea (Forster et al. 2001). From the published full sequences (Ingman and Gyllensten 2003), additional Sahul-specific basal branches of M and N—namely S and O, which were baptized in Palanichamy et al. (2004)—are confirmed. Thus, both basal trunks of the Eurasian mtDNA tree show deep-rooting Sahul-specific branches.

South Asia, with its own specific branches of M and N, represents the third mtDNA domain in Asia (Fig. 1). Haplogroups M2 and R5 and subgroups U2a, U2b, and U2c of U2 (Kivisild et al. 1999a, 2003; Quintana-Murci et al. 2004), which make up more than 15% of the South Asian mtDNAs, each show coalescence times of over 50 000 years (Metspalu et al. 2004). These haplogroups form a set of the most ancient Indian-specific haplogroups identified so far. A number of novel Indian-specific basal N and R lineages (N5, R7, R8, R30, and R31) were recently identified from complete sequences (Palanichamy et al. 2004). The phylogeography of these in South Asia needs further attention, but, significantly, their autochthonous presence in India clearly demonstrates that all the basal trunks—M, N, and R—have diversified in situ. The coding-region-based downstream classification of haplogroup M lineages in South Asia is on the way (Sun et al. 2006), testifying to considerable basal diversity and confirming that the M subhaplogroups of South Asia are different from those os East Asia (Kivisild et al. 1999b; Metspalu et al. 2004).

Overall, then, the South Asian mtDNA pool consists of autochthonous branches of the global mtDNA tree that stem directly from each of the basal trunks M, N, and R (Fig. 1). Note that the only major Indian-specific lineages not stemming directly from the trunk are the Indian subhaplogroups of U2, which may have a sister group U2e in West Eurasia (although we note that this putative sister relationship hinges upon a single transition, at nucleotide position 16051 in the control region, which may not have been a unique event at the base of haplogroup U). The divergence time of these U2 daughters reaches 50 000 years (Kivisild et al. 1999a). Meanwhile, haplogroups R2, U7, and W represent an intriguing link between the West and South Asian mtDNA pools. Their spread and coalescence times suggest pre-LGM gene flows in the area spanning from western India and Pakistan up to the Near East. As judged from the coalescence times of the region-specific subclades of these haplogroups, this genetic continuum was apparently interrupted by the expanding deserts in eastern West Asia during the LGM (Metspalu et al. 2004).

Like Sahul and South Asia, the East Asian mtDNA pool is made up of autochthonous offshoots of M and N, most of which show coalescence times exceeding 50 000 years (Kivisild et al. 2002; Yao et al. 2002). While in South Asia we see a number of basal haplogroups branching from trunk R, in East Asia, only a few haplogroups, B (plus R11) and R9 (including F), spring out from the founder haplotype of haplogroup R (Fig. 1). The putative monophyly of a supergroup R11′B is based solely on transitions at unstable nucleotide positions 16189 and 16519, which could very well have happened in parallel. Furthermore, no more than two East Asian haplogroups, A and N9 (including Y), trace back to MRCAs in the N trunk. As in South Asia, the richest trunk in East Asia in terms of haplogroups stemming from it is M. Complete mtDNA sequencing has indicated that Southeast Asia also harbours some autochthonous M, N, and R lineages apparently not found further north, in East Asia (Macaulay et al. 2005; Merriwether et al. 2005).

As we see, all of the three mtDNA domains along the SCR (South Asia, East Asia, and Sahul) harbour haplogroups that stem directly from the M, N, and R trunks (Fig. 1), are primarily spread only within a single domain, and demonstrate coalescence times comparable to the initial expansion of M and N. They are frequent enough to rule out the possibility of major gene flow between the domains (since these haplogroups are not shared between the domains).

A plausible model for the initial peopling of Asia, one might think, would be a series of nested daughter colonizations, where regions were peopled one by one, with a time lag. In such a case, one would observe East Asian haplogroups to be derived from South Asian haplogroups, and Sahul daughter clades from East Asian haplogroups. This is the case with the later colonization of the Americas, but evidently not along the SCR, where deep-rooting autochthonous branches of the mtDNA tree are present throughout. It suggests that the initial colonization of Asia was not a gradual process, but rather a swift one, spreading the same founder types along the shores of the Indian Ocean as far as Sahul (Fig. 2). The fact that all of the domains show autochthonous basal R haplogroups—for example JT and U in West Asia, R5 (and many more) in South Asia, R9 and B in East Asia, and P in Sahul—suggests that, in addition to the differentiation of M and N, the divergence of R also occurred at the start of the AMH expansion along the SCR. Moreover, autochthonous lineages of haplogroup U2 trace back to approximately 50 000 years in both West Asia and South Asia. This time frame is comparable to the coalescence dates of M and N, possibly placing the spread of haplogroup U within the initial wave of the peopling of Asia. This would, however, demand attributing the absence of haplogroup U east of India to loss through genetic drift in the probably small scout population(s). Alternatively, a later, perhaps Upper Palaeolithic diffusion (starting from approximately 30 000 years ago; Chakrabarti 1999) from the west, ultimately from the Middle East (the most probable source of haplogroup U), might have introduced the U2 lineages into India (Kivisild et al. 1999a, 2000).

Here, it is worth briefly touching upon the 'parallel world' of Y-chromosome variation. As with mtDNA, a diverse set of the basal Eurasian Y-chromosome founder lineages congregate in South Asia—C, F, and K. Further, a number of deep-rooting subclades of F are only distributed in South Asia (Kivisild et al. 2003). The package of Y-chromosome founder lineages in West Eurasia is reduced to F and K. These observations support the idea that the out-of-Africa migration first reached South and Southwest Asia and from there dispersed both east and west, consistent with the single SCR scenario. In addition, more recent migration(s) from Africa, probably following the route over the Sinai, have enriched the West Eurasian gene pool with additional paternal lineages of haplogroup E (Underhill et al. 2001; Cruciani et al. 2004; Luis et al. 2004; Semino et al. 2004). A study of 19 Y-chromosome biallelic markers amongst East Asian populations revealed that the southern populations are more diverse than those from the north, the latter essentially representing

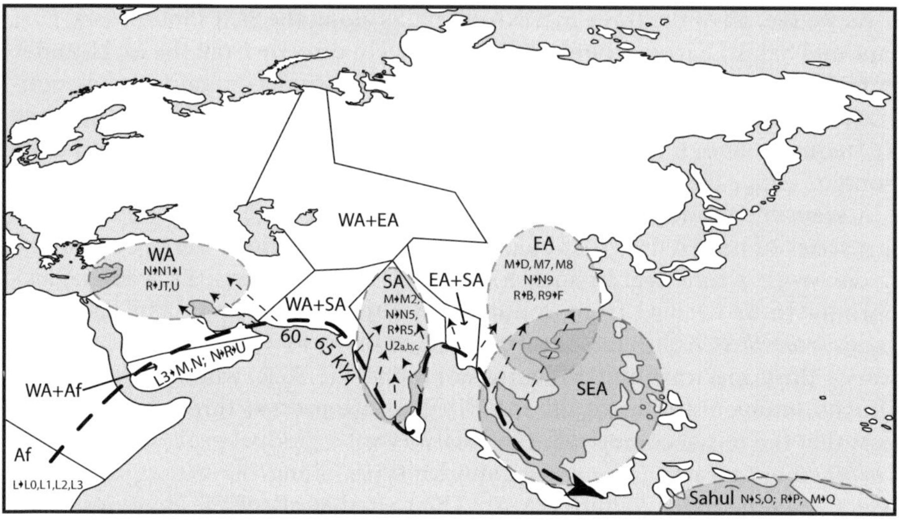

Fig. 2 Map of Eurasia depicting the possible scenario of the pioneer settlement of modern humans in Asia. The *thick dashed arrow* indicates the initial southern (coastal) route of the out-of-Africa event which had taken place by around 60 000–65 000 years ago (*Af* African specific mtDNA variants). During this opening stage, the earliest offshoots of haplogroups M and N were rapidly segregated into West Asian (*WA*; e.g. *JT*, *U*), South Asian (*SA*; e.g. *M2*, *N5*, *R5*, *U2a*, *U2b*, *U2c*), East Asian (*EA*; e.g. *D*, *M7*, M8, *N9*, *R9*, *B*), and further into the Australasia-specific (*Sahul*; *S*, *O*, *P*, *Q*) variants which later became the inocula for the autochthonous mtDNA diversification in the respective regions (*light dashed arrows and ellipses*, Sahul not fully shown). The complete mtDNA sequence data suggest a number of autochthonous Southeast Asian (*SEA*) M and N lineages that are absent from northern East Asia. During later stages of the colonization of Eurasia, modern humans moved further inland (not shown). Admixture between these basic domains of human settlement in Asia has been surprisingly limited on the maternal side ever since. Approximate boundary admixture zones (over 20% of admixture) between the three domains are shown by crude *solid lines* together with an indication to the mixed domains. Note that the whole of Central Asia appears as the biggest admixture zone where the mtDNA pools of West Asia and East Asia, and to a very much lower extent South Asia, intermix.

a subset of the variation present among the former. This is consistent with the SCR scenario, suggesting that mainland Southeast Asia was the starting point for the peopling of East Asia some 60 000 years ago (Su et al. 1999).

6
A Route Through Northern Asia?

Extant mtDNA variation in Asia suggests that the southern route was the major, if not the only, course used by the initial colonizers of Asia. But did some of the initial settlers of Asia populate first Central Asia and reach East Asia via southern Siberia (Maca-Meyer et al. 2001; Wells et al. 2001; Oppenheimer 2003; Tanaka et al. 2004)? The distinction between northern and southern East Asia seen in other biological traits (e.g. the spread of "sinodontic" and "sundadontic" teeth; Scott and Turner 1997) is also evident in the mtDNA variation. The two East Asian haplogroups of the mitochondrial R trunk (B and R9) are predominantly of Southeast Asian provenance, while those with the MRCA in its ancestral N trunk (A and N9) are more frequent in northern East Asia. Unaware of the newly identified Indian-specific branch of N, N5 (Palanichamy et al. 2004), Tanaka et al. (2004) interpreted the apparent lack of basal N lineages in India, and their presence in northern East Asia, as a strong argument for the Northern Asian Route for the peopling of Asia by a supposed 'N population'. However, when the gap in basal N lineages in South Asia is erased, the higher frequencies of some basal N lineages along the Northern Asian Route alone are not enough to corroborate its role in the process of peopling the continent. In addition, Tanaka et al. (2004) overlooked the phylogenetic relatedness of N and R and the abundance of basal R lineages in India.

Similarly, taken as a whole, the M trunk is more frequent in northern than in southern East Asia, but at the subhaplogroup level the picture is more complicated. For example, M7 and E are largely specific to mainland and island Southeast Asia (Ballinger et al. 1992), while others like G, M8 (including C and Z), and the most frequent M subhaplogroup D are much more frequent in northern East Asia. Haplogroups C and D are co-dominant in southern Siberia (Derenko et al. 2003), while haplogroups C and G are more frequent in northeastern Siberia (summarized in Tanaka et al. 2004). However, the mtDNA pools of northern and southern East Asia overlap, and the haplogroups that are most frequent among the Siberian populations also amount to one quarter of the Southeast Asian mtDNA pool. In turn, the southeastern haplogroups (excluding E, which is absent) take a notable share of the East Asian specific mtDNAs in Central Asia (circa 22%) and southern Siberia (circa 13%).

Is this pattern a result of two separate initial migration routes, carrying different founder types, or a unilateral diffusion, followed by genetic drift? Central Asian specific and southern Siberian specific basal branches of M and N would be the 'smoking gun' for the Northern Asian Route. Such branches have, however, not yet been found (Derenko et al. 2003; Comas et al. 2004). In the absence of direct evidence, a more detailed analysis of the phylogeography of the mtDNA haplogroups that make up the East Asian share of the mtDNA pools of Central Asia and southern Siberia ought to help corroborate or rule out the existence of the Northern Asian Route. More specifically, these pools should be checked for haplogroups ancestral to at least some mtDNA lineages in East Asia.

Before we go any further, we should remind ourselves that the extant mtDNA pools of southern Siberia and, especially, Central Asia are mixtures of East and West Eurasian mtDNA domains (Derenko et al. 2003; Bermisheva et al. 2004; Comas et al. 2004). As one would expect, the share of the western haplogroups diminishes as we move eastward (more than 40% in Central Asia, less than 20% in southern Siberia, and roughly 1% in East Asia). This pattern was shaped by admixture along the Steppe Belt long after the initial peopling of Asia and, thus, lies outside the scope of this chapter. In the following discussion we shall, therefore, concentrate on the eastern Eurasian specific share of the maternal lineages in Central Asia and southern Siberia.

Overall, the most frequent subhaplogroup of M in eastern Eurasia is D, which further branches into D4, D5, and D6 (Fig. 1). We see decreasing representation of these as we move westwards from East Asia. Only haplogroup D4 has dispersed into Central Asia (inferred from Comas et al. 2004), whereas the frequency of D5 in southern Siberia (1.5%; Derenko et al. 2003) is fivefold lower than that in China (5–10%; Yao et al. 2002). This pattern itself is best explained by an East Asian origin of haplogroup D, a pattern that recurs for other haplogroups present in both East Asia and Central Asia. Here we illustrate this reasoning by taking a more detailed look at several examples, starting with haplogroup D4.

Haplogroup D4 accounts for a third of the East Asian mtDNA lineages in Central Asia and a quarter of those in southern Siberia. Similarly, the frequency of D4 stays around 25% in northern China, while it drops to about 10% in the south of the country (Yao et al. 2002) and even more as one travels further south (island Southeast Asia) or west (Indochina). This pattern might seem consistent with the spread of D4 from Central Asia. However, the more detailed phylogeographic analysis questions that view.

One out of the myriad of D4 subhaplogroups found in East Asia (Fig. 1), D4c accounts for approximately 40% of D4 in Central Asia (reinferred from the HVS-I data of Comas et al. 2004). By comparison, for example, in southern Siberia the share of D4c in D4 is 10 times smaller. In Japan, where D4c makes up circa 13% of D4 (inferred from Maruyama et al. 2003), we see additional branches of the haplogroup, such as D4c1a (Tanaka et al. 2004), which seems to be absent in Central Asia (as judged from the absence of HVS-II motif 194-207). Other Central Asian D4 (HVS-I) haplotypes have close or exact matches in China, southern Siberia, and tribal populations of eastern India. Hence, the palette of D4 subhaplogroups in Central Asia appears poorer than that in more eastern regions. This is consistent with an eastern origin of the haplogroup.

The situation is similar in several other haplogroups. Like D4c, G2a1a (G2a in the original publication; see http://www.evolutsioon.ut.ee for clarifications) also shows high frequency in Central Asia, and is virtually the only variant of G found there (Comas et al. 2004). In southern Siberia, G2a1a constitutes a third of G while its sister clade G2a2 (3%), mother clade G2a (24%), and two other G subclades G1a1 (36%) and G3 (6%) make up the other two thirds (Derenko et al. 2003) (Fig. 1). G2a1a is, along with its sister clade G2a1b, also present in Japan. Furthermore, there is a subgroup of G2a1a in Japan, defined by transitions at nucleotide positions 16194 and 16195 (Maruyama et al. 2003; Tanaka et al. 2004). As we see, only one subhaplogroup, one bough of the G phylogeny, predominantly represents the dispersal of this haplogroup into Central Asia (Fig. 1). This shows clearly that the phylogeography of neither haplogroup D nor haplogroup G can be interpreted as supporting the origin of major East Asian specific haplogroups in Central Asia.

This conclusion is further supported by the phylogeography of haplogroup M8 (Fig. 1). This is the most frequent haplogroup in southern Siberia, accounting for around half of the East Eurasian mtDNA lineages (Derenko et al. 2003). It is also relatively more frequent in Central Asia than in the east of the continent. However, at the basal level, we see again that in East Asia the presence of the various subhaplogroups of M8 (Fig. 1) is more balanced, as C, Z, and M8a are spread in comparable frequencies (e.g. among Han Chinese; Yao et al. 2002); whilst in southern Siberia, and particularly in Central Asia, subgroups C and Z are predominant. Therefore, an eastern origin and subsequent westwards spread is a more likely history for M8 than the pincer model suggested by Oppenheimer (2003).

Let us have a look at one last example. The frequency of haplogroup A in East Asia is generally between 5 and 10% (Yao et al. 2002). Similarly, in Central Asia, it accounts for less than 10% of the mtDNAs of East Asian origin (Comas et al. 2004). Significantly, only one subclade of A, A4, is present in Central Asia, while A3, A5, A7, and a set of unclassified A* lineages are found alongside A4 in East Asia (see clarification of haplogroup A subgroups classification at http://www.evolutsioon.ut.ee; cf. Tanaka et al. 2004).

Unfortunately, only a fraction of A4 can be assigned to subclades using the HVS-I motifs of the three completely sequenced examples of A4 (excluding the Native American A2) mtDNAs. Available HVS-I data on Asian populations (e.g. summarized in Metspalu et al. 2004) suggest that haplogroup A4 displays further region-specific subclades. The root HVS-I haplotype has been found mainly in Chinese samples, from both tribal and Han people, but also in tribal populations from East India and Thailand and a few Central Asians and southern Siberians. Transitions at nucleotide positions 16124, 16260, and 16274 delineate Thai, East Indian, and Chinese-specific subclades, respectively, while a number of additional minor branches exist. Most Central Asian and southern Siberian A4 lineages group with the Chinese variants. The observation that the spread of haplogroup A in Central Asia is restricted to only one of the subclades and that, within that subclade, the lineages present are generally shared with the East Asians is, again, consistent with the eastern origin of haplogroup A.

Together the haplogroups (M8, D, G, and A) shown here to have radiated out from East Asia cover 70 and 80% of the East Eurasian specific mtDNAs in Central Asia and southern Siberia, respectively. Without going into the details concerning the rest of the maternal lineages of eastern provenance in these regions we add that none bear signs of being ancestral to the respective lineages in East Asia.

In their study of Y-chromosome variation, Wells et al. (2001) argued that Central Asia has been the source area of multiple migrations leading both west and east. The metaphor "important reservoir of genetic diversity", which was used to describe high genetic diversity among the present-day Central Asians (Wells et al. 2001), is indeed true. However, it appears increasingly more likely (Comas et al. 2004; Bermisheva et al. 2004) that this particular reservoir, as it is reflected in its present gene pool, was formed by gene flows *to* Central Asia *from* both east and west, long after the initial settlement of Asia.

7 Conclusion

The SCR of the pioneer phase of the peopling of the vast territories of Asia has gained increasingly strong experimental support, thanks to recently acquired deeper phylogenetic and phylogeographic knowledge about the spread of mtDNA (and Y-chromosomal) variation in this continent. Much, if not all, of the early settlement process can be seen as a 'fast train to Southeast Asia and Australia along the SCR'—indeed, so fast that the founder haplotypes at the base of haplogroups M, N, and R reached all major destinations alongside the route, as far down as Australia. It appears that Central Asia and southern Siberia were not involved in the initial peopling of the continent. It is also evident that the initial fast train phase was followed by a long-lasting freezing of the major geographic pools of maternal lineages in the south and further gene flows northwards from Southeast Asia and subsequently back westwards along the Steppe Belt extending from Manchuria to Europe. At present, western Siberia, the Urals, and Central Asia form a huge continuous admixture zone encompassing East and West Eurasian maternal lineages—a process that has so far had only a minimal influence on the essentially distinct autochthonous patterns of mtDNA variation in most of South Asia, East Asia, Southeast Asia, and Australasia.

Acknowledgement We thank William Davies for a critical reading of parts of this chapter.

References

Adams JM, Faure H (1997) Global land environments during the last interglacial. Oak Ridge National Laboratory, Oak Ridge

Adams JM, Maslin M, Thomas E (1999) Sudden climate transitions during the Quaternary. Prog Phys Geogr 23:1–36

Ballinger SW, Schurr TG, Torroni A, Gan YY, Hodge JA, Hassan K, Chen KH, Wallace DC (1992) Southeast Asian mitochondrial DNA analysis reveals genetic continuity of ancient mongoloid migrations. Genetics 130:139–152

Barker G, Barton H, Beavitt P, Bird M, Daly P, Doherty C, Gilbertson D, Hunt C, Krigbaum J, Lewis H, Manser J, McLaren S, Paz V, Piper P, Pyatt B, Rabett R, Reynolds T, Rose J, Rushworth G, Stephens M (2002) Prehistoric foragers and farmers in Southeast Asia: renewed investigations at Niah Cave, Sarawak. Proc Prehist Soc 68:147–164

Bar-Yosef O (2001) Dating the transition from the Middle to Upper Palaeolithic. Paper presented at XXI[e] Rencontres internationales d'archeologie et d'historie d'Antibes, Antibes

Bermisheva MA, Kutuev IA, Korshunova TY, Dubova NA, Villems R, Khusnutdinova EK (2004) Phylogeographic analysis of mitochondrial DNA in the Nogays: A strong mixture of maternal lineages from eastern and western Eurasia. Mol Biol (Mosk) 38:516–523

Bowler J, Johnston H, Olley J, Prescott J, Roberts R, Shawcross W, Spooner N (2003) New ages for human occupation and climatic change at Lake Mungo, Australia. Nature 421:837–840

Chakrabarti DK (1999) India. An archaeological history. Palaeolithic beginnings to early historic foundations. Oxford University Press, New Delhi

Chappell J (2002) Sea level changes forced ice breakouts in the Last Glacial cycle: new results from coral terraces. Quat Sci Rev 21:1229–1240

Chen TM, Zhang YY (1991) Palaeolithic chronology and possible co-existence of *H. erectus* and *H. sapiens* in China. World Archeol 23:147–154

Comas D, Plaza S, Wells RS, Yuldaseva N, Lao O, Calafell F, Bertranpetit J (2004) Admixture, migrations, and dispersals in Central Asia: evidence from maternal DNA lineages. Eur J Hum Genet 12:495–504

Cruciani F, La Fratta R, Santolamazza P, Sellitto D, Pascone R, Moral P, Watson E, Guida V, Colomb EB, Zaharova B, Lavinha J, Vona G, Aman R, Calì F, Akar N, Richards M, Torroni A, Novelletto A, Scozzari R (2004) Phylogeographic analysis of haplogroup E3b (E-M215) Y chromosomes reveals multiple migratory events within and out of Africa. Am J Hum Genet 74:1014–1022

Derenko MV, Grzybowski T, Malyarchuk BA, Dambueva IK, Denisova GA, Czarny J, Dorzhu CM, Kakpakov VT, Miścicka-Śliwka D, Woźniak M, Zakharov IA (2003) Diversity of mitochondrial DNA lineages in south Siberia. Ann Hum Genet 67:391–411

Dolukhanov PM, Shukurov AM, Tarasov PE, Zaitseva GI (2002) Colonization of northern Eurasia by modern humans: radiocarbon chronology and environment. J Arch Sci 29:593–606

Etler DA (1996) The fossil evidence for human evolution in Asia. Annu Rev Anthropol 25:275–301

Foley R (1998) The context of human genetic evolution. Genome Res 8:339–347

Forster P, Harding R, Torroni A, Bandelt H-J (1996) Origin and evolution of Native American mtDNA variation: a reappraisal. Am J Hum Genet 59:935–945

Forster P, Torroni A, Renfrew C, Röhl A (2001) Phylogenetic star contraction applied to Asian and Papuan mtDNA evolution. Mol Biol Evol 18:1864–1881

Gilead I (1991) The Upper Palaeolithic period in the Levant. J World Prehist 5:105–154

Ingman M, Gyllensten U (2003) Mitochondrial genome variation and evolutionary history of Australian and New Guinean aborigines. Genome Res 13:1600–1606

Jobling MA, Tyler-Smith C (2003) The human Y chromosome: an evolutionary marker comes of age. Nat Rev Genet 4:598–612

Joshi RV (1996) South Asia in the period of Homo sapiens neanderthalensis and contemporaries (Middle Palaeolithic), vol I. UNESCO

Kennedy KA, Deraniyagala SU, Roertgen WJ, Chiment J, Disotell T (1987) Upper Pleistocene fossil hominids from Sri Lanka. Am J Phys Anthropol 72:441–461

Kennedy KAR, Deraniyagala SU (1989) Fossil remains of 28 000-year-old hominids from Sri Lanka. Curr Anthropol 30:394–399

Kivisild T, Bamshad MJ, Kaldma K, Metspalu M, Metspalu E, Reidla M, Laos S, Parik J, Watkins WS, Dixon ME, Papiha SS, Mastana SS, Mir MR, Ferak V, Villems R (1999a) Deep common ancestry of Indian and western-Eurasian mitochondrial DNA lineages. Curr Biol 9:1331–1334

Kivisild T, Kaldma K, Metspalu M, Parik J, Papiha SS, Villems R (1999b) The place of the Indian mitochondrial DNA variants in the global network of maternal lineages and the peopling of the Old World. In: Papiha SS, Deka R, Chakraborty R (eds) Genomic diversity. Kluwer/Plenum, Dordrecht, pp 135–152

Kivisild T, Papiha SS, Rootsi S, Parik J, Kaldma K, Reidla M, Laos S, Metspalu M, Pielberg G, Adojaan M, Metspalu E, Mastana SS, Wang Y, Gölge M, Demirtas H, Schnekenberg E, Stefano GF, Geberhiwot T, Claustres M, Villems R (2000) An Indian ancestry: a key for understanding human diversity in Europe and beyond. In: Renfrew C, Boyle K (eds) Archaeogenetics: DNA and the population prehistory of Europe. McDonald Institute for Archaeological Research University of Cambridge, Cambridge, pp 267–279

Kivisild T, Tolk H-V, Parik J, Wang Y, Papiha SS, Bandelt H-J, Villems R (2002) The emerging limbs and twigs of the East Asian mtDNA tree. Mol Biol Evol 19:1737–1751 (erratum 20:162)

Kivisild T, Rootsi S, Metspalu M, Mastana S, Kaldma K, Parik J, Metspalu E, Adojaan M, Tolk H-V, Stepanov V, Gölge M, Usanga E, Papiha SS, Cinnioğlu C, King R, Cavalli-Sforza L, Underhill PA, Villems R (2003) The genetic heritage of the earliest settlers persists both in Indian tribal and caste populations. Am J Hum Genet 72:313–332

Kivisild T, Reidla M, Metspalu E, Rosa A, Brehm A, Pennarun E, Parik J, Geberhiwot T, Usanga E, Villems R (2004) Ethiopian mitochondrial DNA heritage: tracking gene flows across and around the Strait of Tears. Am J Hum Genet 75:752–770

Klein R (1999) The human career. University of Chicago Press, Chicago

Kong Q-P, Yao Y-G, Sun C, Bandelt H-J, Zhu C-L, Zhang Y-P (2003) Phylogeny of East Asian mitochondrial DNA lineages inferred from complete sequences. Am J Hum Genet 73:671–676

Kumar V, Reddy M (2003) Status of Austro-Asiatic groups in the peopling of India: An exploratory study based on the available prehistoric, linguistic and biological evidences. J Biosci 28:507–522

Kuzmin YV, Keates SG (2004) Comment on "Colonization of northern Eurasia by modern humans: radiocarbon chronology and environment" by P.M. Dolukhanov, A.M. Shukurov, P.E. Tarasov and G.I. Zaitseva. Journal of Archaeological Science 29, 593–606 (2002). J Archaeol Sci 31:141–143

Lahr M, Foley R (1994) Multiple dispersals and modern human origins. Evol Anthropol 3:48–60

Luis JR, Rowold DJ, Regueiro M, Caeiro B, Cinnioğlu C, Roseman C, Underhill PA, Cavalli-Sforza LL, Herrera RJ (2004) The Levant versus the Horn of Africa: evidence for bidirectional corridors of human migrations. Am J Hum Genet 74:532–544 (erratum 74:788)

Maca-Meyer N, González AM, Larruga JM, Flores C, Cabrera VM (2001) Major genomic mitochondrial lineages delineate early human expansions. BMC Genet 2:13

Macaulay V, Hill C, Achilli A, Rengo C, Clarke D, Meehan W, Blackburn J, Semino O, Scozzari R, Cruciani F, Taha A, Shaari NK, Raja JM, Ismail P, Zainuddin Z, Goodwin W, Bulbeck D, Bandelt H-J, Oppenheimer S, Torroni A, Richards M (2005) Single, rapid coastal settlement of Asia revealed by analysis of complete mitochondrial genomes. Science 308:1034–1036

Maruyama S, Minaguchi K, Saitou N (2003) Sequence polymorphisms of the mitochondrial DNA control region and phylogenetic analysis of mtDNA lineages in the Japanese population. Int J Legal Med 117:218–225

Merriwether DA, Hodgson JA, Friedlaender FR, Allaby R, Cerchio S, Koki G, Friedlaender JS (2005) Ancient mitochondrial M haplogroups identified in the Southwest Pacific. Proc Natl Acad Sci USA 102:13034–13039

Metspalu M, Kivisild T, Metspalu E, Parik J, Hudjashov G, Kaldma K, Serk P, Karmin M, Behar DM, Gilbert MTP, Endicott P, Mastana S, Papiha SS, Skorecki K, Torroni A, Villems R (2004) Most of the extant mtDNA boundaries in South and Southwest Asia were likely shaped during the initial settlement of Eurasia by anatomically modern humans. BMC Genet 5:26 (erratum 6:41)

Mishmar D, Ruiz-Pesini E, Golik P, Macaulay V, Clark AG, Hosseini S, Brandon M, Easley K, Chen E, Brown MD, Sukernik RI, Olckers A, Wallace DC (2003) Natural selection shaped regional mtDNA variation in humans. Proc Natl Acad Sci USA 100:171–176

Olszewski DI, Dibble HL (1994) The Zagros Aurignacian. Curr Anthropol 35:68–75

Oppenheimer S (2003) Out of Eden: the peopling of the world. Constable, London

Pakendorf B, Stoneking M (2005) Mitochondrial DNA and human evolution. Annu Rev Genomics Hum Genet 6:165–183

Palanichamy M, Sun C, Agrawal S, Bandelt H-J, Kong Q-P, Khan F, Wang C-Y, Chaudhuri T, Palla V, Zhang Y-P (2004) Phylogeny of mtDNA macrohaplogroup N in India based on complete sequencing: implications for the peopling of South Asia. Am J Hum Genet 75:966–978

Quintana-Murci L, Semino O, Bandelt H-J, Passarino G, McElreavey K, Santachiara-Benerecetti AS (1999) Genetic evidence of an early exit of *Homo sapiens sapiens* from Africa through eastern Africa. Nat Genet 23:437–441

Quintana-Murci L, Chaix R, Wells S, Behar D, Sayar H, Scozzari R, Rengo C, Al-Zahery N, Semino O, Santachiara-Benerecetti A, Coppa A, Ayub Q, Mohyuddin A, Tyler-Smith C, Mehdi Q, Torroni A, McElreavey K (2004) Where West meets East: the complex mtDNA landscape of the Southwest and Central Asian corridor. Am J Hum Genet 74:827–845

Ray N, Adams JM (2001) A GIS-based vegetation map of the world at the Last Glacial Maximum (25 000–15 000 BP). Internet Archaeol 11

Richards M, Macaulay V, Hickey E, Vega E, Sykes B, Guida V, Rengo C et al. (2000) Tracing European founder lineages in the Near Eastern mtDNA pool. Am J Hum Genet 67:1251–1276

Roberts RG, Jones R, Smith MA (1990) Thermoluminescence dating of a 50 000-year-old human occupation site in northern Australia. Nature 345:153–156

Roberts RG, Jones R, Spooner NA, Head MJ, Murray AS, Smith MA (1994) The human colonisation of Australia: optical dates of 53 000 and 60 000 years bracket human arrival at Deaf Adder Gorge, Northern Territory. Quat Sci Rev 13:575–583

Sauer C (1962) Seashore—primitive home of man? Proc Am Philos Soc 106:41–47

Scott GR, Turner CGI (1997) The anthropology of modern human teeth. Dental morphology and its variation in recent human populations. Cambridge University Press, Cambridge

Semino O, Magri C, Benuzzi G, Lin AA, Al-Zahery N, Battaglia V, Maccioni L, Triantaphyllidis C, Shen P, Oefner PJ, Zhivotovsky LA, King R, Torroni A, Cavalli-Sforza LL, Underhill PA, Santachiara-Benerecetti AS (2004) Origin, diffusion, and differentiation of Y-chromosome haplogroups E and J: inferences on the neolithization of Europe and later migratory events in the Mediterranean area. Am J Hum Genet 74:1023–1034

Shen G, Wang W, Wang Q, Zhao J, Collerson K, Zhou C, Tobias PV (2002) U-Series dating of Liujiang hominid site in Guangxi, southern China. J Hum Evol 43:817–829

Stoneking M, Jorde LB, Bhatia K, Wilson AC (1990) Geographic variation in human mitochondrial DNA from Papua New Guinea. Genetics 124:717–733

Stringer C (2000) Coasting out of Africa. Nature 405:24–25, 27

Su B, Xiao J, Underhill P, Deka R, Zhang W, Akey J, Huang W, Shen D, Lu D, Luo J, Chu J, Tan J, Shen P, Davis R, Cavalli-Sforza L, Chakraborty R, Xiong M, Du R, Oefner P, Chen Z, Jin L (1999) Y-Chromosome evidence for a northward migration of modern humans into Eastern Asia during the last ice age. Am J Hum Genet 65:1718–1724

Sun C, Kong Q-P, Palanichamy Mg, Agrawal S, Bandelt H-J, Yao Y-G, Khan F, Zhu C-L, Chaudhuri TK, Zhang Y-P, (2006) The dazzling array of basal branches in the mtDNA macrohaplogroup M from India as inferred from complete genomes. Mol Biol Evol 23:683–690

Sykes B, Leiboff A, Low-Beer J, Tetzner S, Richards M (1995) The origins of the Polynesians: an interpretation from mitochondrial lineage analysis. Am J Hum Genet 57:1463–1475

Tanaka M, Cabrera VM, González AM, Larruga JM, Takeyasu T, Fuku N, Guo L-J, et al. (2004) Mitochondrial genome variation in eastern Asia and the peopling of Japan. Genome Res 14:1832–1850

Thorne A, Grun R, Mortimer G, Spooner NA, Simpson JJ, McCulloch M, Taylor L, Curnoe D (1999) Australia's oldest human remains: age of the Lake Mungo 3 skeleton. J Hum Evol 36:591–612

Torroni A, Schurr TG, Cabell MF, Brown MD, Neel JV, Larsen M, Smith DG, Vullo CM, Wallace DC (1993a) Asian affinities and continental radiation of the four founding Native American mtDNAs. Am J Hum Genet 53:563–590

Torroni A, Sukernik RI, Schurr TG, Starikorskaya YB, Cabell MF, Crawford MH, Comuzzie AG, Wallace DC (1993b) mtDNA variation of aboriginal Siberians reveals distinct genetic affinities with Native Americans. Am J Hum Genet 53:591–608

Torroni A, Neel JV, Barrantes R, Schurr TG, Wallace DC (1994) Mitochondrial DNA "clock" for the Amerinds and its implications for timing their entry into North America. Proc Natl Acad Sci USA 91:1158–1162

Turney CSM, Bird MI, Fifield LK, Roberts RG, Smith M, Dortch CE, Grun R, Lawson E, Ayliffe LK, Miller GH (2001) Early human occupation at Devil's Lair, southwestern Australia 50 000 years ago. Quat Res 55:3–13

Underhill PA, Passarino G, Lin AA, Shen P, Mirazon Lahr M, Foley R, Oefner PJ, Cavalli-Sforza LL (2001) The phylogeography of Y chromosome binary haplotypes and the origins of modern human populations. Ann Hum Genet 65:43–62

van Andel T, Davies W, Weninger B (2003) The human presence in Europe during the last glacial period I: human migrations and the changing climate. In: van Andel T, Davies W (eds) Neanderthals and modern humans in the European landscape during the last glaciation. McDonald Institute for Archaeological Research, Cambridge, pp 31–56

Vasil'ev SA, Kuzmin YV, Orlova LA, Dementiev VN (2002) Radiocarbon-based chronology of the Palaeolithic in Siberia and its relevance to the peopling of the New World. Radiocarbon 44:503–530

Wells RS, Yuldasheva N, Ruzibakiev R, Underhill PA, Evseeva I, Blue-Smith J, Jin L, et al. (2001) The Eurasian heartland: a continental perspective on Y-chromosome diversity. Proc Natl Acad Sci USA 98:10244–10249

Yao Y-G, Kong Q-P, Bandelt H-J, Kivisild T, Zhang Y-P (2002) Phylogeographic differentiation of mitochondrial DNA in Han Chinese. Am J Hum Genet 70:635–651

Nucleic Acids and Molecular Biology, Vol. 18
Hans-Jürgen Bandelt, Vincent Macaulay, Martin Richards (Eds.)
Human Mitochondrial DNA and the Evolution of *Homo sapiens*

Ancient DNA and the Neanderthals

William Goodwin[1] (✉) · Igor Ovchinnikov[2]

[1]University of Central Lancashire, Preston PR1 2HE, UK
whgoodwin@uclan.ac.uk

[2]University of Connecticut, Storrs, CT 06269-2131, USA

1 Introduction

The relationship between the Neanderthals and modern humans has been vigorously debated ever since the first Neanderthal remains were recognised as being distinct from those of modern humans. The main thrust of the discussion can be summarised as: "To what extent did various populations of Ancients such as the Neanderthals contribute to the evident physical and cultural diversity of anatomically modern humans?" (Stringer and Gamble 1993). Until recently the subject was in the realm of physical anthropologists and archaeologists. In 1987 molecular geneticists entered the arena with a seminal paper by Cann et al. (1987) that made a major contribution to the theories on the origins of modern humans. This paper examined current populations and by analysing the diversity within the extant mitochondrial DNA (mtDNA) pool concluded that modern humans had originated in Africa within the last few hundred thousand years. Multiple studies, examining a large number of loci, have built on this work, and the vast majority of them have come to similar conclusions (Underhill et al. 2001; Cavalli-Sforza and Feldman 2003; Watkins et al. 2003). All these reports used the diversity that exists in current populations to extrapolate back in time.

It was not possible in the early days of molecular biology to examine the DNA of early hominins directly. However, the development of DNA amplification using the polymerase chain reaction (PCR) made such analysis a realistic proposition. Most hominin fossils are still outside the scope of molecular genetics as specimens greater than 100 000 years old do not contain enough DNA to analyse. Luckily, the Neanderthals are an exception amongst extinct hominins: some of them are young enough to contain endogenous DNA. Ten years after the first paper by Cann et al., a paper was published entitled "Neanderthal DNA sequences and the origin of modern humans" (Krings et al. 1997); this was an attempt to assess directly the genetic relationship of Neanderthals and modern humans by analysing Neanderthal DNA. This along with subsequent DNA analysis of Neanderthal remains will be the subject of this chapter.

2 Origins of the Neanderthals

It is widely accepted that Neanderthals were descended from early *Homo* populations that left Africa and moved into Europe and western Asia. The model depicted in Figs. 1 and 2 shows them as descendants of a migration of *H. heidelbergensis* which occurred around 600 000 years ago (YBP) – defining the abbreviation YBP used in Fig. 1. *H. heidelbergensis* is an example of an archaic hominin (sometimes referred to as 'archaic *H. sapiens*'), an intermediate or transitional fossil group between *H. erectus* and modern *H. sapiens*. Whether Neanderthals are direct descendants of *H. heidelbergensis* or are descended from an earlier population of *H. erectus* which had evolved towards an archaic *H. sapiens* state *in situ* in Europe

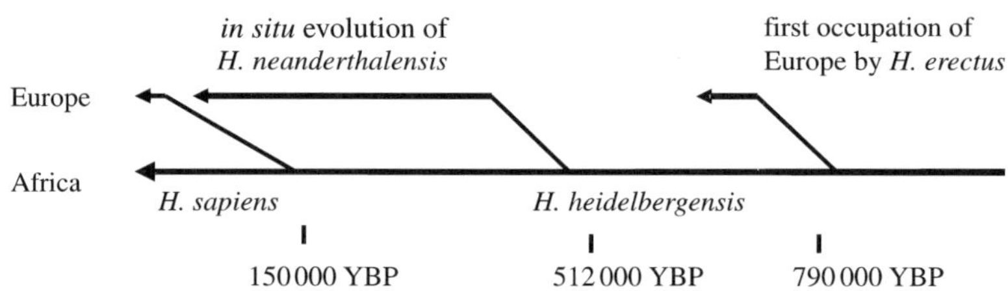

Fig. 1 Key later stages in the evolution of the *Homo* genus (based on Lahr and Foley 1998)

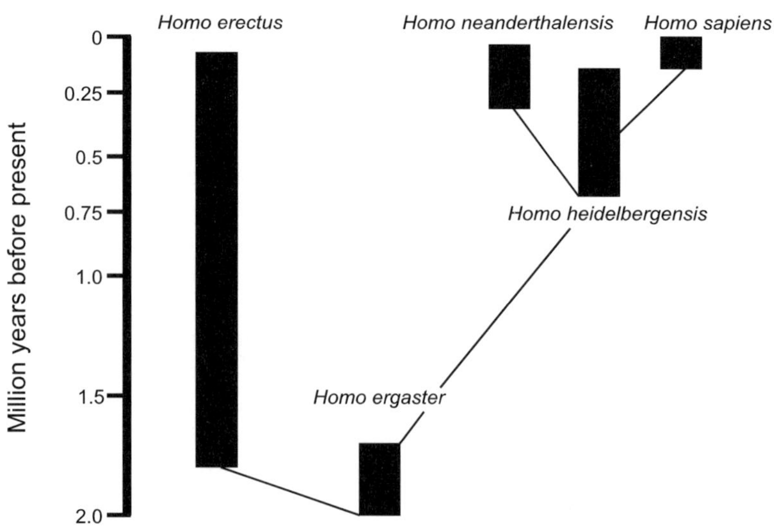

Fig. 2 A phylogenetic tree showing the relationship of the different members of the *Homo* genus that are believed to be the ancestors of *Homo sapiens*, dating back two million years (based on Conroy 1997)

(or even a later African hominin species, *Homo helmei*) is unresolved—the representation given here is just one interpretation of the fossil record (Conroy 1997).

What is clear from the archaeological and anthropological evidence is that one group of archaic hominins came to dominate the European scene for around 200 000 years; these were the Neanderthals. Characteristic Neanderthal features made their first appearance in the fossil record around 230 000 years ago. 'Classic' Neanderthals, with the full suite of Neanderthal characteristics, were present by around 130 000 years ago and the best examples have been found in western Europe (Stringer and Gamble 1993).

The gradual evolution of Neanderthals in Europe over a period of 200 000 years came to an end around 30 000 years ago. Neanderthals became extinct and were replaced by anatomically modern humans—the Cro-Magnons. The replacement started in the east around 45 000 years ago in Europe and ended in several refugia 30 000–28 000 years ago. The areas of Europe where Neanderthals lived are shown in Fig. 3.

There are several principal theories that explore the potential contributions of archaic hominins to the origin of modern humans. The theories can be broadly classified into multiregional evolution, hybridisation and replacement by an African population. The key features of these theories are:

- *Multiregional evolution*, also known as regional continuity, hypothesises that the Neanderthals and their regional contempories in other parts of the world are the direct ancestors of modern populations; therefore modern humans evolved simultaneously in Africa, Europe and Asia. The model relies on sufficient gene flow between these regional populations to maintain the different populations as a single species (Wolpoff 1989).
- *Replacement-out of Africa* hypothesises that all modern humans originated from a single ancestral population which is dated to around 150 000 years ago. As the name of the hypothesis suggests, the proposed homeland for this population is Africa (Cann et al. 1987; Stringer and Andrews 1988).

Between these two extremes of high and zero continuity lies the possibility of medium-to-low continuity. If anatomically modern humans did evolve in Africa and did migrate into Europe then there is also the possibility that there was genetic exchange between the new population and the Neanderthals, resulting in a hybrid population.

3
DNA Analysis of Neanderthal Specimens

What questions about the relationship can DNA address? Analysis of contemporary populations can be used to infer some events in the evolutionary

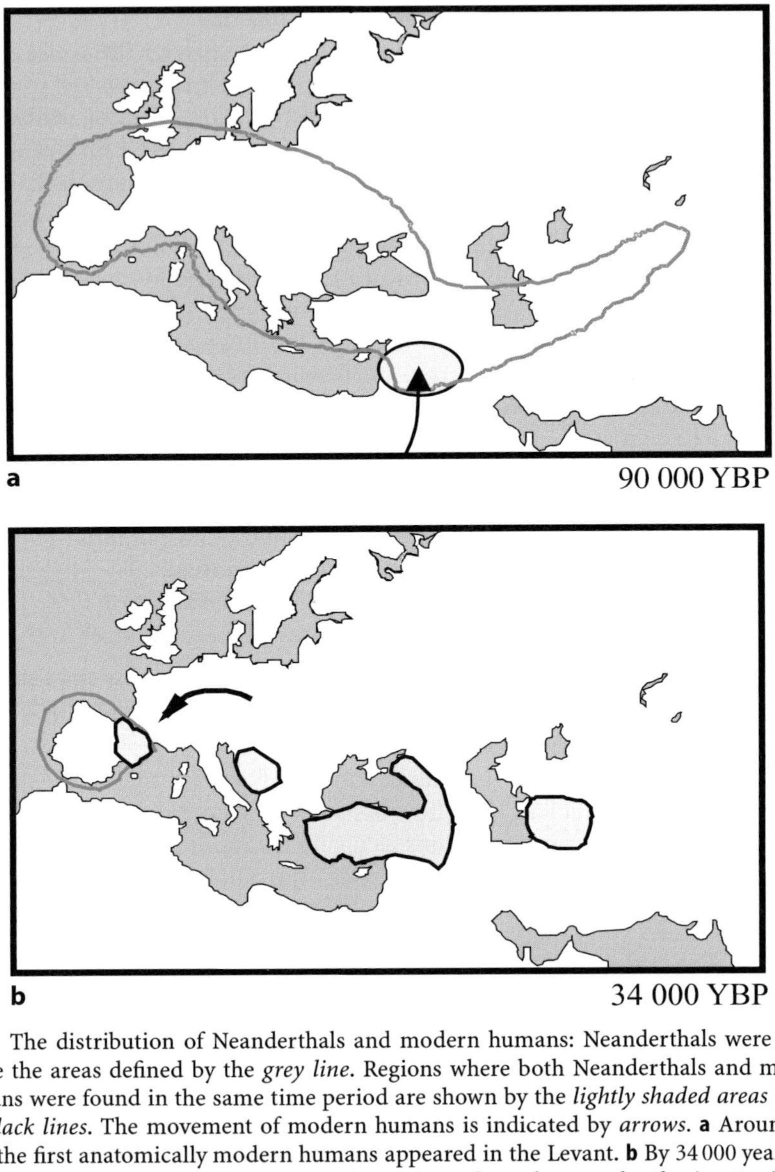

Fig. 3 The distribution of Neanderthals and modern humans: Neanderthals were found inside the areas defined by the *grey line*. Regions where both Neanderthals and modern humans were found in the same time period are shown by the *lightly shaded areas within the black lines*. The movement of modern humans is indicated by *arrows*. **a** Around this time the first anatomically modern humans appeared in the Levant. **b** By 34 000 years ago, Neanderthals could only be found in a limited number of areas; the Iberian Peninsular was last preserve of the Neanderthals

history of *H. sapiens*. The direct analysis of Neanderthal remains allowed the direct comparison of Neanderthal and modern human mtDNA pools.

In 1997, the first successful analysis of Neanderthal mtDNA was published. Following this, seven additional Neanderthal specimens have been success-

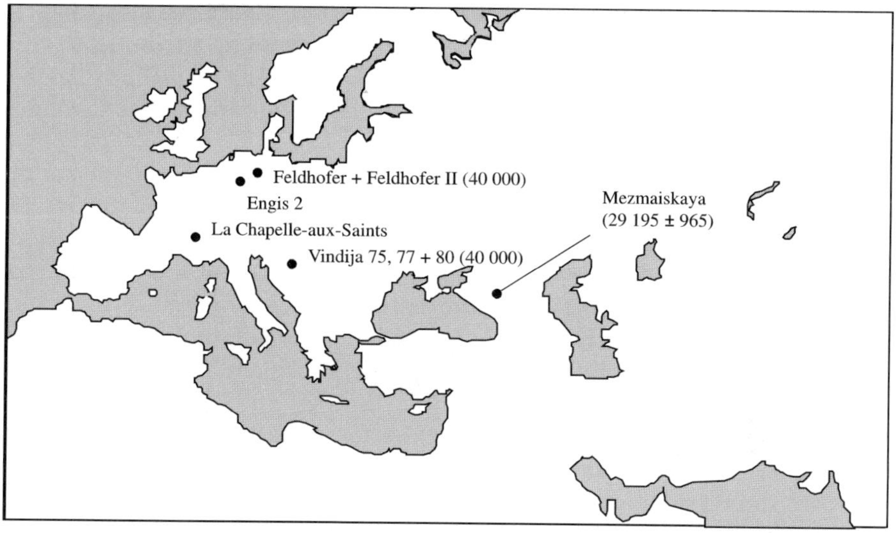

Fig. 4 The locations of the Neanderthal remains that have yielded DNA. The radiocarbon age estimates of specimens that have been dated are shown in *parentheses*

fully analysed. In addition to these Neanderthals, several anatomically modern human remains from the Pleistocene have also been subject to molecular analysis (Caramelli et al. 2003; Serre et al. 2004); although, as discussed later (see also Chaps. 5, 6), these results need to be corroborated by additional investigations (Gilbert et al. 2005). The locations of the Neanderthals that have been successfully analysed are shown in Fig. 4.

Analysis of the composition and diversity of the Neanderthal gene pool in comparison with that of modern humans allows questions regarding the date of the split between modern humans and the Neanderthals and admixture between the two populations to be addressed. However, as with all ancient DNA (aDNA) analysis, before interpreting the data, great care must be taken to ensure that the DNA that is being examined is really aDNA and not the result of exogenous contamination.

4 The Limitations of Ancient DNA Analysis

Soon after the advent of PCR, a series of spectacular reports appeared in respected scientific journals that described the recovery of DNA from a diverse range of material, including Miocene plants that were 17–20 million years old (Golenberg et al. 1990), insects that had been embedded in amber for 120–135 million years (Desalle et al. 1992; Cano et al. 1993) and Cretaceous dinosaurs (Woodward et al. 1994; Wang et al. 1997); see also Chap. 3. These early results, while reported in good faith, can all be attributed to contamination. The erroneous claims damaged the emerging discipline of aDNA studies, and many people became very sceptical about any claims involving the recovery of aDNA.

Over the last 10–15 years, great effort has been spent on characterising the preservation of DNA and developing more robust protocols for the extraction and analysis of aDNA. The sentiments of anybody undertaking serious work involving aDNA are well summarised by the title of a comment that was published following the Fifth International DNA Conference by two of the pre-eminent researchers in the field, urging the community to improve the standard of research: "Ancient DNA: do it right or not at all" (Cooper and Poinar 2000).

A well-defined set of criteria now exist for undertaking the extraction, amplification and analysis of aDNA, and any reports that are to be taken seriously should adhere to these criteria. These break down into two main groups. The first fall under the heading of preliminary assessment: they include thermal age, morphological preservation, amino acid racemisation and collagen composition. The second concern the extraction, amplification and analysis processes, and include cleaning of bone surface, multiple extractions in an environment free of contaminating DNA, multiple PCR amplifications, subcloning of PCR products and sequencing of several individual products, and independent analysis in a second laboratory. Because aDNA studies almost exclusively use the sequence of the mitochondrial HVS-I (composed of overlapping shorter fragments)—and occasionally a few restriction fragment length polymorphism sites from the coding region in addition—the presence of artefacts can often be detected through the appearance of hybrid or mosaic mtDNAs and the presence of mosaic molecules is strong evidence that the sequence data that have been presented are an artefact (Bandelt 2005). In addition to satisfying the aforementioned criteria, the results from any analysis of aDNA should be considered in the context of the particular project and evaluated to assess the likelihood that the DNA that has been analysed is truly aDNA and not a modern contaminant or artefact (Gilbert et al. 2005).

Unfortunately, and much to the detriment of aDNA research, peer-reviewed papers still manage to be published in high-impact journals when there is little or no possibility of the recovered DNA being from the fossil remains. It may seem arrogant to dismiss some papers out of hand, especially when in some cases they are published by respected scientists. However, studies of the preservation of DNA should inform the critical reader that DNA survival has its limits.

5
DNA Degradation

In most environments, following death, an organism's soft tissue rapidly degrades through the processes of autolysis, putrefaction and scavenging. Therefore, when analysing fossil samples, bones and teeth are usually the only type of material available for DNA analysis. There are some exceptions; in particular, coprolites have proven to be a rich source of aDNA (Poinar et al. 1998, 2001, 2003).

Bones and teeth act as a harbour for DNA, providing a physical barrier against attack by bacteria and fungi; teeth are afforded additional protection through their enamel. In addition to the physical protection, the hydroxylapatite mineral, which is present at high levels in bones and teeth, helps to stabilise DNA molecules (Lindahl 1993).

Even when the endogenous DNA is in a relatively stable environment provided by bones and teeth, it will continue to break down over time largely through the processes of hydrolysis and oxidation (Lindahl 1993; Willerslev et al. 2004). Hydrolytic damage results in the removal of bases, in particular purines; this process of depurination is one of the main routes of DNA degradation. Oxidative damage, which is mediated through the effects of both direct and indirect ionising radiation, and which can lead to lesions in the sugar–phosphate backbone of the DNA molecule and chemical alterations of the bases, is another source of DNA degradation (Lindahl 1993).

With an increased understanding of the process of DNA degradation, it is now generally accepted that it is very unlikely that endogenous DNA will be recovered from any samples older than 50 000–100 000 years, even with extremely favourable environmental conditions. Fortunately, the last Neanderthals fall into this window of potential DNA preservation.

6
The Extraction and Analysis of Neanderthal DNA: Assessing the Preservation

One of the key criteria for determining the authenticity of aDNA is that some assessment should be made of the fossil material to determine if there is any reasonable possibility that endogenous DNA could exist. There are no absolute methods available for this, but there are some key indicators that can be used; these are assessment of the environmental history and the molecular preservation of a sample. If any given sample displays poor molecular preservation or has been exposed to high temperatures for extended periods there is little merit in undertaking destructive analysis.

Consideration of the environment that the sample has been found in, along with a chemical assessment, can virtually exclude the possibility of finding aDNA. However, it is important to realise that a positive assessment is no guarantee that endogenous DNA will be recovered from a given sample; rather it is an indication that the sample could potentially harbour aDNA and will improve confidence in any results if aDNA appears to have been recovered.

7 Environmental Factors

The environment plays a big role in the rate of DNA degradation. Low temperature is generally considered to be the most important single factor in the preservation of aDNA. It is therefore not surprising to find that most of the successful analyses involving aDNA have been with samples that are from cooler climates. Furthermore, the effect of temperature on the chemical preservation of DNA has been demonstrated directly in studies where higher levels of chemical damage in aDNA have been correlated directly with higher environmental temperatures (Höss et al. 1996).

One theoretical method that has been employed to estimate the potential for a given fossil sample to contain DNA is the estimate of the thermal age (Smith et al. 2001). The thermal age is defined as "the time taken to produce a given degree of DNA degradation when temperature is held as a constant 10 °C". Calculating the thermal age takes into consideration current temperature, which is assumed to be relatively constant through the Holocene (the last 10 000 years or so) and the cooler temperatures during the proceeding glacial period. For example, taking the Neanderthal location, this has an estimated mean air temperature of 10 °C, the thermal age at 10 000 years will be 10 000 years, while the thermal age at 32 000 years will be 15 000 years; the rate of decay during 22 000 years in the cooler glacial periods is comparable to that for just 5000 years at 10 °C. A survey of a number of fossils recovered from Pleistocene sites demonstrated a positive correlation between the thermal age of the fossils and the recovery of endogenous DNA (Smith et al. 2001, 2003). Table 1 shows the thermal ages of some of the sites where the Neanderthals that have yielded DNA were discovered (Smith et al. 2003). The thermal age of the Feldhofer Neanderthal is estimated to be towards the upper limit for DNA preservation. Neanderthals from sites such as La Chapelle-aux-Saints are thought unlikely to yield DNA on the basis of their thermal ages (Smith et al. 2001, 2003).

While temperature is the most important single factor, other environmental conditions also have to be taken into consideration when estimating if DNA could be present in a sample, including air and soil humidity, soil pH, average temperatures in different earth layers and microbial-mediated decay (Ovchinnikov et al. 2001; Smith et al. 2001). The interplay of these and other factors makes predicting the preservation of material from a site based on en-

Table 1 Estimated thermal ages of European sites where Neanderthals have been discovered that have yielded DNA

Site	Age (KY) of Neanderthal	Thermal age (at 10 °C)		
		10 KY	32 KY	50 KY
La Chapelle-aux-Saints	NA	17	28	35
Neanderthal	40	10	16	19
Engis	NA	9	14	16
Mezmaiskaya[a]	30	4	7	NA

KY thousand years, *NA* not applicable
[a]Based on estimates as no accurate meteorological data are available (data from Smith et al. 2003)

vironmental information complex and the information can act only as a guide rather than being definitive. The variations in microenvironments make precise predictions of aDNA preservation very difficult.

8 Molecular Preservation

Proteins in biological material are more stable and easier to analyse than the DNA, and therefore provide a good proxy for assessing DNA degradation in a given sample. Assessment of changes in the proteins allows a measure of diagenetic change, which in turn provides an estimation of the amount of DNA degradation and modification that is likely to have occurred. The techniques used do have the disadvantage that they are destructive, although using less material than DNA extraction, and careful consideration has to be given before valuable samples are analysed.

The most widely used method has been the measurement of different forms of amino acid. With the exception of glycine, amino acids can exist in the form of two optical isomers, D and L. In living organisms the L enantiomer is exclusively used in protein biosynthesis; however, upon death, when the amino acids are no longer part of a living organism they undergo racemisation to the D enantiomer. Eventually, the two forms will reach equilibrium, when they will be present at equal levels. Poinar et al. (1996) measured the racemisation of aspartic acid and found that when the D-to-L ratios were below 0.08, DNA could be extracted, and that DNA extracted from samples which had lower D-to-L ratios was less degraded, containing longer fragments. Samples with D-to-L ratios above 0.08 yielded no endogenous aDNA.

The Neanderthal from La Chapelle-aux-Saints illustrates the value of direct molecular assessment. On the basis of the thermal age of this sample there would have been little point in subjecting it to destructive analysis: its thermal age is approximately 1.8 times older than the Feldhofer Neanderthal which is predicted to be towards the upper limit where DNA retrieval is possible (Smith et al. 2001, 2003). However, the amino acid racemisation analysis indicated that it was well preserved; and while the method has limitations (Collins et al. 1999), it has often been successful in identifying samples with good molecular preservation. In the case of the Mezmaiskaya Neanderthal, we used the amount of collagen-type debris and its associated levels of nitrogen and carbon as an indicator of molecular preservation (Ovchinnikov et al. 2000).

Other Neanderthal samples from southern Europe have been examined using the overall levels of amino acids and nitrogen, the racemisation of aspartic acid and alanine and histological preservation to infer molecular preservation. These specimens were from warmer areas and displayed too much diagenetic change to justify destructive DNA analysis (Cooper et al. 1997).

9 Isolation of Neanderthal DNA

When a fossil displays good molecular and morphological preservation there is a possibility that it will also contain some DNA. The first step in any molecular analysis is DNA extraction. While there are slight variations in the extraction protocols, all the successful Neanderthal DNA extractions have followed very similar protocols (Krings et al. 1997). At all times during the extraction and amplification of aDNA, rigorous precautions must be followed in order to minimise the possibility of contamination with exogenous DNA, and controls should be included to maximise the possibility of detecting any potential sources of contamination. Ideally a laboratory that is dedicated for aDNA analysis should be used.

10 DNA Extraction

The extraction of DNA from fossilised bone is a destructive process. Using bone samples for aDNA analysis has the advantage over using other material that the surfaces can be cleaned using abrasive physical and chemical agents, which helps to reduce contaminating exogenous DNA and potential PCR inhibitors. After cleaning, the surfaces are exposed to UV light; this causes thymine dimers to form which render the DNA inert in subsequent PCR reactions, as the DNA is unable to denature. Once clean, the bone sample is ground in a freezer mill under liquid nitrogen and the resulting powder is incubated with 0.5 M EDTA and proteinase K. Further developments, such as the silicone embedding technique, may reduce the problems of contamination with exogenous DNA (Gilbert et al. 2003b, 2004).

The EDTA helps to break down the mineral structure of the bone and the proteinase K digests much of the protein material in the matrix; the effect is to release the DNA into solution. The addition of the chemical compound *N*-phenacylthiazolium bromide (PTB), which is a reagent that cleaves glucose-derived protein cross-links, to the DNA extraction has also been shown to help the release of DNA from ancient bone material (Poinar et al. 1998; Krings et al. 2000). Once the bone is in solution, protein is precipitated using the chaotropic agents phenol and chloroform, and then the DNA is concentrated and washed using a filter column. The DNA extract can then be added to the PCR reaction, although in some cases it requires further cleaning to remove PCR inhibitors (Höss and Pääbo 1993).

The DNA extracts from fossil samples cannot be easily quantified as the levels of DNA are normally very low. Large amounts of DNA will sometimes be present, but the source of this DNA is always bacterial or fungal rather than endogenous DNA from the sample. Real-time PCR provides a sensitive method for quantifying the target DNA (von Wurmb-Schwark et al. 2002).

11 Amplification and Sequence Analysis of Neanderthal DNA

In any samples from the Pleistocene (older than 10 000 years), even if the sample displays a remarkable preservation status, the DNA will be in a highly degraded state and only a small number of chemically modified molecules can normally be recovered (Höss et al. 1996). While early studies attempted to analyse aDNA directly without an amplification phase (Higuchi et al. 1984; Pääbo 1985) the low number of starting molecules made such analysis extremely difficult and the technique was of limited scope. PCR provided a possible solution to the low number of starting molecules; in theory, one single molecule can be amplified several billion times and can generate enough DNA to subject to analysis. The drawback of PCR is that any contaminating homologous DNA that gets into the DNA extract or PCR reaction will also be amplified, and the exogenous DNA may well be amplified preferentially as the endogenous DNA will be chemically modified to some degree; hence the need for extreme caution to avoid contamination with exogenous DNA.

Multiple PCR amplifications from aDNA extracts followed by cloning and sequencing of individual PCR products are important steps to undertake when the number of target molecules in an aDNA extract is very low. Sequenced PCR products can vary slightly. The sequence heterogeneity is caused by a combination of DNA damage leading to the wrong nucleotides being inserted into the nascent DNA strand and the limited fidelity of the *Taq* polymerase. When analysing fossil DNA this problem is exacerbated by

DNA damage. It is therefore very important to carry out at least two separate PCR reactions, followed by subcloning and sequencing so that any errors introduced by the PCR reaction can be detected (Handt et al. 1994; Krings et al. 1997, 1999, 2000; Ovchinnikov et al. 2000). Producing an accurate sequence can still be problematic. The original Neanderthal sequence probably contains four errors (Schmitz et al. 2002). These were incorporated into the consensus sequence based on all 17 PCR products that were from a single PCR reaction containing the substitutions; six out of nine PCR products generated using a separate primer set only detected two of the substitutions—deriving a consensus sequence in such circumstances is intrinsically problematic.

Another critical aspect of the subcloning of the PCR products is that it allows mixtures to be detected. Even with all the precautions and controls in place, it is still not uncommon to detect products from more than one source: endogenous DNA along with contaminating sequences. The problems of contamination are particularly acute when working with hominin remains as contamination from modern humans is very difficult to eliminate (Handt et al. 1996; Krings et al. 1997, 1999, 2000; Serre et al. 2004).

(a)

```
Ref    ACAGCAATCAACCCTCAACTATCACACATCAACTGCAACTCCAAAGCCACCCCT -CACCCAG
17.1   .............T.....G...T.........A...........A.G...T.A.......
17.2   .............T.....G...T.......T.AT..........A.G...T.A.......
17.3   .T...........T.....G...T........ .A...........A.G...T.A.......
17.4   .............T.....G...T.......T.AT..........A.G...T.A.......
17.5   .............T.....G...T.........A...........A.G...T.A.......
17.6   .............T.....G...T.........A...........A.G...T.A.......
17.7   .............T.....G...T.........A...........A.G...T.A.......
17.8   .............T.....G...T.........A...........A.G...T.A.......
17.9   .............T...T.G...T.........A...........A.G...T.A..T....
17.10  .............T.....G...T.........A.....T.....A.G...T.A.......
```

(b)

	16037	16078	16086	16093	16107	16108	16111	16112	16118	16129	16139	16148	16154	16156	16169	16182	16183	16189	16209	16223	16230	16234	16243	16244	16250	16256	16261	16262	16263.1	16278	16299	16311	16320	16344	16362	16365	16393
CRS	A	A	T	T	C	C	C	C	G	G	A	C	T	G	C	A	A	T	T	C	A	C	T	G	C	C	C	C	-	C	A	T	C	C	T	C	C
Direct 1		.	C	.	.	.	.	.	.	A	T	T	.	A	T	C	C	C	C	T	G	T	.	A	.	A											
P1		.	C	.	.	.	.	.	.	A	T	T	.	A	T	C	C	C	C	T	G	T	.	A	.	A	.	T	A	T							
P2		.	C	.	.	.	.	.	.	A	T	T	.	A	T	C	C	C	C	T	G	T	.	A	.	A	.	T	A	T							
P3		.	C	.	.	.	.	.	A	A	T	T	.	A	T	C	C	C	C	T	G	T	.	A	.	A	.	T	A	T							
Direct 2																				T	G	T	.	A	.	A	.	T	A	T	G	C	T	T	C	.	.
577.1																				T	G	T	.	A	.	A	T	T	A	T	G	C	T	T	C	.	T
557.2																				T	G	T	.	A	T	A	.	T	A	T	G	C	T	T	C	.	.
581.2																				T	G	T	.	A	.	A	.	T	A	T	G	C	T	T	C	.	.
581.3																				T	G	T	C	A	.	A	.	T	A	T	G	C	T	T	C	T	.
Mezmaiskaya		.	C	.	.	.	.	.	.	A	T	T	.	A	T	C	C	C	C	**T**	**G**	**T**	.	**A**	.	**A**	.	**T**	**A**	**T**	**G**	**C**	**T**	T	C	.	.
Feldhofer	G	G	.	C	T	T	T	T	.	A	T	T	C	.	T	.	C	C	C	T	G	T	.	A	.	A	.	T	A	T	G	C	T	.	C	.	.

Fig. 5 The sequence of related PCR products in comparison to the Cambridge reference sequence (*CRS*; Anderson et al. 1981) derived from **a** the Feldhofer Neanderthal using the primers L16209 and H16271, the numbers refer to different PCR clones (Krings et al. 1997), and **b** the Mezmaiskaya Neanderthal between positions 16056 and 16399. *Direct 1* and *Direct 2* refer to PCR products that were sequenced without cloning; the other sequences were generated from cloned fragments. The composite Mezmaiskaya and Feldhofer sequences are shown and the nucleotides in *boldface* indicate the region of the sequence that was independently sequenced in a second laboratory (Ovchinnikov et al. 2000)

When the first Neanderthal extract was analysed, Krings et al. (1997) used primers that produced a short 105-bp amplicon. They carried out two separate PCR reactions, and cloned the PCR products into a plasmid vector for sequencing. Out of 30 clones that were analysed, 27 represented a sequence that was quite different from that of any extant human sequence; three clones appeared to be very similar to modern human mtDNA, and are likely to have been due to low-level contamination. They repeated the process with a number of overlapping amplicons until the HVS-I region had been fully sequenced between positions 16023 and 16400. Subsequent analyses of Neanderthals have been able to use this first sequence to create 'Neanderthal-specific' primers, which are based on regions that differ between the Neanderthal and modern human DNA.

Figure 5a shows the sequences of ten PCR products generated using one set of primers for the first Feldhofer Neanderthal, and Fig. 5b shows the complete published sequence of the Mezmaiskaya Neanderthal. Both of the sets of cloned PCR products display nucleotide differences in some of the PCR products, but there are also substitutions and insertions that occur in all of the PCR products, which allow a consensus sequence to be derived. It is interesting to note that the number of non-consensus substitutions in the Mezmaiskaya Neanderthal is much lower than that in the Feldhofer Neanderthal—probably a reflection of the lower levels of DNA degradation/diagenesis in the Mezmaiskaya Neanderthal.

12
Ancient DNA Artefacts

In addition to the difficulties associated with retrieving DNA from fossil material there is also a problem with distinguishing between the actual endogenous DNA sequence from sequences that have been created in the PCR tube. DNA degradation and damage, in addition to limiting the length of any fragment of DNA that might reasonably be expected to be found in a fossil sample, also complicates the PCR amplification process. The DNA damage can lead to processes that are capable of producing erroneous and, at times, misleading results.

Jumping PCR, for example, can knit partial PCR products together, resulting in hybrid molecules (Pääbo et al. 1989, 1990). If extension occurs from a primer but is terminated prematurely owing to DNA damage, the resulting truncated PCR product can then act as primer in the next round of PCR. The priming now starts further downstream from the original primer site and can lead to the formation of a hybrid molecule. This can create problems for interpreting data, particularly from loci that may well have two different alleles within any given individual or when dealing with a mixture.

Damaged DNA is also more problematic for the *Taq* DNA polymerase to copy: the purine bases, guanine and adenine, are particularly prone to hydrolytic attack, leaving the DNA template with gaps in the sequence of bases. If the damage is severe enough PCR-mediated amplification may prove impossible: in one study oxidation-mediated chemical changes in the pyramiding bases were shown to be positively correlated with the inability to amplify endogenous DNA (Höss et al. 1996). Deamination, in particular of the cytosine residue, has been shown to be common in both the DNA of living organisms and in that of fossils (Hofreiter et al. 2001). The deamination of cytosine to uracil will lead to transitions being detected in the PCR products that are artefacts of the DNA damage rather than reflections of the endogenous sequence: the modified deoxycytidine leads to C-to-T and G-to-A substitutions in the copied DNA. Treatment of the template with *N*-glycosylase removes the deaminated cytosine from the template, a strand break then occurs through a hydrolysis reaction and the errors are therefore not incorporated into the template (Hofreiter et al. 2001; Gilbert et al. 2003a, b; Chap. 5).

The postmortem damage that occurs in DNA, in particular the hydrolytic deamination and depurination, has been shown to act in a non-random manner, with some sites being much more prone to postmortem damage than others: in fact, sites that show rapid rates of mutation *in vivo* tend to be prone to postmortem modification (Gilbert et al. 2003b). This is a potential problem when assigning aDNA types to haplogroups. With the Neanderthal samples there is little possibility that this has had a major impact on the deduced sequences: only one out of the five informative HVS-I sites identified by the discriminating approach of Knight (2003) has been identified as being prone to high levels of postmortem damage (Gilbert et al. 2003b). There is no indication that the highly informative insertion between nucleotide positions 16263 and 16264, which is common to all Neanderthals sequences is the result of postmortem damage. In addition, that enough PCR product could be generated from the Mezmaiskaya Neanderthal to allow direct sequencing reduces the likelihood of postmortem damage being a significant problem.

13 Authentication

The measures that are required in order to satisfy researchers that they have in fact analysed aDNA, rather than some exogenous contamination or PCR artefact, are numerous. For the Neanderthal samples, it should be stressed that these criteria have been met. The molecular preservation of the analysed samples has been shown to be good. The key factor in accepting the data is that two of the Neanderthals, both the Feldhofer and Mezmaiskaya specimens, have been analysed independently in different laboratories with the appropriate negative controls and appropriate sequence analysis. The Mezmaiskaya Neanderthal was also analysed in completely separate laboratories to all the other specimens. With all of the analysed Neanderthals the extracted DNA has shown appropriate molecular behaviour—only short PCR products could be amplified. Finally, the fact that all the Neanderthal sequences form a distinct phylogenetic clade also supports their validity.

14
Evaluation of Neanderthal DNA

Despite some early concerns about the validity of the first Neanderthal mtDNA sequence, virtually the entire mtDNA community is now satisfied that endogenous DNA has been successfully extracted from Neanderthal fossils. The different Neanderthals have been analysed to different degrees. The HVS-I and HVS-II regions of the Feldhofer and Vindija 75 Neanderthals have been analysed, whereas only HVS-I of the Mezmaiskaya, Feldhofer II and Vindija 80 Neanderthals was sequenced. Only 31 bp have been sequenced from the Vindija 77, La Chapelle-aux-Saints and Engis 2 Neanderthals—just enough to show that their mtDNA is similar to that of the other Neanderthals. The details of the published sequence information are summarised in Table 2.

The differences between the five Neanderthal sequences for which more than 300 bp of HVS-I have been sequenced are illustrated in Fig. 6. Even a cursory examination of the sequence data from the Neanderthals reveals that they are much more similar to each other than to modern human sequences. However this superficial comparison needs to be corroborated by more rigorous phylogenetic analysis.

Table 2 Regions of the mitochondrial HVS-I and HVS-II that have been characterised from different Neanderthal specimens

Neanderthal	Region sequenced	
	HVS-I	HVS-II
Feldhofer	16023–16400	57–396
Feldhofer II	16023–16378	–
Mezmaiskaya	16056–16399	–
Vindija 75	16023–16378	57–343
Vindija 77	16231–16261	–
Vindija 80	16023–16378	–
La Chapelle-aux-Saints	16231–16261	–
Engis 2	16231–16261	–

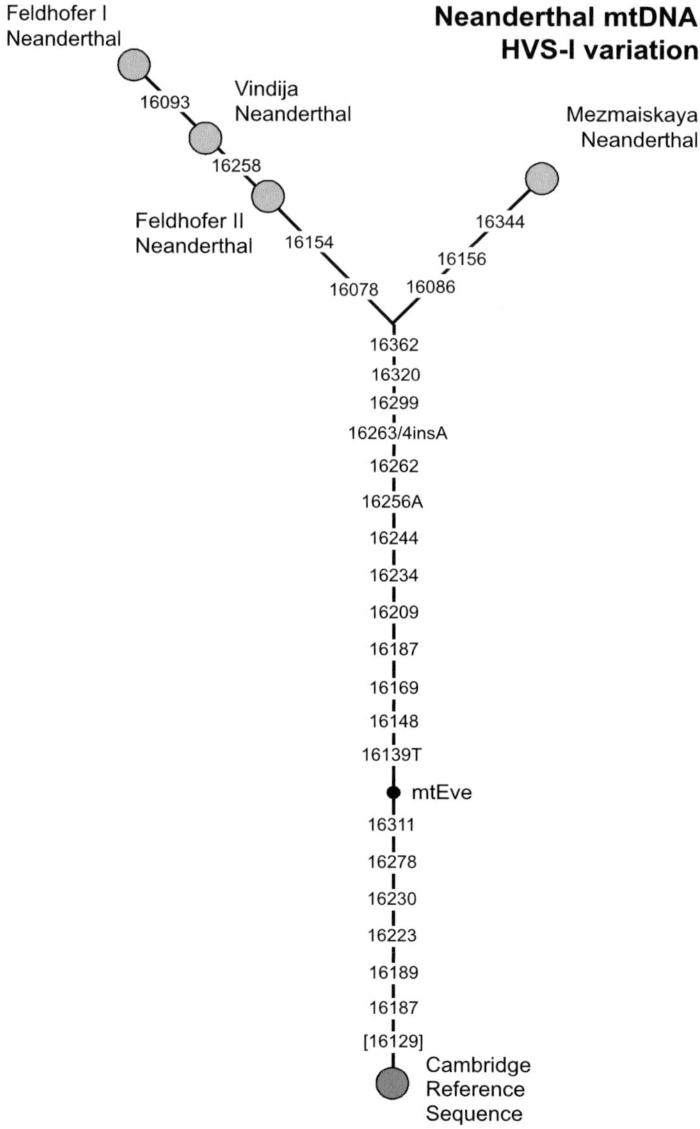

Fig. 6 The relationship of Neanderthal sequences. The substitutions that differentiate the different Neanderthal HVS-I sequences between positions 16056 and 16378 are shown, with reference to the CRS

15 Phylogenetic Analysis

In order to assess the relationship between Neanderthal mtDNA and modern human mtDNA, various kinds of phylogenetic analyses have been carried out.

The common aim of the different methods employed has been to determine whether the Neanderthal sequences fall within the diversity of modern human sequences or whether their mtDNA belongs to a sister clade of modern human mtDNA.

The results of four types of analysis are shown in Fig. 7. Figure 7a shows a tree constructed using the neighbour-joining algorithm, rooted with chimpanzee and bonobo sequences; 663 contemporary human sequences were compared with sequences from two Neanderthals. The support values on the tree refer to the quartet-puzzling support values (Krings et al. 2000), although there are caveats with this method (Chap. 4). Figure 7b shows a neighbour-joining tree with the Feldhofer and Mezmaiskaya Neanderthals. These have been compared to 5846 modern human sequences collected from HvrBase (Burckhardt et al. 1999), which were, however, not free of copying errors (Árnason 2003); the values on the branches are the original bootstrap frequencies (percent) obtained from 1000 replicates (Ovchinnikov et al. 2000). Figure 7c shows again the Feldhofer and Mezmaiskaya Neanderthals compared with ten modern humans; this analysis uses a maximum parsimony branch and bound search (Ovchinnikov et al. 2000).

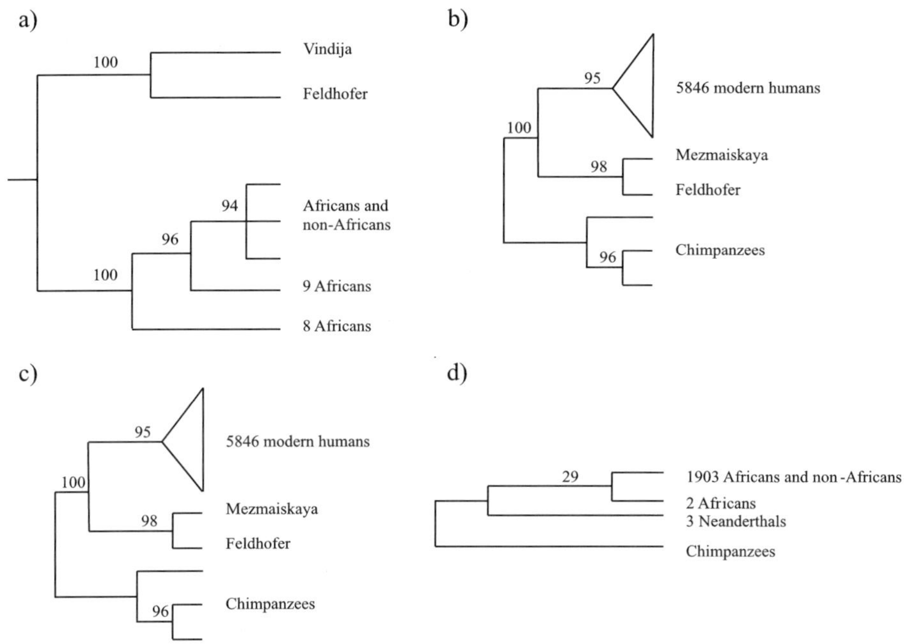

Fig. 7 Four different analyses examining the relationship of Neanderthal to modern human mtDNA. Three of the analyses, **a–c**, come to similar conclusions (Krings et al. 2000; Ovchinnikov et al. 2000). Analysis **d** does not find the same level of support for the Neanderthal–modern human split (Gutiérrez et al. 2002)

The three analyses all indicate that the Neanderthal and modern human mtDNA sequences fall into separate clades. Different phylogenetic analyses have not always supported the separation: a maximum-likelihood analysis by Gutiérrez et al. (2002) resulted in insignificant support for the Neanderthal–modern human split (Fig. 7d). In this case, the high level of recurrent mutation in the human mtDNA control region appears to have drowned out the phylogenetic signal (Tamura and Nei 1993; Excoffier and Yang 1999). However, Knight (2003) provides a more nuanced analysis by focusing on the most phylogenetically informative characters (positions 16037, 16139, 16244 and 16262, and the insertion of an A between 16263 and 16264 in HVS-I, and 243, 245 and 391 in HVS-II), and discarding those that are highly recurrent (positions 16093, 16148, 16256, 16278 and 16362 in HVS-I, and 236 in HVS-II; as well as the two cytosine-rich segments prone to slipped-strand mispairing between positions 16181–16194 and 302–322). Knight (2003) has shown that eight sites indeed consistently split Neanderthals from set of modern sub-Saharan Africans.

The conclusion of the analysis is that even when the very conservative value of 0.5 is used for the probability of each of the highly informative sites having been selected by error, the observation of eight apparently stable sites differentiating modern humans and Neanderthals occurring by chance is very low (0.4%). This approach provides strong evidence that the deep split between Neanderthals and modern humans is real and not a phylogenetic artefact.

16 Admixture between Modern Humans and Neanderthals

If we accept that the Neanderthal and modern human mtDNA lineages are separated by a deep split, then a separate issue can be addressed which lies at the centre of much of the debate concerning the relationship between modern humans and the Neanderthals: Did the Neanderthals and early modern humans interbreed?

What can be concluded from the mtDNA evidence from Neanderthals and the modern humans is that there is no signal visible for Neanderthals contributing to the mtDNA gene pool of modern humans. However, this does not mean that there can have been no possible contribution: several explanations can be offered to explain the lack of Neanderthal mtDNA in the modern mtDNA pool.

Genetic drift acts on all loci, and the haploid mtDNA and the Y chromosome are particularly susceptible, the process occurring 4 times faster than in diploid loci. As Neanderthals became extinct approximately 30 000 years ago, this date marks the end of the period of potential admixture. However, since this time, there have been 1000–1500 generations, which provides a lot of scope for genetic drift. The decrease in diversity would have been even more pronounced during the likely population constrictions caused by global cooling during the last glacial maximum, around 20 000 years ago.

Under two very simple (not to say, simplistic) demographic models, Nordborg (1998) suggested that the presence of Neanderthal sequences in an admixed ancient population would have had a relatively low probability of being represented in the modern gene pool. Although the possibility of modern humans and Neanderthals randomly interbreeding on a large scale could be excluded, smaller contributions by the Neanderthals could not be ruled out. This was simply because modern mtDNAs coalesced to relatively few lineages 30 000 years ago, and many more may have been lost to drift. Analysis of modern populations bears out the effect of genetic drift empirically. In an analysis of modern European populations, the majority of lineages can be dated to a period after the early Upper Palaeolithic, the period when admixed populations may have existed (Richards et al. 2000). A further analysis of the Neanderthal and modern human mtDNA data by Currat and Excoffier (2004) which used a 'realistic' model of the range expansion of modern humans into Europe concluded that if there was any admixture the levels must have been very low (less than 0.1%); this is based largely on the assumption that the modern human population would be undergoing a population expansion at the time of the range expansion—this would lead to any admixed Neanderthal mtDNA becoming relatively common in a short period of time.

Another approach to examining the question of admixture has been to examine early modern human populations using aDNA, with the aim of directly increasing the number of lineages sampled; however, this type of analysis has major limitations. If the endogenous sequences of the mtDNA of the early modern humans are similar to those of contemporary mtDNA then it is almost impossible to determine the sequence confidently because of the risk of contamination by modern DNA (Serre et al. 2004). Therefore reports describing modern mtDNA sequences recovered from archaic humans (Adcock et al. 2001; Caramelli et al. 2003) must be treated with much scepticism (Cooper et al. 2001; Chap. 6).

A different approach to this problem was taken by Serre et al. (2004). Early modern human remains that displayed good molecular preservation were identified, DNA was extracted and then Neanderthal-specific primers were used in an attempt to amplify Neanderthal-'type' sequences. Five early modern humans displayed good preservation but the Neanderthal-specific primers failed to produce an amplicon, indicating that the endogenous sequence was not similar to that of the Neanderthals. This approach is itself limited by the small number of early modern human remains that have been discovered, of which only a small number are suitable for molecular analysis. In order to exclude a Neanderthal contribution of 10% to the ancient modern human gene pool, using this approach would require 50 early modern human remains (Serre et al. 2004)—which is, to say the least, unlikely.

17 Neanderthal Diversity

Neanderthal mtDNA has been recovered from three distinct locations, providing an indication of the diversity that is present in the Neanderthal lineage. Using only four specimens, the Feldhofer and Feldhofer II, Mezmaiskaya and Vindija 75, Rosenberg and Nordborg (2002) estimated that there is a 60% probability that the deepest split in the Neanderthal lineage has been detected (probability of sampling the deepest split is $(n-1)/(n+1)$, where n is the number of sampled specimens). It could be concluded that it is unlikely that a Neanderthal specimen will be found that is highly diverse from the specimens analysed to date. However, this probability is only valid if dealing with an unstructured population (Rosenberg and Nordborg 2002). This may well not have been true for the Neanderthals, given their large geographic and temporal range. The relative homogeneity of the western Neanderthals compared with the one eastern Neanderthal is an indication that the Neanderthals did not comprise a single homogeneous population. Further sampling from different geographic areas may provide some insight into this.

The average number of pairwise sequence differences amongst the Feldhofer and Feldhofer II, the Mezmaiskaya and the Vindija 75 Neanderthals in HVS-I is 4.17 ± 2.64, which is similar to the levels of diversity that are found in modern humans (the three western Neanderthals show considerably less diversity, differing at 2 ± 1 positions). This contrasts with the high levels of diversity found in chimpanzees and gorillas (Gagneux et al. 1999; Krings et al. 2000). The low levels of diversity found in modern humans have been interpreted as reflecting a rapid growth from a small founder population in the past (Harpending et al. 1998). The similar levels of diversity in sampled Neanderthals may hint that the demography of the Neanderthals could have been similar to that of modern humans in this respect.

18 The Age of Divergence

Mutations in mtDNA have been used as a molecular clock to estimate the divergence time of lineages, although there are well-documented problems with measuring the rate of change in the mitochondrial genome, particularly when dealing with the hypervariable regions (Chap. 4). It is nevertheless therefore possible to estimate the date of the modern human–Neanderthal split, albeit with large confidence intervals.

The genetic distance between modern human and Neanderthal mtDNA has been used both to date the split between the modern humans and the Neanderthal mtDNAs and to estimate the age of the Neanderthal mtDNA lineage (Tamura and Nei 1993). Ovchinnikov et al. (2000) estimated the age of the split between the modern human and Neanderthal lineages, on the basis of the analysis of the Feldhofer and Mezmaiskaya specimens, as approximately 600 000 years (365 000–853 000 years); they estimated the age of the most recent common ancestor of the Neanderthals to 151 000–352 000 years ago. The common ancestor for modern humans was calculated to be between 106 000 and 246 000 years ago using the same algorithm and assumptions. Other comparisons, using both HVS-I and HVS-II, have produced similar estimates (Krings et al. 1999).

While it is accepted that these figures offer only a broad indication of the age of the lineages, and are based on only one locus, they do correlate with archaeological evidence for the scenario depicted in Figs. 1 and 2, in which the Neanderthals evolved from a population of *H. heidelbergensis* that migrated into Europe around 600 000 years ago while modern humans evolved from the *H. heidelbergensis* population that stayed in Africa. Similarly, the ranges calculated for the Neanderthal and modern human most recent common ancestor corroborate the archaeological evidence for the first appearance of these lineages in the archaeological records. This may indicate that the mtDNAs of both species coalesce at around the time of, or a little before, the appearance of their defining morphological characteristics in the fossil record, possibly indicating a speciation bottleneck in both cases.

19 Conclusions

Fragments of the non-coding portion of mtDNA of various lengths have been successfully isolated from a total of eight Neanderthal specimens. This has provided an insight into the mtDNA gene pool and has enabled some aspects of the diversity and age of the Neanderthal lineage to be assessed. No admixture between modern humans and Neanderthals has been detected, but the limited number of samples available for molecular analysis limit the conclusions that can be made with respect to potential admixture. Other explanations for the lack of Neanderthal lineages in the modern mtDNA gene pool, in particular genetic drift, can also explain the results, especially as the conclusions are based on the analysis of one haploid locus, the mtDNA. Further analysis will provide a better view of the Neanderthal gene pool, but the number of potential samples is limited: in total 70 sites have yielded Neanderthal bones (Klein 2003). Many of the sites, particularly those from southern Europe, do not show good molecular preservation (Cooper et al. 1997; Smith et al. 2003).

References

Adcock GJ, Dennis ES, Easteal S, Huttley GA, Jermiin LS, Peacock WJ, Thorne A (2001) Mitochondrial DNA sequences in ancient Australians: implications for modern human origins. Proc Natl Acad Sci USA 98:537–542

Anderson S, Bankier AT, Barrell BG, Debruijn MHL, Coulson AR, Drouin J, Eperon IC, Nierlich DP, Roe BA, Sanger F, Schreier PH, Smith AJH, Staden R, Young IG (1981) Sequence and organization of the human mitochondrial genome. Nature 290:457–465

Árnason E (2003) Genetic heterogeneity of Icelanders. Ann Hum Genet 67:5–16

Bandelt H-J (2005) Mosaics of ancient mitochondrial DNA: positive indicators of nonauthenticity. Eur J Hum Genet 13:1106–1112

Burckhardt F, von Haeseler A, Meyer S (1999) HvrBase: compilation of mtDNA control region sequences from primates. Nucleic Acids Res 27:138–142

Cann RL, Stoneking M, Wilson AC (1987) Mitochondrial-DNA and human-evolution. Nature 325:31–36

Cano RJ, Poinar HN, Pieniazek NJ, Acra A, Poinar GO (1993) Amplification and sequencing of DNA from a 120–135-million-year-old weevil. Nature 363:536–538

Caramelli D, Lalueza-Fox C, Vernesi C, Lari M, Casoli A, Mallegni F, Chiarelli B, Dupanloup I, Bertranpetit J, Barbujani G, Bertorelle G (2003) Evidence for a genetic discontinuity between Neandertals and 24,000-year-old anatomically modern Europeans. Proc Natl Acad Sci USA 100:6593–6597

Cavalli-Sforza LL, Feldman MW (2003) The application of molecular genetic approaches to the study of human evolution. Nature Genet 33:266–275

Collins MJ, Waite ER, van Duin ACT (1999) Predicting protein decomposition: the case of aspartic-acid racemization kinetics. Philos Trans R Soc Lond Ser B 354:51–64

Conroy G (1997) Reconstructing human origins. Norton, London

Cooper A, Poinar HN (2000) Ancient DNA: do it right or not at all. Science 289:1139

Cooper A, Poinar HN, Pääbo S, Radovcic J, Debenath A, Caparros M, Barroso-Ruiz C, Bertranpetit J, Nielsen-Marsh C, Hedges REM, Sykes B (1997) Neandertal genetics. Science 277:1021–1024

Cooper A, Rambaut A, Macaulay V, Willerslev E, Hansen AJ, Stringer C (2001) Human origins and ancient human DNA. Science 292:1655–1656

Currat M, Excoffier L (2004) Modern humans did not admix with Neanderthals during their range expansion into Europe. PLoS Biol 2:2264–2274

Desalle R, Gatesy J, Wheeler W, Grimaldi D (1992) DNA-sequences from a fossil termite in Oligomiocene amber and their phylogenetic implications. Science 257:1933–1936

Excoffier L, Yang ZH (1999) Substitution rate variation among sites in mitochondrial hypervariable region I of humans and chimpanzees. Mol Biol Evol 16:1357–1368

Gagneux P, Wills C, Gerloff U, Tautz D, Morin PA, Boesch C, Fruth B, Hohmann G, Ryder OA, Woodruff DS (1999) Mitochondrial sequences show diverse evolutionary histories of African hominoids. Proc Natl Acad Sci USA 96:5077–5082

Gilbert MTP, Hansen AJ, Willerslev E, Rudbeck L, Barnes I, Lynnerup N, Cooper A (2003a) Characterization of genetic miscoding lesions caused by postmortem damage. Am J Hum Genet 72:48–61

Gilbert MTP, Willerslev E, Hansen AJ, Barnes I, Rudbeck L, Lynnerup N, Cooper A (2003b) Distribution patterns of postmortem damage in human mitochondrial DNA. Am J Hum Genet 72:32–47

Gilbert MTP, Cuccui J, White W, Lynnerup N, Titball RW, Cooper A, Prentice MB (2004) Absence of Yersinia pestis-specific DNA in human teeth from five European excavations of putative plague victims. Microbiology 150:341–354

Gilbert MTP, Bandelt H-J, Hofreiter M, Barnes I (2005) Assessing ancient DNA studies. Trends Ecol Evol 20:541–544

Golenberg EM, Giannasi DE, Clegg MT, Smiley CJ, Durbin M, Henderson D, Zurawski G (1990) Chloroplast DNA-sequence from a Miocene magnolia species. Nature 344:656–658

Gutiérrez G, Sánchez D, Marín A (2002) A reanalysis of the ancient mitochondrial DNA sequences recovered from Neandertal bones. Mol Biol Evol 19:1359–1366

Handt O, Richards M, Trommsdorff M, Kilger C, Simanainen J, Georgiev O, Bauer K, Stone A, Hedges R, Schaffner W, Utermann G, Sykes B, Pääbo S (1994) Molecular-genetic analyses of the Tyrolean Ice Man. Science 264:1775–1778

Handt O, Krings M, Ward RH, Pääbo S (1996) The retrieval of ancient human DNA sequences. Am J Hum Genet 59:368–376

Harpending HC, Batzer MA, Gurven M, Jorde LB, Rogers AR, Sherry ST (1998) Genetic traces of ancient demography. Proc Natl Acad Sci USA 95:1961–1967

Higuchi R, Bowman B, Freiberger M, Ryder OA, Wilson AC (1984) DNA-sequences from the quagga, an extinct member of the horse family. Nature 312:282–284

Hofreiter M, Jaenicke V, Serre D, von Haeseler A, Pääbo S (2001) DNA sequences from multiple amplifications reveal artifacts induced by cytosine deamination in ancient DNA. Nucleic Acids Res 29:4793–4799

Höss M, Pääbo S (1993) DNA Extraction from Pleistocene bones by a silica-based purification method. Nucl Acids Res 21:3913–3914

Höss M, Dilling A, Currant A, Pääbo S (1996) Molecular phylogeny of the extinct ground sloth Mylodon darwinii. Proc Natl Acad Sci USA 93:181–185

Klein RG (2003) Whither the Neanderthals? Science 299:1525–1527

Knight A (2003) The phylogenetic relationship of Neandertal and modern human mitochondrial DNAs based on informative nucleotide. J Hum Evol 44:627–632

Krings M, Stone A, Schmitz RW, Krainitzki H, Stoneking M, Pääbo S (1997) Neanderthal DNA sequences and the origin of modern humans. Cell 90:19–30

Krings M, Geisert H, Schmitz RW, Krainitzki H, Pääbo S (1999) DNA sequence of the mitochondrial hypervariable region II from the Neandertal type specimen. Proc Natl Acad Sci USA 96:5581–5585

Krings M, Capelli C, Tschentscher F, Geisert H, Meyer S, von Haeseler A, Grossschmidt K, Possnert G, Paunovic M, Pääbo S (2000) A view of Neandertal genetic diversity. Nat Genet 26:144–146

Lahr MM, Foley RA (1998) Towards a theory of modern human origins: geography, demography, and diversity in recent human evolution. Yearb Phys Anthropol 41:137–176

Lindahl T (1993) Instability and decay of the primary structure of DNA. Nature 362:709–715

Nordborg M (1998) On the probability of Neanderthal ancestry. Am J Hum Genet 63:1237–1240

Ovchinnikov IV, Götherstrom A, Romanova GP, Kharitonov VM, Liden K, Goodwin W (2000) Molecular analysis of Neanderthal DNA from the northern Caucasus. Nature 404:490–493

Ovchinnikov IV, Götherström A, Romanova GP, Kharitonov VM, Lindén K, Goodwin W (2001) Not just old but old and cold?—Reply. Nature 410:772–772

Pääbo S (1985) Molecular-cloning of ancient Egyptian mummy DNA. Nature 314:644–645

Pääbo S, Higuchi RG, Wilson AC (1989) Ancient DNA and the polymerase chain-reaction—the emerging field of molecular archaeology. J Biol Chem 264:9709–9712

Pääbo S, Irwin DM, Wilson AC (1990) DNA damage promotes jumping between templates during enzymatic amplification. J Biol Chem 265:4718–4721

Poinar H, Kuch M, McDonald G, Martin P, Pääbo S (2003) Nuclear gene sequences from a late Pleistocene sloth coprolite. Curr Biol 12:1150–1152

Poinar HN, Höss M, Bada JL, Pääbo S (1996) Amino acid racemization and the preservation of ancient DNA. Science 272:864–866

Poinar HN, Hofreiter M, Spaulding WG, Martin PS, Stankiewicz BA, Bland H, Evershed RP, Possnert G, Pääbo S (1998) Molecular coproscopy: dung and diet of the extinct ground sloth Nothrotheriops shastensis. Science 281:402–406

Poinar HN, Kuch M, Sobolik KD, Barnes I, Stankiewicz AB, Kuder T, Spaulding WG, Bryant VM, Cooper A, Pääbo S (2001) A molecular analysis of dietary diversity for three archaic Native Americans. Proc Natl Acad Sci USA 98:4317–4322

Richards M, Macaulay V, Hickey E, Vega E, Sykes B, Guida V, Rengo C et al. (2000) Tracing European founder lineages in the near eastern mtDNA pool. Am J Hum Genet 67:1251–1276

Rosenberg NA, Nordborg M (2002) Genealogical trees, coalescent theory and the analysis of genetic polymorphisms. Nat Rev Genet 3:380–390

Schmitz RW, Serre D, Bonani G, Feine S, Hillgruber F, Krainitzki H, Pääbo S, Smith FH (2002) The Neandertal type site revisited: Interdisciplinary investigations of skeletal remains from the Neander Valley, Germany. Proc Natl Acad Sci USA 99:13342–13347

Serre D, Langaney A, Chech M, Teschler-Nicola M, Paunovic M, Mennecier P, Hofreiter M, Possnert G, Pääbo S (2004) No evidence of Neandertal mtDNA contribution to early modern humans. PLoS Biol 2:313–317

Smith CI, Chamberlain AT, Riley MS, Cooper A, Stringer CB, Collins MJ (2001) Not just old but old and cold? Nature 410:771–772

Smith CI, Chamberlain AT, Riley MS, Stringer C, Collins MJ (2003) The thermal history of human fossils and the likelihood of successful DNA amplification. J Hum Evol 45:203–217

Stringer C, Gamble C (1993) In search of the Neanderthals. Thames and Hudson, London

Stringer CB, Andrews P (1988) Genetic and fossil evidence for the origin of modern humans. Science 239:1263–1268

Tamura K, Nei M (1993) Estimation of the number of nucleotide substitutions in the control region of mitochondrial-DNA in humans and chimpanzees. Mol Biol Evol 10:512–526

Underhill PA, Passarino G, Lin AA, Shen P, Lahr MM, Foley RA, Oefner PJ, Cavalli-Sforza LL (2001) The phylogeography of Y chromosome binary haplotypes and the origins of modern human populations. Ann Hum Genet 65:43–62

von Wurmb-Schwark N, Higuchi R, Fenech AP, Elfstroem C, Meissner C, Oehmichen M, Cortopassi GA (2002) Quantification of human mitochondrial DNA in a real time PCR. Forensic Sci Int 126:34–39

Wang HL, Yan ZY, Jin DY (1997) Reanalysis of published DNA sequence amplified from cretaceous dinosaur egg fossil. Mol Biol Evol 14:589–591

Watkins WS, Rogers AR, Ostler CT, Wooding S, Bamshad MJ, Brassington AME, Carroll ML, Nguyen SV, Walker JA, Prasad BVR, Reddy PG, Das PK, Batzer MA, Jorde LB (2003) Genetic variation among world populations: Inferences from 100 Alu insertion polymorphisms. Genome Res 13:1607–1618

Willerslev E, Hansen AJ, Poinar HN (2004) Isolation of nucleic acids and cultures from fossil ice and permafrost. Trends Ecol Evol 19:141–147

Wolpoff M (1989) Multiregional evolution: the fossil alternative to Eden. In: Mellars P, Stringer C (eds) The human revolution. Edinburgh University Press, Edinburgh, pp 62–108

Woodward SR, Weyand NJ, Bunnell M (1994) DNA sequence from Cretaceous period bone fragments. Science 266:1229–1232

Nucleic Acids and Molecular Biology, Vol. 18
Hans-Jürgen Bandelt, Vincent Macaulay, Martin Richards (Eds.)
Human Mitochondrial DNA and the Evolution of *Homo sapiens*

A Model for the Dispersal of Modern Humans out of Africa

Martin Richards (✉) · Hans-Jürgen Bandelt · Toomas Kivisild ·
Stephen Oppenheimer

Institute of Integrative & Comparative Biology, Faculty of Biological Sciences,
University of Leeds, Leeds LS2 9JT, UK
m.b.richards@leeds.ac.uk

The World was all before them, where to choose
Their place of rest, and Providence their guide:
They hand in hand with wand'ring steps and slow
Through Eden took their solitary way.
Milton, *Paradise Lost* (1667)

1 Introduction: Setting the Agenda

As the above quotation cited by Gamble (1993a, b) indicates, interest in human dispersals has a long pedigree. Eight years before the publication of Darwin's *Origin of Species*, Latham's *Man and His Migrations* highlighted three great problems facing anthropology (Latham 1851):

1. The unity or non-unity of the human species
2. Its antiquity
3. Its geographical origin

As Gamble points out, this particular agenda for human origins research has been followed closely to this day. It has certainly been central to the genetic debate about the origin of anatomically modern humans. The debate between 'multi-regional evolution' on the one hand, and 'replacement', the 'garden of Eden', 'Noah's ark' or 'out of Africa' on the other, has determined the argument since the first major contributions from geneticists in late 1980s. This debate has focused precisely on the place and time of origin of modern humans, and has claimed as one of its major implications (from both sides) that this reveals something important about the unity or lack of unity of humanity as a whole.

However, Gamble suggests that, from his archaeologist's perspective, there are good reasons now for updating this agenda. He proposes replacing Latham's 'big three' questions with the following two:

1. Why were humans everywhere in prehistory?
2. What is the purpose of a world prehistory?

Gamble (1993b) has his own answers to these questions, and they clearly go beyond the role of human mitochondrial phylogeography, which is primarily to track the appearance of genetic variants in time and space and organize them into a narrative about dispersals. Nevertheless, we suggest that, whilst the debate concerning the original agenda seems to rumble on interminably, geneticists too should be attempting to look forward. In particular, in an attempt to get away from the never-ending dispute between the multiregional and replacement models, we shall take the latter as our starting point. Whilst some analyses of certain autosomal marker systems remain equivocal, the patterns of variation in human mitochondrial DNA (mtDNA) and the non-recombining portion of the Y-chromosome (NRY) now clearly support the out-of-Africa replacement model. We shall argue that by taking the detailed phylogeographic information that has become available to us from these genetic marker systems into account, it is now possible to reconstruct the dispersals of anatomically modern humans from Africa in some detail. And Gamble's tale provides a possible twist to our own, which we shall return to at the end. Firstly, though, some background.

2
Genetics and the Traditional Debate over Human Origins

As early as 1980, Wesley Brown's pioneering study of 21 humans from diverse ethnic and geographic backgrounds indicated that mtDNA restriction-enzyme cleavage patterns could be used to trace human genetic history (Brown 1980). On the basis of the observed diversity in a worldwide sample, he obtained a surprisingly recent coalescence age estimate of around 180 000 years (taking 1% per million years as an estimated rate of base substitution) for the global mtDNA variation. Shortly after the publication of the first complete sequence of human mtDNA (Anderson et al. 1981), Denaro et al. (1981) showed that a polymorphic *Hpa*I recognition site at nucleotide positions 3592–3597 separated most modern sub-Saharan Africans from Europeans and Asians. Curiously, with hindsight, this key polymorphism in the human mtDNA phylogeny stood in a derived state among Africans with respect to other primate species, meaning it was possible that non-African human populations could share an ancestor that had founded African populations.

The observation that the ancestral state of a *Hinc*II site (at nucleotide positions 12406–12411) is frequent in some Asian populations likewise suggested a similar conclusion (Denaro et al. 1981; Blanc et al. 1983). The 'Oriental' origins of human mtDNA variation were further deduced from a study of Tharus from Nepal (Brega et al. 1986), and high mtDNA diversity was found also to be present among Japanese populations (Horai et al. 1984; Horai and Matsunaga 1986). Besides Asian populations and Africans, Cann (1984) reported that native Australians displayed equally high diversity estimates. Johnson et al. (1983) observed the highest diversity in Africans, but suggested that this could be due to differences in evolutionary rates among different mtDNA lineages.

Therefore, mtDNA researchers (as well as others: Wainscoat et al. 1986) had been quietly working on human origins for a number of years when Allan Wilson's group launched their support for the out-of-Africa replacement model (Cann et al. 1987). They refocused the debate around the opposition between a recent African origin versus multiregional evolution, as indeed palaeontologists were also doing at the time (Stringer and Andrews 1988).

Human genetics hit human origins research (and the headlines), then, on New Year's Day, January 1987, with the publication from Wilson's laboratory of a high-resolution restriction analysis of mtDNAs from a sample representative of worldwide variation from living people. Cann et al. (1987) extracted mtDNA from 147 individuals from European Americans, African Americans, East Asians, New Guineans and Aboriginal Australians. They then digested the samples with 12 restriction enzymes, enabling them to construct fairly high resolution restriction maps of the mtDNA genome and infer the genealogical history of the molecule. This early attempt of high-resolution restriction fragment length polymorphism analysis suffered, however, as we now know (Forster et al. 2001), from several misscorings of sites. The phylogenetic tree they presented, constructed using the maximum-parsimony package PAUP (phylogenetic analysis using parsimony; Swofford 1993), was an estimate of the maternal genealogy (since mtDNA is maternally inherited, without recombination), and was divided into two basal clades. One of these clades was found only in African Americans, whereas the other was found not only in African Americans but throughout the other populations sampled.

Cann et al. (1987) argued that this pattern implied that the most recent common ancestor (MRCA) of human mtDNAs lived in Africa. Furthermore, they used the molecular clock to date the time of the MRCA to between 140 000 and 290 000 years ago. Therefore, their interpretation of the results directly addressed Latham's points 2 and 3: the ancestor of the modern human maternal lineage had been African, and lived relatively recently, in the Late Pleistocene. This suggested that the expansion out of Africa had been recent (dated to between 13 000 and 180 000 years ago), involving anatomically modern *Homo sapiens*.

These genetic analyses contradicted the prevailing 'multiregional' view of modern human origins, which argued that anatomically modern human populations were the descendants of regional *H. erectus* or archaic human groups who had dispersed from Africa in the Middle Pleistocene, perhaps a million years ago. Instead, mtDNA suggested that anatomically modern traits had evolved solely in tropical Africa, and had been dispersed quite recently by emigration. Hence the debate also brought in Latham's first point, because researchers who supported the out-of-Africa scenario could now argue that their perspective clearly indicated the very recent origin and therefore relative unity of modern humanity (Stringer and McKie 1996). This contrast in dates implied that, with its longer timescale for modern human evolution, multiregionalism might be more likely to give succour to racialist thinking. Modern multiregionalism, however, unlike its antecedents (Coon 1962), postulated extensive gene flow throughout the world following the initial dispersals of *H. erectus*, and in more recent versions the evolving model has also allowed for further dispersals from Africa (and yet more gene flow). Modern proponents of the multiregional view, therefore, were able to counter that this reasoning was fallacious and that the replacement view itself implied a genocidal tendency in modern humans that was inimical to modern liberalism (Wolpoff and Caspari 1997).

Although similar conclusions concerning an African origin for modern humans had been arrived at in the previous year (Wainscoat et al. 1986), the publication by Cann et al. (1987) led to a protracted period of debate. This intensified after a supporting paper, using sequence data from the non-coding and fast-evolving mtDNA control region, was published (Vigilant et al. 1991)—again not free of problems (Penny et al. 1995; Bandelt 2005). In this latter case, the tree that was presented was rooted using a chimpanzee outgroup, and the phylogeographic argument was much stronger. Now, instead of merely an African (American) clade and a worldwide clade at the base of the tree, the first 14 deepest branches all led to lineages that were exclusively sub-Saharan African. This indicated long-term evolution within Africa before a dispersal around the globe. Although these conclusions came under serious scrutiny as the quality of the phylogenetic analyses was questioned (Templeton 1992; Hedges et al. 2002), further analyses (Penny et al. 1995; Watson et al. 1997), improved rooting using resurrected Neanderthal sequences as an outgroup (Krings et al. 1997), and better resolved genealogical reconstructions using complete mtDNA sequences (Ingman et al. 2000) eventually confirmed the picture. Moreover, Y-chromosome single-nucleotide polymorphisms (SNPs), most of which are stable markers that have evolved only once during the course of modern human evolution, clearly displayed a similar pattern (Underhill et al. 2000; Ke et al. 2001).

2.1 The Debate Continues

It is now widely accepted by geneticists that both the mtDNA and the Y chromosome have a recent ancestry in sub-Saharan Africa. The hopes of multiregionalists that mtDNA coalescence times would become older and older as data accumulated (Hawks et al. 2001) have not been fulfilled and, in retrospect, were never likely to be: they were fuelled by ill-applied models as well as insufficient data, and complete mtDNA sequences show that earlier dating attempts were by and large accurate (Mishmar et al. 2003). One well-studied X-chromosome locus with low recombination has also been viewed as supporting a recent African origin (Kaessmann et al. 1999), whereas another has been interpreted differently (Yu et al. 2002)—but, we would argue, fallaciously (see later). There has been on-going discussion as to the extent to which the various autosomal markers that have been studied in any detail support the model. Few have attempted, so far, to test the model against multiple genetic systems. Templeton (2002) used nested clade analysis—a formal, ill-based paraphyletic analysis (Weale et al. 2003)—to suggest that, whilst complete mtDNA sequence and well-resolved Y-chromosome data now supported a recent African origin (in marked contrast to his earlier, highly publicized, interpretations), some autosomal loci indicated earlier dispersals.

A number of logical quirks come along with claims such as "mtDNA data ... of course contains no information whatsoever about the validity of the replacement hypothesis and hence is irrelevant to any test of the replacement hypothesis" (Templeton 2004). The ingredients of such views are (1) misdating coalescence times (Chap. 4) and (2) the belief that "mtDNA is informative about human evolution only to about 100–125 KYA" (Templeton 2003), that is, half of the assumed coalescence time. This misconception reads into real data expected values of random variables gleaned from a simplistic coalescence model of a constant-size panmictic population and tacitly assumes that the evolution of human mtDNA conformed to this expectation by a rule of thumb.

On the other hand, the analysis of Takahata et al. (2001) suggested that a suite of ten molecular marker systems was largely concordant with the replacement model. In fact, the evidence from most autosomal systems is difficult to interpret because of a poverty of informative variation. This is because nuclear base variants arise much more slowly than base variants on the mtDNA, and since the X chromosome and the autosomes undergo recombination, which destroys the genealogical patterns, it has been difficult to find and sequence a sufficiently long stretch that has not recombined over the timescale of modern human evolution. A recent example is seen in Garrigan et al. (2005), where the claimed Asian root is not supported by the phylogeographic distribution, and the supposedly ancient Asian-specific clade is subject to enormous uncertainty as to both its origin and its age.

As a consequence of this poor genealogical resolution, and poorly tested (and very imprecise) age estimates, it seems likely that Templeton's (2002) analysis of autosomal markers is encountering problems similar to his earlier analyses of mtDNA, which have been corrected by the improved resolution afforded by complete mtDNA sequences. This weakness of the evidence from other genetic systems renders them compatible with a number of perspectives. In rare cases where the level of variation is higher, the evidence supports a recent African ancestry, without any signals from the earlier dispersal events from Africa (of *H. erectus*, *H. heidelbergensis*, and perhaps *H. helmei* or archaic *H. sapiens*) which almost certainly took place (Alonso and Armour 2001). The distribution of *Alu* insertion elements, which have a simple ancestral-derived polarity, in the human genome further supports a recent ancestry in Africa for non-African populations (Stoneking et al. 1997; Watkins et al. 2001). Their status as neutral, multilocus markers means that they provide some of the strongest autosomal support for the out-of-Africa model.

Yet, the overall relative paucity of the genetic evidence deriving both from the sequencing of autosomal loci and from classical markers, along with the weakness of traditional methods of analysis, has allowed population geneticists and anthropologists with an interest in the multiregionalist perspective to argue that the issue is not resolved (Relethford 2001). From a technical point of view this may be correct (since most of the human genome has not been analysed, and some markers that have been analysed give equivocal results) and, as we have already noted, indeed some autosomal sequencing studies have been explicitly interpreted as supporting if not exactly full-blooded multiregionalism, then at least a high level of interbreeding between modern and archaic humans in Eurasia (e.g. Harding et al. 1997; Zietkiewicz et al. 2003; Garrigan et al. 2005). However, such studies rarely provide a clear test of the alternatives, because it is usually possible to postulate several founder types, and, as importantly, the dating of Eurasian-specific autosomal lineages has been sufficiently imprecise to date not to be able to rule out the replacement hypothesis.

Whether there was some interbreeding of anatomically modern humans with more archaic forms of human along the way, as some insist, has still to be tested against the similarly detailed evidence that will soon become available from autosomal and X-chromosomal haplotypes. However, the desire of some population geneticists working on X-chromosome and autosomal systems to stress the complexity of our evolutionary past should not be accepted uncritically. The emphasis on complexity seems unarguable: who could deny that human evolution over the last few hundred thousand years must have been complex? Yet statements such as "both the out-of-Africa and the multiregional models appear to be too simple to explain the evolution of modern humans" (Yu et al. 2002) warrant some scrutiny. In the case of Yu et al. (2002), for example, they are based on two lines of argument. One is a premature (and inappropriate) coalescent modelling and dating of X-chromosomal mutations, which could equally (we would argue better) be interpreted as indicating an African origin and a number of Eurasian founder clades diversifying over the last 65 000 years or so.

The second is the suggestion that "it is important to recognize that each locus in the human genome can capture only a fraction of the human history, and different loci can have rather different genealogies. Thus, some conclusions from different loci are *necessarily conflicting*. Only after a sufficient number of studies have been conducted, can we gradually reach a consensus about the history of modern humans" (Yu et al. 2002) (our italics). This would surely be the case only if multiregionalism had actually been shown to be correct. In fact, if we accept the evidence of the mtDNA and the Y chromosome, the default inference is that the non-African human population has undergone dramatic founder effects and range expansions within the last 70 000 years or so. We see no reason to think, under these circumstances, that every locus would have a completely separate evolutionary history: unlike later movements, such as those enforced by slavery, it is simply not plausible to imagine the out-of-Africa migration being carried out only by one sex, for example. Haplotypes come bundled together in individuals and populations, and, given a history such as is indicated by the non-recombining markers, there is every reason to think that other (neutral or nearly neutral) loci would have shared their fate. In particular, those who keep the prayer wheel revolving ("mtDNA is only one locus, and only reflects the maternal history of a population; the history of a single locus may not accurately reflect the history of a population because of chance [drift] effects or because of selection acting on that locus"; Pakendorf and Stoneking 2005) have yet to come up with a single example where complete mtDNA variation would be misleading in regard to modern human dispersals in the Stone Age. What is clear, however, is that we cannot expect that every migration or major expansion event left its decipherable trace in every stretch of the genome, so that a number of loci could neither support nor reject the reconstructed event.

It is also worth mentioning that the presence of some deep-rooting lineages in non-African populations would not necessarily invalidate the replacement model. Ancestral diversity has been preserved in sub-Saharan Africa since the exodus, and may be detected in non-African populations as a result of recent gene flow. We will see later, for example, that the African-derived Y-chromosome haplogroup E3b underwent a limited spread into the Near East in the late Pleistocene/early Holocene, and from there may have been dispersed into southern Europe during the Neolithic as well as directly from North Africa in the medieval period (Cruciani et al. 2004; Semino et al. 2004). It is certainly possible that autosomal markers from Africa experienced the same fate. Therefore, the simple detection of deep-rooting diversity in non-African populations, without a thorough phylogeographic analysis, may be misleading.

We can hope for more information in the future, as haplotype blocks, thought to have been unaffected by recombination, are now being identified within the autosomes, and sequencing technology has progressed sufficiently that it is starting to become feasible to analyse very long stretches of such DNA in many individuals (Phillips et al. 2003). In the meantime, suggestions such as that of Eswaran (2003), which attempt to reconcile the multiregional and out-of-Africa models, seem to be based on a false premise: namely, that the genetic and fossil evidence is not compatible with the latter (Pearson and Stone 2003).

2.2
Moving on

It is not obvious, though it is possible, that amongst the small human population sizes of the Late Pleistocene there would have been much opportunity for interbreeding between archaic and anatomically modern individuals. In any case, our reconstruction will follow the routes of our anatomically modern ancestors from Africa through Eurasia, traced out by mtDNA and the Y chromosome. We believe that, given the strength of the evidence from the two most powerful genetic markers systems, the non-recombining mtDNA (Ingman et al. 2000) and the NRY (Underhill et al. 2000), backed by a number of autosomal marker systems that do give a clear result (Tishkoff et al. 1996; Labuda et al. 2000; Alonso and Armour 2001), it now seems reasonable to attempt to move the discussion forward. In what follows, we attempt to use the non-recombining marker systems to plot out the course of the dispersals around the world, and leave any narratives of contact and interaction with other hominins for others to elaborate. In this sense, we follow the strategy pursued in anthropology by Lahr and Foley (1994, 1998), Lahr (1996), Foley and Lahr (1997), and Foley (1998). A particular contribution of theirs has been to emphasize the climatological and ecological framework of glacial and interglacial faunal expansions and contractions pumping large mammals, including hominins, out of Africa on a regular basis, throughout the Pleistocene. We will be led to rather different conclusions, at least so far as modern human dispersals are concerned, but we make our suggestions in a similar spirit.

We should not persuade ourselves, however, that the genetic evidence alone will allow us to chart the patterns of human prehistory; this can only be achieved in the context of archaeology, physical geography, palaeontology and palaeoclimatology. The problem then is that simply combining different lines of evidence risks the kind of circularity sometimes seen in the phylogeographic literature, as well as in attempts to combine archaeology and historical linguistics, where two lines of weak evidence become mutually reinforcing. We can try to avoid this by using the genetic evidence to distinguish hypotheses within a plausible model built from other disciplines (Richards et al. 2002)—whilst not ignoring the possibility that the genetic evidence can itself also suggest useful hypotheses for further exploration.

Several authors, using mainly Y-chromosome evidence with the mtDNA in a largely supporting role, but in some cases based primarily on the distribution of mtDNA, have indeed attempted such a project already (Maca-Meyer et al. 2001; Underhill et al. 2001). Our main line of argument is only substantially prefigured, however, by Forster (2004) (see also Forster et al. 2001; Oppenheimer 2003), although we differ in the details of the subsequent dispersals through Eurasia. A detailed consideration of the mtDNA evidence will lead us to rather different conclusions from the present consensus (cf. Lewin and Foley 2004) that the out-of-Africa settlements were achieved through multiple dispersals. In particular, we believe that recent work on the genetic variation in the Indian subcontinent should lead to a major reassessment of the dispersal process.

3
How Many Dispersals of Modern Humans from Africa?

As it stands, the replacement hypothesis recognizes a number of distinct dispersals out of Africa in prehistoric times, beginning with *H. erectus* about 1.8 million years ago, but in its strict form contends that the most recent emigration, of anatomically modern *H. sapiens* within the last 100 000 years, led to the supplanting of all earlier *H. erectus* and archaic human populations (such as the Neanderthals). This may or may not have involved some interbreeding between modern humans and Neanderthals, or earlier archaic human species such as *H. heidelbergensis*, *H. antecessor* or even *H. erectus* (Lahr and Foley 1994; Foley and Lahr 1997, 1998), but in any case no lineages on either the male or the female lines of descent survive in the human gene pool today.

However, there has been debate about how many distinct dispersals of *modern* humans contributed to the modern pool of *H. sapiens*. This was fuelled by the dating of early modern human remains at Skhul and Qafzeh caves in the Levant to the Eemian interglacial, early in oxygen isotope stage 5, sometime between ~ 90 000 and 120 000 years ago (Grün and Stringer 1991). This indicated the appearance of anatomically modern humans in the Near East, in a Middle Palaeolithic context, well before the widespread appearance of the Upper Palaeolithic in Eurasia about 50 000 years ago, and also preceding the appearance of modern humans in Australia, thought to be at most ~ 70 000 years ago (Thorne et al. 1999). One possibility was that the Skhul and Qafzeh remains were the traces of an earlier dispersal than that which gave rise to the Upper Palaeolithic of Eurasia. However, the Levant can be regarded ecologically as an extension of Ethiopia during stage 5, with a whole range of fauna expanding into the region at that time, and contracting subsequently as the Ice Age regained its grip and Neanderthals took hold of the region (Tchernov 1992; Lahr and Foley 1994). Therefore, it may well be the case that the Skhul and Qafzeh specimens left no descendants in the modern human population, at least outside Africa.

An alternative argument was mounted by Lahr and Foley (1994), when it was realized that dates for the settlement of Greater Australia were moving back beyond the point at which it seemed that they could be attributed to the same dispersal event that led to the settlement of western Eurasia ~ 50 000 years ago. Some dates for early Australian sites, whilst remaining controversial, have been put at 50 000–60 000 years or even slightly earlier (Roberts et al. 1994; Thorne et al. 1999). This suggested that at least two dispersals out of Africa were likely: an earlier 'southern route' from the Horn of Africa, perhaps ~ 70 000 years ago, and one through North Africa and the Levantine corridor ~ 50 000 years ago (Lahr and Foley 1994). Some genetic evidence from classic marker frequencies in modern populations was also cited in favour of this view (Nei and Roychoudhury 1993; Cavalli-Sforza et al. 1994).

Further arguments for multiple dispersals have come from physical anthropology. Lahr and Foley (1994) argued that it was necessary to postulate multiple dispersals out of Africa in order to satisfactorily explain diversity in Upper Pleistocene human remains, in particular the retention of ancestral features in early Australians—although other assessments of the skeletal evidence have disagreed (Rayner and Bulbeck 2001). More generally, it has often been suggested that there is an extant tropical belt of human populations that anatomically resemble sub-Saharan Africans (with 'racial' features such as very dark skin, curly hair and so on). They include some southern Indians, the Andamanese, the so-called Negritos of the Philippines (Aeta/Agta) and the Malay Peninsula (Semang), Papuans and Aboriginal Australians. These people, it was suggested, might be the survivors of a 'southern coastal route', from the Horn of Africa along the tropical coastline through to Southeast Asia and Australasia (Nei and Roychoudhury 1993). The bulk of Eurasian populations were then suggested to be the survivors of a 'northern route': out of Egypt into the 'Levantine corridor', and thence into both Europe and Asia (Lahr 1996). A version of this model has been pursued in what is perhaps the most comprehensive interdisciplinary study to date, using Y-chromosome data and archaeological and palaeoanthropological evidence (Underhill et al. 2001). It has also been suggested that mtDNA data (on haplogroup M) support an early southerly migration from Africa ~ 60 000 years ago to India and beyond (Quintana-Murci et al. 1999), thus leaving the possibility open for a later northern route into the rest of Eurasia. Lewin and Foley (2004) even present an (imaginary) sketch of the mtDNA tree skewed to appear to support an early migration to Southeast Asia and Australia.

3.1 The mtDNA Evidence and Multiple Dispersals

A closer look at the mtDNA evidence forces us to question this picture. The number of mtDNA lineages moving out of Africa had not been clear in the earlier analyses of Cann et al. (1987) and Vigilant et al. (1991). It was cleared up by using more African data and focusing only on the commoner mtDNA types, thus cutting out any rare back-migrants from Eurasia into Africa (Watson et al. 1997). This showed that all non-Africans (with the exception of a few much more recent emigrants from Africa) carry variants of a single mtDNA clade, haplogroup L3, which traces back to sub-Saharan Africa—most probably East Africa. Modern non-Africans all trace their maternal line of descent back to a single mtDNA type, which arose in Africa around 83 500 (±8400) years ago (Macaulay et al. 2005). This made a single exodus from Africa highly likely.

There was a complication. The analysis of Watson et al. (1997) was based largely on control-region sequence data, which fails to resolve many mtDNA haplogroups. By targeting newly identified coding-region variants, Quintana-Murci et al. (1999) distinguished two major clades in non-Africans, within haplogroup L3. One of these had already been identified as the Asian super-haplogroup M (Torroni et al. 1994a); Quintana-Murci et al. (1999) showed that all other non-African L3 lineages fell into a second major clade, later named haplogroup N. Haplogroups M and N appear to be almost identical in age, estimated from complete coding-region sequences to be about 63 000–69 000 (±5000) years old (Macaulay et al. 2005; see also Forster et al. 2001; Kong et al. 2003; Forster 2004). Moreover, haplogroup M was present at high frequency in Ethiopia and Somalia, in addition to its known distribution from the Near East to East Asia and Australasia—and amongst Native Americans (Torroni et al. 1993, 1994b). This dichotomy in non-Africans raised the possibility that there may have been two migrations out of Africa, one carrying haplogroup M along the 'southern coastal route' from East Africa to India and on to Southeast Asia and Australia, and one carrying haplogroup N along the putative northern route via northeast Africa and the Levant (Maca-Meyer et al. 2001; Tanaka et al. 2004).

However, persuasive phylogeographic arguments militate against this view. Haplogroup M is found in East Africa and the Near East (including Arabia) only in one rare variety, a subclade splinter referred to as M1. Although very diverse in East Africa (Quintana-Murci et al. 1999), M1 is not present in either the Indian subcontinent or East Asia. Both India and East Asia each have a particular set of diverse haplogroup M subclades with virtually no overlap between the two regions (Chap. 8). A further M clade, now known as haplogroup Q (Forster et al. 2001) is found in New Guinea, and there is at least one clade further present in Aboriginal Australians (Huoponen et al. 2001; Ingman and Gyllensten 2003). It is clear from this picture that the greatest diversity of haplogroup M subclades is found not in Africa, but rather in the Indian subcontinent and Southeast Asia (Forster et al. 2001; Edwin et al. 2002; Kivisild et al. 2002, 2003; Kong et al. 2003; Metspalu et al. 2004; Macaulay et al. 2005; Thangaraj et al. 2005). This suggests that the Indian subcontinent may be close to the origin of haplogroup M and that Africa and the Near East are peripheral, the result of secondary dispersals, perhaps from the Gulf area.

The suggestion that haplogroup M—specifically haplogroup M1—is most likely the result of a back-migration into East Africa is supported by several further lines of evidence (Richards et al. 2003; Forster 2004; Kivisild et al. 2004). Despite its high diversity, within sub-Saharan Africa, M1 is only present in Ethiopia and Somalia: it is entirely absent elsewhere. Furthermore, Ethiopia and Somalia also show the presence of other diverse haplogroups of clear Eurasian origin, in particular haplogroup (pre-HV)1. Finally, haplogroup M's Eurasian sister clade, haplogroup N, which has a very similar age to haplogroup M, is almost entirely non-African, extending to all parts of the non-African world, including Australasia, with no indication of an African origin in its present distribution. The only members of haplogroup N in Africa are all most likely immigrants, belonging to derived branches of haplogroups such as Near Eastern U (predominantly found in North Africa as its descendant U6, and in Ethiopia as U9) or J and T (Richards et al. 2003; Kivisild et al. 2004). A possible exception, haplogroup X1, although found occasionally in Ethiopia, is largely restricted to North African and Near Eastern populations (Reidla et al. 2003).

Since the work of Quintana-Murci et al. (1999), complete mtDNA sequences have begun to become available from Africa, Eurasia and Australia (Ingman et al. 2000; Maca-Meyer et al. 2001; Herrnstadt et al. 2002). These indicate that the root of L3 gives rise to a multifurcation from a single haplotype producing a number of distinct subclades. L3b′d, L3e and L3f, for instance, are clearly of African origin, whereas haplogroup N is of apparently Eurasian origin. Unfortunately, there is no evidence that haplogroups M and N are more closely related to each other than they are to other clades within L3 (although they could have been linked in the maternal genealogy without us now being able to detect this—that is to say, by lines of descent not affected by any mutations in the mtDNA). Since the complete-sequence tree represents the limit of the resolution of the maternal genealogy, this is something we may never know.

The simplest explanation for this geographical distribution, however, is an expansion of the root type within East Africa, where several independent L3 branches thrive, including a sister group to L3, christened L4 (Kivisild et al. 2004; Chap. 7), followed by divergence into haplogroups M and N somewhere between the Horn of Africa and the Indian subcontinent. Since neither the L3 root type nor any other descendants survive outside Africa, the root type itself must have become extinct during a period of genetic drift in the founder population as it diversified into haplogroups M and N, if the diversification was outside Africa. If on the other hand the diversification was indeed within East Africa, then haplogroups M and N must either have been carried out of Africa in their entirety or subsequently have become extinct within Africa, with the singular exception of the derived M1. Parsimony argues against this latter model, particularly since we have independent evidence for later migrations back into the Horn of Africa.

This picture strongly suggests a single dispersal out of Africa, since it implies that all modern non-Africans carry the descendants of a single African mtDNA type. In the complete mtDNA sequence tree, M and N diverge by four and five mutations respectively from the ancestral L3 type. Although it is not impossible that the root type of L3 might have been carried out of Africa more than once, perhaps diversifying separately into M on one occasion and N on the other, the very high diversity of mtDNA lineages in East Africa makes this very unlikely—particularly since the two-dispersals model suggests that the two source populations were different: the Horn of Africa for the 'southern route' and northeast Africa for the 'northern route', respectively (Lahr 1996). The more-or-less identical ages of haplogroups M and N add further weight to a single exit, which took place somewhere between the estimated ages of L3 ($\sim$ 85 000 years ago) and M and N ($\sim$ 65 000 years ago). A separate, more recent migration out of Africa into Eurasia, most likely around the time of the Last Glacial Maximum, is, however, evidenced in Y-chromosome haplogroup E3b distribution, as discussed in more detail later. In contrast to the widely co-spread mtDNA basic founder haplogroups M, N and R, and Y-chromosomal C, D, F and K, the recent immigrant haplogroup E3b in Eurasia has a narrow geographic distribution, restricted to the Near East and Europe, and is also associated with significantly lower background short tandem repeat (STR) variation than the 'ancient' haplogroups in Eurasia (Cruciani et al. 2004; Luis et al. 2004; Semino et al. 2004).

Furthermore, nested in haplogroup N sits haplogroup R, which has sent its descendants into nearly all non-African parts of the world. The non-overlapping distribution of the derived haplogroups within M and N and its subclade R in South Asia, eastern Asia and Australasia indicates that the only founder types in each of these regions were the ancestral types of M, N and R, strongly implying a very rapid dispersal. This contrasts with the situation when there has been a known pause before a subsequent expansion, for example in the case of the Americas, where East Asian haplogroups A–D are recruited (Torroni et al. 1993), and the Pacific islands, where derived subclades of haplogroups B and Q are involved (Sykes et al. 1995; Richards et al. 1998). The movement of the founders inland from the coastal settlements, and the evolution of region-specific haplogroups, may have taken place more gradually (cf. Chap. 8).

This perspective, which argues in rather traditional fashion for a rapid replacement model, also rules out the 'weak garden of Eden' suggestion. Harpending et al. (1993) argued that demographic history was best inferred from a plot of pairwise comparisons between sequences (which they called the 'mismatch distribution'). Simulations suggested that the wave in the distribution that they found, signalling a star-shaped phylogeny, indicated a population expansion or a selective sweep (Slatkin and Hudson 1991; Rogers and Harpending 1992). They estimated the timing of the expansion to be $\sim$ 40 000–60 000 years ago.

However, Harpending et al. (1993) also postulated the existence of a prior bottleneck, a time when they suggested that modern humans had been on the verge of extinction. They furthermore used the 'intermatch distribution' (using the distribution of sequence pairs from separate populations) to argue that the expansions had been distinct in different continents. This led them to the 'weak garden of Eden' hypothesis, which suggests that modern humans dispersed out of Africa into different parts of the world ~ 100 000 years ago, forming the distinct gene pools of the modern 'racial groups', long before the expansions. One possible explanation for this pattern was that the regional groups were decimated by the Mount Toba volcanic eruption on Sumatra ~ 74 000 years ago, and then re-expanded, generating the star phylogenies we see today (Rogers and Jorde 1995; Ambrose 1998).

Although this suggestion has been widely publicized (e.g. Lewin and Foley 2004), there seems to be almost no evidence in its favour. The populations compared in the intermatch analysis were Africans, Europeans and Native Americans—which undoubtedly did expand at different times—but of course these are not the appropriate populations with which to test a rapid expansion through Eurasia and Australasia. The mtDNA may signal an early range expansion within Africa, perhaps more than 100 000 years ago (perhaps involving haplogroups L0, L1, L2 and L5), and more clearly signals a second expansion (involving haplogroup L3) after ~ 85 000 years ago, which subsequently spilled out of Africa (Watson et al. 1997; Salas et al. 2002). This may support a 'weak garden of Eden' scenario *within* Africa (Labuda et al. 2000)—but it offers no support for it outside Africa, or for the involvement of the Toba eruption in human evolution.

3.2 Y-Chromosome Founders

Can this picture be reconciled with the Y-chromosome evidence? Y-chromosome variation has hitherto been interpreted as suggesting multiple dispersals out of Africa (Underhill et al. 2001; Luis et al. 2004). The argument is as follows. The tree of the Y-chromosome variation is similar to that of mtDNA in many respects: the root is within African lineages (Y-chromosome haplogroups A and B), and there is a trifurcation leading to two clades that are non-African (C and F) and one (DE) which is both African and non-African (Underhill et al. 2000, 2001); see Fig. 1. Haplogroup DE, defined by the YAP^+ marker, was the subject of debate for a number of years, presenting a puzzling phylogeographic pattern. It falls into two subclades: haplogroup D, which is entirely southern, eastern and Central Asian, and haplogroup E which is largely African but also occurs in the Near East and southern Europe. Hammer et al. (1998) argued, using a poorly resolved tree, that haplogroup E was derived with respect to haplogroup D and therefore represented a 'back-to-Africa' migration, ~ 40 000 years ago. The much better resolved system of Underhill et al. (2000) clearly shows that D and E are in fact sister clades, rendering the postulated back-to-Africa migration superfluous (Underhill and Roseman 2001). In addition, Weale et al. (2003) have since discovered several underived DE* lineages in Nigeria, indicating an origin for the YAP^+ marker in west Africa. Haplogroups E2 and E3a were dispersed east and south within

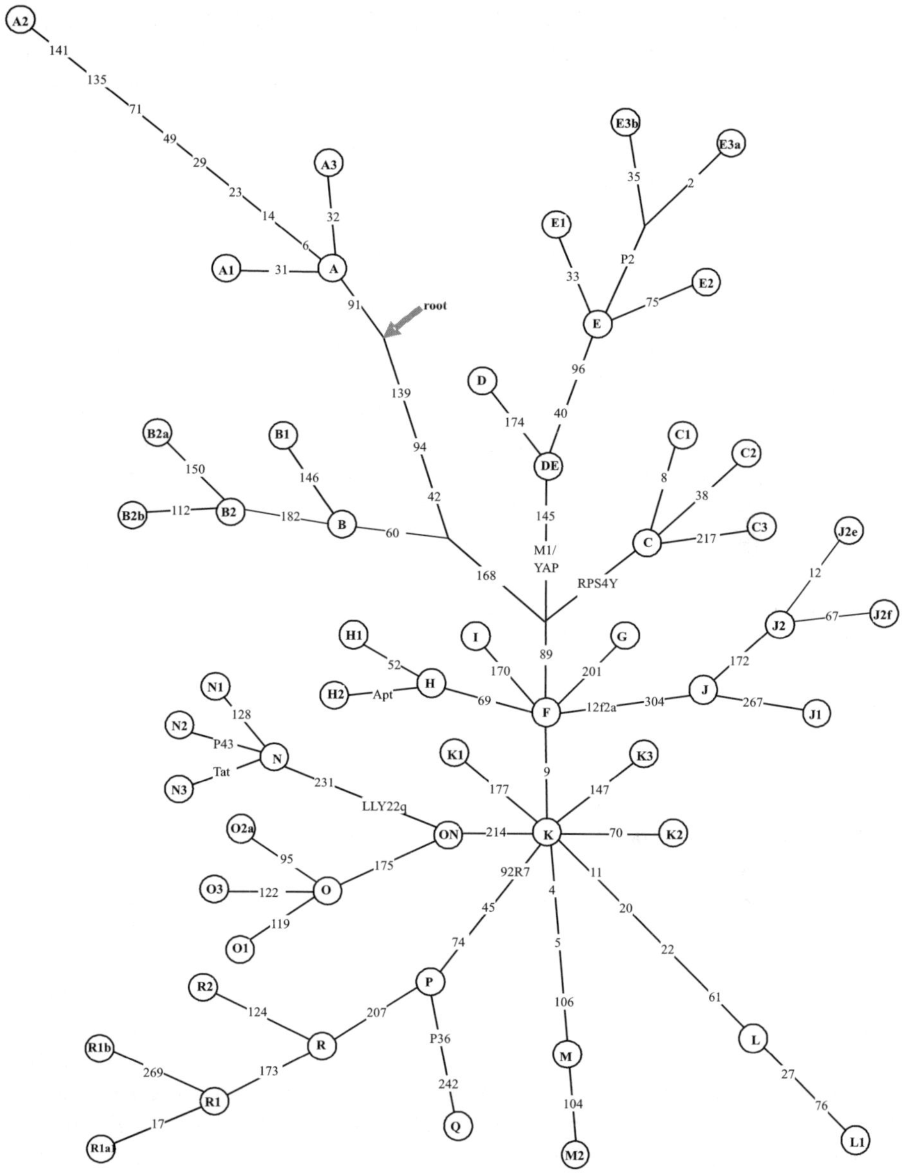

Fig. 1 Schematic outline of the human Y-chromosome tree, with major branches and useful markers indicated (based on YCC 2002)

the past ~ 4000 years by the Bantu dispersals (Passarino et al. 1998; Scozzari et al. 1999; Cruciani et al. 2002).

Technically, haplogroup E3b probably also represents an out-of-Africa movement into the Levant, and thence both west to Anatolia and Europe and south into Arabia, but this is much more recent (probably well within the last ~ 15 000 years) and does not therefore substantiate the use of the 'Levantine corridor' ~ 50 000 years ago (Cruciani et al. 2004; Luis et al. 2004). (Those who prefer their genetics served with historical linguistics might, however, like to associate this with the spread of Afro-Asiatic languages from East Africa into the Near East.) E3b also predominates in North Africa (Bosch et al. 2001), but probably only spread through this region in the last approximately 5000 years (Arredi et al. 2004; Cruciani et al. 2004). YAP^+ chromosomes in Europe also mainly belong to various subclades of E3b, some of which may have spread into Europe with the Neolithic ~ 8000 years ago, and some of which probably entered directly from North Africa more recently, for example in medieval times (Cruciani et al. 2004). There is no indication that E3b was transmitted from the Horn of Africa into Arabia (Luis et al. 2004), whereas the low levels of E3a in Arabia most likely result from the action of the Arab slave trade within the last 2000 years (Richards et al. 2003; Luis et al. 2004).

It seems likely, therefore, that the ancestors of haplogroup D left Africa for South and East Asia without the ancestors of haplogroup E. This rules out the possibility that extant Eurasian Y chromosomes might trace to a single Eurasian founder lineage, as mtDNAs probably do (i.e. that there is a Y-chromosome equivalent of a dispersing mtDNA haplogroup L3). The equivalent in the Y-chromosome tree of mtDNA L3 is the clade CR defined by the derived state at the M168 marker and giving rise to three haplogroups C, DE and F. Haplogroups C and F are non-African, while DE (defined by YAP^+) holds both African and Asian branches.

So a minimum of either two or three Y-chromosome types probably participated in the dispersals out of Africa. It does not necessarily follow, however, that there were two or three migrations. The two or three ancestral Y-chromosome types, which are closely related (at least to the extent of all being derived with respect to M168), could equally well have been members of the same population, and could have been carried together in the exodus (Kivisild et al. 2003). The distribution of haplogroup D—present at low levels in Central, East and Southeast Asians as well as in Andaman islanders (Underhill and Roseman 2001; Thangaraj et al. 2003) supports this proposition, since these groups would, on a multiple-dispersal model, not normally be considered likely to be part of the same process (Endicott et al. 2003b). Andamanese would rather be thought likely to be part of the supposed 'southern route', and most East Asians part of the supposedly later northern Upper Palaeolithic expansion. Conversely, haplogroup F, which Underhill et al. (2001) identify with the later expansion, is found at rather high levels in isolated populations that they would classify with the earlier southern expansion (Kayser et al. 2001; Kivisild et al. 2002; Thangaraj et al. 2003). We discuss this in more detail later.

We conclude that, like the mtDNA, Y-chromosome variation is best explained by a single exodus. This is further supported by studies of STRs and an *Alu* element at the autosomal *CD4* locus (Tishkoff et al. 1996). These show a drastic reduction in diversity to just three main variants in western Eurasians, two of which are carried into eastern Eurasians. Again, resampling for a second exodus would be likely to have resulted in a greater distinction between the gene pools of east and west. All non-African copies of the 16p13.3 locus (Alonso and Armour 2001) also coalesce to a single founder sequence, again supporting a single exodus.

A population tree and principal coordinates (PC) analysis of *Alu* insertion elements dispersed throughout the genome suggested that Aboriginal Australian and Papuan populations were more closely related to sub-Saharan Africans than Eurasians (Stoneking et al. 1997). This was due to an excess of ancestral alleles in these populations, suggesting an earlier tropical coastal migration. However, analysis of larger numbers of elements, whilst not including Aboriginal Australian and Papuan samples, failed to support this view by indicating that Indian tribal populations did not possess an excess of ancestral alleles (Watkins et al. 2001); and inspection of the PC plot in Stoneking et al. (1997) suggests that the link, if it exists, is a tenuous one, which could equally be explained by lineage sorting during the single process of expansion. Other loci, such as Xq13.3 (Kaessmann et al. 1999), *MX1* (Jin et al. 1999) and β-globin (Harding et al. 1997), are consistent with a single exodus with a variable number of founders. More founders are indeed to be expected for X-chromosome and autosomal loci, where the effective populations sizes are respectively 3 and 4 times larger than those for mtDNA and the Y chromosome, so that the impact of founder effects is correspondingly less. However, these loci have been variously interpreted, and do not provide strong evidence either for or against a single dispersal. For example, the short sequence stretch (only $\sim$ 100 bp) of the *MX1* locus seems to exhibit six founder types in Eurasia, with almost no additional diversity accumulated since the spread outside Africa (Jin et al. 1999). Whilst this has been interpreted in terms of multiple dispersals (Jin et al. 1999), there is no good reason to do so, given the limited evidence available from such a system (Kivisild et al. 2003).

4
Which Way out of Africa?

The evidence of Y-chromosome variation has recently been used in a throughgoing attempt to map out the modern human dispersal process in detail (Underhill et al. 2001). As the authors point out, this is best achieved within a framework supplied by archaeology, palaeoanthropology and palaeoclimatology, which mitigate the reliance on a single locus or a few loci (that may, for example, also be affected by non-demographic forces such as natural selection). However, we would argue that their model is overdetermined by the prior information fed in from previous anthropological studies (Lahr and Foley 1994, 1998). We suggest that the Y-chromosome evidence (perhaps as the result of very high levels of genetic drift) is sufficiently flexible to be compatible with a number of distinct exit models, and that insufficient attention has been paid to testing the model with mtDNA data. We aim to rectify that here and to modify the model accordingly.

The model of Underhill et al. (2001), although a variant of the out-of-Africa replacement hypothesis, involves multiple dispersals of anatomically modern humans. All non-African Y-chromosome lineages are derived at the M168 position, which leads to a trifurcation resulting in haplogroups C, DE and F. Underhill et al. suggest that small groups of emigrants carrying the M168 mutation left Africa several times. One group left via the Horn of Africa 45 000–50 000 years ago, dispersing to southern Asia where the M130/RPS4Y mutations defining haplogroup C arose. Haplogroup C was then carried into Southeast Asia, with subgroups dispersing further into both central and eastern Asia and Australasia (and ultimately North America).

They argue, as mentioned already, against an Asian origin for the YAP mutation defining haplogroup DE (see also Underhill and Roseman 2001), and that the YAP mutation was dispersed from the Horn of Africa to southern Asia. They suggest that this may have been alongside the haplogroup C dispersal, but conclude that two dispersals were likely to have been involved, albeit at a similar time. The M174 mutation defining haplogroup D then occurred in eastern Asia, leaving haplogroup D concentrated mainly in parts of East Asia. They regard these dispersal events to have been the same ones that carried mtDNA haplogroup M from East Africa to India, according to Quintana-Murci et al. (1999).

Underhill et al. (2001) thus regard the distributions of both Y-chromosome haplogroups C and D as relict distributions. The main out-of-Africa dispersal they believe occurred via the Levantine corridor, ~ 45 000 years ago, by members of haplogroup F, alongside members of mtDNA haplogroup N. This would correspond to the archaeological record for the expansion of Upper Palaeolithic cultures in Eurasia (Mellars 1992). The Y-chromosome haplogroup F is defined by the mutation M89, which they argue evolved in northeast Africa. They suggest that it was carried into the Near East ~ 45 000 years ago and dispersed west, north and east from there from ~ 40 000 years ago Four major clades are derived from the base of haplogroup F. Haplogroup J would have remained largely concentrated in the Near East. The expansions into Europe involved F* and subsequently haplogroup I lineages. The eastwards expansion into India involved haplogroup H lineages. The northwards expansion into the Caucasus or Central Asia involved the major clade defined by the M9 marker, haplogroup K. This diverged into haplogroup O, which filled much of East Asia, haplogroup L, haplogroup M and haplogroup P, as well as several further minor clades. Haplogroup P, defined by M45 and M74, spread across Asia and (as haplogroup R, with the M173 variant) into Europe, probably ∼ 30 000 years ago. They suggest that this can be recognized archaeologically as the Aurignacian, the Gravettian or both. A second derivative, haplogroup Q (with the M3 variant) became the main contributor to Native American Y-chromosome lineages.

4.1 mtDNA and the Indian Staging Post

As we have argued already, the mtDNA data strongly suggest a single main dispersal out of Africa, rather than multiple dispersal events. In general, we favour a 'lumping' rather than a 'splitting' approach to the genetic composition of dispersing groups, on the basis of parsimony: if two groups of lineages could either have been dispersed by the same population or by two populations, the former explanation is more parsimonious. But, more to the point, it seems extremely unlikely that a single mtDNA lineage (haplogroup L3) or two very closely related lineages (the ancestors of M and N) could have been dispersed from two completely different source regions (the Horn of Africa and northeast Africa). If the mtDNA data suggest rather a single dispersal, which route was taken?

The most widely influential mtDNA work on the route out of Africa suggests a 'southern route' to India on the basis of the distribution of haplogroup M (Quintana-Murci et al. 1999). Our reassessment of this evidence has implied that it does not provide a basis for the early migration routes, because the M1 haplogroup trail from Asia to Africa most likely runs in the opposite direction, and at a more recent time (see the discussion earlier). However, the evidence of Kivisild et al. (1999a, b, 2003) and Metspalu et al. (2004) suggests that the proposed route from East Africa to India may, nevertheless, have been the route that the early humans took. As already noted, mtDNA haplogroup M is very widely distributed in southern and eastern Eurasia, reaching its highest frequencies in India (more than 60% in many groups, especially tribal populations). Although Underhill et al. (2001) suggest that mtDNA haplogroup M matches the distribution of Y-chromosome haplogroups C and D, the situation is more complicated. It is true that Y-chromosome haplogroups C and D resemble mtDNA haplogroup M in being present only in South Asia and eastern Eurasia, and not in western Eurasia. However, far from being relict in eastern Eurasia, the mtDNA haplogroup M is common and widespread there, whereas Y-chromosome haplogroups C and D have a much more limited distribution in eastern Eurasia—although both appear in widely separated parts of central and eastern Eurasia. A major counterargument for that simple distinction of the two waves is that most Australian Aboriginal mtDNAs derive directly from the ancestral nodes of N and R (not M) via their region-specific founders (Ingman and Gyllensten 2003), whilst the Upper Palaeolithic culture of the Levant never reached Australia. Moreover, so do some of the 'relict' populations on the route, such as the Andamanese and the Semang (Macaulay et al. 2005; Thangaraj et al. 2005).

There is even a suggestion, based largely on control-region sequences, that the diversity of basal clades within mtDNA haplogroup M in India may exceed that in eastern Eurasia; and numerous so-called M* lineages (sequence types unassigned to any described subclade within M) occur in India, but not in East Asia. These include the most recent ancestor sequence of haplogroup M, found much more frequently in Indian data sets than anywhere else yet sampled (Endicott et al. 2003b; Kivisild et al. 2003). This may well be a consequence of insufficiently detailed phylogenetic investigation, as sufficient complete sequence data on haplogroup M from India are not yet available, but at face value this evidence suggests, as Kivisild et al. (1999b, 2003) argue, that India was the first major locus of modern human expansion outside Africa. Furthermore, since the only haplogroup M lineages found in substantial numbers to the west of the subcontinent are members of the M1 fragment, it also seems unlikely that haplogroup M originated much further west. The coalescence time of haplogroup M mtDNAs based on control-region data in Indian tribal populations is 62 000 ($\pm$12500) years (Kivisild et al. 2003), closely in line with the complete-sequence estimates of the age of haplogroup M sequences (from East Eurasia) of 63 000 ($\pm$5000) years, cited earlier.

We should emphasize that the coalescence times for haplogroup M in South Asia and eastern Eurasia, to the extent that they are correctly estimated, strongly imply a Middle Palaeolithic entry into one or other of these regions (or, at least, a pre-Upper Palaeolithic entry—since the term 'Middle Palaeolithic' is used rather loosely for much of Eurasia). Although there have been many previous attempts to explain the rationale for this (e.g. Richards and Macaulay 2000; Richards et al. 2002; Endicott et al. 2003b), it has recently been challenged again by Cordeaux and Stoneking (2003), citing the oft-quoted dictum that "the divergence of genes predates that of populations". Despite its elevation to a mantra amongst some human population geneticists (e.g. Barbujani et al. 1998), this idea emerged in the context of the application of molecular data to the reconstruction of species trees (Nei 1987), and only applies if there is genetic diversity carried over from the source population of interest. In our current example, however, as we have explained already, there are no reasons to speculate that M and N would have carried their sub-branches, extant now only outside Africa, in the African source population.

Provided that our argument about the back-migration of M1 is sound, we have to suppose that the four mutations defining haplogroup M evolved somewhere in South or West Asia. In that case, the coalescence time estimated from the accumulation of sequence variation *on top of* these four mutations in haplogroup M in Asia is a minimum, not a maximum, estimate for the human antiquity in the region. Similarly, contra Cordeaux and Stoneking (2003), the age of mtDNA haplogroup U in Europe/the Near East, assuming (reasonably, given the distribution of haplogroup U lineages) that that is where it evolved, approximates to the age of modern human settlement there. This age (approximately 45 000–72 000 years according to complete genome data; Achilli et al. 2005) overlaps with the error range for the earliest archaeological evidence for Early Upper Palaeolithic settlement in Europe and the Near East (van Andel et al. 2003). These points are well made by Endicott et al. (2003b).

This argument applies with even more force to mtDNA haplogroup N, and its immediate descendant and major subclade, haplogroup R, since neither shows any sign whatsoever of an African origin. The Indian subcontinent also includes numerous lineages within haplogroup R that are not found elsewhere, estimated at $\sim$ 65 000 (64200 $\pm$ 6300) years old from the complete sequence data of Palanichamy et al. (2004)—very similar to estimates for haplogroup M (Kivisild et al. 1999b). These include the HVS-I founder type (at the resolution level of control-region sequences and diagnostic coding-region sites) that is the MRCA of both the predominant east Eurasian (B, R9—including haplogroup F—and P) and west Eurasian (pre-HV, JT and U) haplogroups.

This suggests that the roots of haplogroups M, N and R probably arrived at the same time in the Indian subcontinent, most probably $\sim$ 60 000–70 000 years ago. A founder analysis of complete mtDNA genomes from India suggests a date of $\sim$ 66 000 (66100 $\pm$ 5700) years, with founder ages for East Asia and Australasia progressively younger at 64 500 ($\pm$3800) years and 63 400 ($\pm$5, 200) years (Macaulay et al. 2005). Incidentally, a founder age for European lineages is estimated at 66 300 ($\pm$5600) years, approximately similar to the age for India, strongly implying that the split took place close to the Indian subcontinent. Therefore, we agree with Quintana-Murci et al. (1999) that the earliest migration out of Africa was a southern route to India, even though we differ in the detail of our interpretation of the mtDNA evidence. It also seems likely, as suggested by Kivisild et al. (1999b) and Endicott et al. (2003b), that it was the Indian subcontinent that provided the inocula for further dispersals to the east. Moreover the founder age estimates suggest that the dispersal was extremely rapid, at roughly 4 km/year between the Indian subcontinent and Australasia, with an approximate lower bound of 0.7 km/year.

4.2
A Far Eastern Ancestry?

Both haplogroups M and N, and N's subclade R, are also found further east, in northeast Asia, East Asia, Southeast Asia and Australasia, with distinct haplogroup compositions in each region. Papuans and Australian Aborigines are distinct, both from Southeast Asians and, largely, also from each other in regard to mtDNA—although they do share one large and ancient clade, within R, haplogroup P. The simplest explanation for this pattern is that the ancestral population moved on east along the coast, with successive founder effects amplifying the founder lineages of haplogroups M, N and R as they moved, diverging in mainland Southeast Asia and the prehistoric continent of Sundaland (now the Indo-Malaysian archipelago following sea-level rises in the late Pleistocene/early Holocene) and spreading onwards into eastern Eurasia and Australasia from the coast.

The sharing of haplogroup P between Papuans, where one major branch is extant, and Aboriginal Australians, where at least four major branches seem to have survived (Ingman and Gyllensten 2003) suggests that the early Australasians may have entered the continent of Greater Australia, or Sahul—which formed one landmass until the end of the Ice Age—as a single group. Sea level in the region fell with the onset of oxygen isotope stage 4 ~ 70 000 years ago, rising again slightly for the duration of stage 3, and falling again drastically as the world headed for the Last Glacial Maximum after ~ 30 000 years ago (Lambeck and Chappell 2001; Chappell 2002). The glacial maxima, when the sea would have been at its narrowest as well as lowest, appear to be ruled out for the time of the crossing, given the ages of haplogroups M, N* and R. It is therefore possible that the easier 'northern route' from Sundaland, via Wallacea (rather than the 'southern route' via the Nusa Tenggara) was the more likely path taken, although there is no clear archaeological evidence for human settlement before ~ 30 000 years ago. However, the water crossing would have been considerably shorter by either route at any time during this period than it is today.

An alternative scenario has provided a large role for the volcanic eruption of Toba in Sumatra, ~ 74 000 years ago (Ambrose 1998, 2003). This issue was much discussed in the context of mtDNA mismatch distributions popular with some researchers in the early 1990s (Rogers and Jorde 1995). The Toba ash may have had a major effect to the northwest, especially the Indian subcontinent, with a relatively minor impact on much of Southeast Asia (Acharyya and Basu 1993; but see Song et al. 2000; Gathorne-Hardy and Harcourt-Smith 2002; Oppenheimer 2002). The question here is whether the initial peopling of these regions by modern humans had already taken place when Toba erupted. If so, it is possible that Indian subcontinental populations were decimated ~ 74 000 years ago, and that the region was repopulated from the east (e.g. Burma) and/or the west, e.g. the Gulf (Oppenheimer 2003). The relict distribution of Y-chromosome haplogroups C and D might be considered consistent with a resettlement of the subcontinent from the east.

There is insufficient confidence in genetic dating to finally resolve this issue at present, and there is no indication of derived deep branching in the topology of either the mtDNA or the Y-chromosome tree at the level of resolution so far achieved that may be used to test the hypothesis. However, the most recent estimates for the ages of mtDNA haplogroups M and N in eastern Eurasia, cited earlier, suggest that the settlement there was most probably post-Toba, and that therefore the effect of the eruption on modern human evolution is likely to have been minimal. An intriguing intermediate possibility is that the high drift that led to the geographical separation of haplogroups M and N from their ancestor haplogroup L3, somewhere between the Horn of Africa and South Asia, might be explained as an effect of Toba, if the exodus took place more than 74 000 years ago.

4.3 Opening up the West

By contrast with the situation in South Asia, which includes highly diverse mtDNA lineages of haplogroups M, N and R, the Near East (although also very diverse) harbours mainly lineages within the nested haplogroups N and R (with low frequencies of M1, and some recently arrived eastern Eurasian and African lineages). These might be derived directly from populations expanding northwest from the western part of the Indian subcontinent, or there may have been an earlier, more westerly, divergence located between East Africa and India that led to west Eurasian mtDNAs, such as the Persian/Arabian Gulf (Oppenheimer 2003). The elevation of haplogroup R at the expense of other clades within haplogroup N, and the loss of the ancestral L3 types, indicates that haplogroup R arose during the period of drift that took place somewhere between East Africa and India before the geographical spread started.

Unfortunately, there is no trail surviving in either the mtDNA or the Y-chromosome evidence in these regions of this period, as they seem to have been repopulated much more recently from the Fertile Crescent by derived lineages within mtDNA haplogroup R and Y-chromosome haplogroup J (Richards et al. 2003; authors' unpublished mtDNA data from Dubai). There are, however, mtDNA clades specific to both western Eurasia (N1, N2), India (N5) and eastern Eurasia (A, N9) belonging to haplogroup N (minus R), which we can refer to as N* for convenience. Haplogroup X, which is found in both western Eurasia and the Americas, may even have had a pan-Eurasian distribution after the Last Glacial Maximum (Reidla et al. 2003).

The Fertile Crescent was occupied by modern humans by $\sim$ 50 000 years ago, according to the archaeological record (Gilead 1991; Bar-Yosef 1998; Kuhn 2002), and the oldest western Eurasian clades (JT and U, within haplogroup R) date to this time and appear to have differentiated largely in this region (Richards et al. 2000). North Africa was settled (or resettled) by western Eurasian (probably Near Eastern) people carrying mtDNA haplogroup U6 sometime between $\sim$ 20 000 (if diversified into subclades U6a, U6b and U6c) and $\sim$ 45 000 years ago (if arriving undiversified), with the remaining majority of extant mtDNA and Y-chromosome lineages arriving from Europe within the last 10 000 years or so (Rando et al. 1998, 1999; Macaulay et al. 1999; Maca-Meyer et al. 2003). North Africa therefore holds no extant genetic evidence that could test its proposed status as a possible source of a northern-route dispersal (Lahr 1996). However, the mtDNAs extant in the Near East are highly diverse, trace back to the settlement ~ 50 000 years ago, and form the main source for the peopling of Europe, North Africa and much of Central Asia (Richards et al. 2000). Therefore, the fact that Near Eastern mtDNAs belong overwhelmingly to derived forms of haplogroup N (mainly R) only—and not to haplogroup M—clearly indicates that the Near East was not the source of most continental East Eurasian mtDNAs, but received mtDNAs from elsewhere, earlier in the Middle Palaeolithic. This argues forcefully against the view of Cordeaux and Stoneking (2003) that modern humans, including Southeast Asians and Australasians, owe their genetic lineages to dispersals only through North Africa and the Levant, with no contribution from a southern route.

There is one caveat to this picture. The most common mtDNA haplogroup extant in Europe, haplogroup H, seems to have a more recent origin, arriving from the Near East, Caucasus or eastern Europe ~ 20 000–25 000 years ago, alongside mtDNA haplogroup pre-V (Richards et al. 2000; Torroni et al. 2001; Achilli et al. 2004). As selection seems an unlikely explanation in this instance, the fact that it arose only ~ 20 000–30 000 years ago, yet rose to such high frequencies, testifies to the ongoing impact of genetic drift around the Last Glacial Maximum of these regions, and warns that, despite the high mtDNA diversity of the Near East, our reconstructions may only be telling us part of the story.

4.4 Y-Chromosome Passage to India and Beyond

Is this sketch compatible with the Y-chromosome evidence? Before we attempt to answer this question, it is worth mentioning that, despite the great advances in understanding of Y-chromosome variation in the last few years, gaps still remain that hinder its use as a phylogeographic marker. Whilst the outline of the Y-chromosome tree, based on SNP variants, is very robust, not to say bulletproof (Underhill et al. 2000, 2001; YCC 2002), the detail of variation within haplogroups utilizes relatively few SNPs, and is largely based on variation in between six and ten STRs, typically reconstructed using the median-joining (MJ) network algorithm. It is perhaps insufficiently recognized how inaccurate these reconstructions are likely to be, and that inferences on them must therefore be provisional until more STRs (preferably more than 30) or more SNPs are routinely employed. The other important weakness is dating: STR evolution is poorly understood and the rate varies by more than a factor of 10 not only from locus to locus but also depending on the length of the repeat (Brinkmann et al. 1998). Moreover time depth estimates from STRs have often been based on the measured pedigree rate (Heyer et al. 1997; Kayser and Sajantila 2000), which is most likely at least 2–3 times faster than the evolutionary rate. Here we rely on the evolutionary rate calibration of Zhivotovsky et al. (2004), which probably provides somewhat more reliable estimates.

As with the mtDNA, Y-chromosome variation in the Indian subcontinent has only recently been assessed in any detail (Kivisild et al. 2003). Until recently, it had seemed that the Indian subcontinent lacked one of the three main Y-chromosome founder clades (haplogroups C, D and F), since haplogroup D had not been found there. However, Thangaraj et al. (2003) have shown that haplogroup D is present in Andaman islanders, alongside other members of mtDNA haplogroup M (Endicott et al. 2003a, b). India therefore has the full complement of founder Y chromosomes and can be considered a potential source for the peopling of Eurasia along the southern route. Similarly, Y-chromosome haplogroups C and D are present only in central, southern and eastern Eurasia, rather like mtDNA haplogroup M (though at much lower frequencies), and are virtually absent from western Eurasia or North Africa (Underhill et al. 2001; Wells et al. 2001; Endicott et al. 2003b; Kivisild et al. 2003). Intriguingly, haplogroup C is present at low levels in Oman, but whether this represents survival of an early trail along the southern route or backflow from Asia is unclear (Luis et al. 2004).

To rehearse the argument outlined before: the Y-chromosome haplogroup F is widely distributed through Eurasia. This suggests two plausible interpretations of the Y-chromosome evidence: (1) an early southern route (involving C and D), followed by a second, northern dispersal via the Levantine corridor that carried haplogroup F and populated both western and eastern Eurasia (Underhill et al. 2001); and (2) a single southern route carrying the founders of all three haplogroups. The relict distribution of haplogroups C and D in Central, South and East Asia only makes a single northern route, as suggested, for example, by Cordaux and Stoneking (2003), less plausible, if not ruling it out—at least, the single northern route would not explain that distribution. The usual interpretation to date has been interpretation 1. By contrast, we have argued that *only* a single southern dispersal can satisfactorily explain the mtDNA evidence.

An obvious starting point for looking at the composition of southern-route Y chromosomes is Australia/New Guinea. Here we find that sampled Aboriginal Australians Y chromosomes do indeed include a majority of haplogroup C (mainly in a uniquely deleted form not found elsewhere; Kayser et al. 2001, 2003). There is, furthermore, a tenuous link back to Indian haplogroup C lineages. On the basis of STR analyses, Redd et al. (2002) have postulated that the dispersal from the Indian subcontinent to Australia was a Holocene, not a Pleistocene, event. An estimate of the divergence using the programme BATWING gave 95% confidence intervals of 1300–13 000 years using the fast (pedigree) rate. Since this is also an estimate of the MRCA of haplogroup C, however, it seems likely that this is a rather drastic underestimate for C, which diverges directly from one of the out-of-Africa founders, and is certainly at least as old as the other major globally distributed haplogroups; haplogroup C has the highest variance of any Y-chromosome haplogroup in India (Kivisild et al. 2003). A Holocene dispersal and divergence for *all* modern human Y chromosomes would seem implausible. A Pleistocene divergence seems rather more likely, at least potentially supporting a southern out-of-Africa migration route from India to Australia.

However, as well as haplogroup C, Australia and New Guinea also include a substantial amount of undifferentiated haplogroup K (with the markers used, which were ascertained using Eurasians and Africans). K is itself a subclade of F, just as mtDNA haplogroup R is a subclade of N. The composition of New Guinea, both east and west, is different again, with two locally derived forms of haplogroup K (M and K-M230) predominating in most populations, and a derived form of haplogroup C related to eastern Indonesian and Pacific lineages predominating in the northwest (Capelli et al. 2001; Kayser et al. 2003). Haplogroup M predominates amongst central and most western populations and K-M230 predominates in the eastern and southern highlands. Age estimates based on STR diversity are similar to those of Southeast Asian haplogroups (Kayser et al. 2003).

It is therefore clear that Australasia is dominated by Y-chromosome haplogroups C and K (within F), just as on the mtDNA side both M and N/R are present. It is possible that haplogroup C was the first to arrive and that haplogroup K moved in later, via island Southeast Asia/Sundaland, particularly taking over the populations of New Guinea, from an ultimate source in the Levant. Such an interpretation, however, would have to contend with the fact that most of the indigenous Indian subcontinental lineages are also members of haplogroups F and K (in particular, F, H, L and P), and that some of them are ancestral to the main haplogroups now present in Europe (such as R1a and R1b).

The distribution of Y-chromosome haplogroup P and its main subclade haplogroup R have presented a particular conundrum. Haplogroup R1b is the most common haplogroup in western Europe, reaching almost fixation in some Atlantic populations such as northern Iberia, Ireland and western Britain (Hill et al. 2000; Semino et al. 2000; Wilson et al. 2001). Its sister clade, haplogroup R1a, is also found in Europe but has a completely different distribution, being common in the north and east. It has been suggested that these distributions were the result of late-glacial expansions from refugial zones, in Iberia and eastern Europe respectively, after the Last Glacial Maximum (Semino et al. 2000). In this respect they resemble mtDNA haplogroups pre-V (Torroni et al. 1998, 2001) and H (Achilli et al. 2004).

R1a and R1b are both present in Indian populations, in both tribal and caste groups, between them amounting to a third of Indian Y-chromosome lineages. The distinction between tribal and caste groups is usually drawn because it has often been assumed that the non-tribal caste groups have received substantially more 'Indo-Aryan' genetic input from the north in the last few thousand years, whereas tribal groups are thought of as 'aboriginal'. However, both mtDNA and the Y chromosome show only minor introgression in caste groups, largely restricted to the northwest (Bamshad et al. 2001; Kivisild et al. 2003). Furthermore, R1a and R1b have similar variances in India, and the sister clade of R1, R2, is present only in the Indian subcontinent and the adjacent regions of Iran and Central Asia, with a very similar variance. Further R* lineages and lineages from the ancestor of R one step further back, P*, have also both been found in India, albeit at very low frequencies. This indicates a plausible origin for haplogroup R in India, or possibly somewhat to the west in Pakistan or Iran, where the variance of R1a is somewhat higher—although R1b and R2 have not been found there (Kivisild et al. 2003).

This pattern has been taken to suggest a major role for Central Asia in the peopling of Europe, particularly focusing on the distribution of Y-chromosome haplogroup R lineages (Wells et al. 2001). This contradicts the mtDNA evidence, which strongly implies that Central Asia was secondarily populated fairly recently from both East Asia and western Eurasia, since only fragments of East Asian and western Eurasian mtDNA haplogroups are present in Central Asian populations (Comas et al. 1998, 2004; Richards et al. 2000; Quintana-Murci et al. 2004). Furthermore, most of the Central Asian Y-chromosome haplogroups are also present at higher diversity in the Indian subcontinent (Kivisild et al. 2003). Therefore, a south Asian (Indian or Iranian) rather than Central Asian, origin for the spread of the Y-chromosome haplogroup R1 into Europe seems the most plausible explanation. It may have spread first into the vicinity of the Caucasus or eastern Europe, and from there into Europe—possibly with the Gravettian material culture complex, or perhaps with the Badegoulian, around the time of the Last Glacial Maximum, alongside the dispersal of members of mtDNA haplogroups pre-V and H (Gamble et al. 2004).

So all of the major non-African Y-chromosome founders (C, D, F and K) are present in the Indian subcontinent, together with a number of diverse subclades that are unique to the region. On the Y chromosome as well as the mtDNA, the package was reduced by drift both to the west (losing C and D) and to the east (losing the non-K parts of haplogroup F). An early offshoot from southwest Asia, moving up the Persian Gulf and thence into the Levant, perhaps in tandem with mtDNA haplogroup N, would fit the distribution of Y-chromosome haplogroup F (minus K; including F*, I and J) within Europe and the Levant. Haplogroup H diversified in the Indian subcontinent and has remained largely restricted to that region. The main Eurasian Y-chromosome clade, haplogroup K, has diversified in South Asia, East Asia (and amongst Native Americans), and Australasia.

Haplogroups C and D would have been carried by the same dispersing groups beyond the Indian subcontinent into eastern Eurasia (and in the case of haplogroup C, also Central Asia). Their relict representation in the extant populations of eastern Eurasia suggests the strong early action of drift, which is, in general, more powerful on the male than on the female line of descent, to the extent of making problematic the phylogeographic reconstruction of ancient dispersals (Weale et al. 2003). Haplogroup E3b, by contrast, represents a more recent dispersal from Africa into the Near East, and thence into Europe (Kivisild et al. 2003; Cruciani et al. 2004). This would not constitute what has usually been meant by the term 'northern route out of Africa' however, since the extent of its spread beyond the Near East (in part as a result of Neolithic dispersals, and in part due to more recent processes; Semino et al. 2000; Cruciani et al. 2004) is geographically very limited. Its predominance in North Africa has been attributed to either late Palaeolithic (Bosch et al. 2001) or Neolithic (Arredi et al. 2004; Cruciani et al. 2004) dispersals.

4.5 In Context

The evidence of the non-recombining genetic markers therefore suggests that a single, small group of people dispersed from the Horn of Africa along the 'southern route' between ∼ 60 000 and 80 000 years ago. Somewhere to the west of the Indian subcontinent, within East Africa, on the coast of Yemen or around the Persian Gulf, there was a period of isolation and extensive drift. Following this, groups dispersed east, with a parting of the ways somewhere within or to the immediate west of the Indian subcontinent. One group headed northwest, ultimately into the Fertile Crescent (by ∼ 50 000 years ago), and another expanded around the coast of the subcontinent (more than 60 000 years ago). In a succession of founder events, this latter group splintered and dispersed further, giving rise to populations expanding into Southeast Asia, into Australasia, and around the Pacific Rim into northeast Asia, whilst populations remaining on the coast of the subcontinent gradually moved inland.

Colour can be added to this sketch by taking a look at the climatological and archaeological evidence. From the geographical perspective, the usual crossing point for the southern route has been thought to be an 18-km wide, shallow-bottomed isthmus, at the mouth of the Red Sea, known as the Strait of Bab el Mandab (or 'Gate of Grief') (Sirocko 2003). The strait has been flooded since the end of the Pliocene, but narrows markedly during glacial periods, potentially allowing humans to cross. Since modern humans crossed both the great Asian rivers and the sea between Timor and Australia sometime prior to ∼ 50 000 years ago, it seems likely that they would not have found the ∼ 11 km crossing at the Bab el Mandab an insurmountable challenge.

The earliest transitional or anatomically modern human fossils have been found in Africa, for example dating from ∼ 125 000 years ago at Omo-Kibish in Ethiopia (Stringer 2002) and, more recently, ∼ 160 000 years ago at Herto, also in Ethiopia (White et al. 2003), although some argue that the important anatomical transition was to 'archaic *H. sapiens*' (or *H. helmei*)

$\sim$ 260 000 years ago (McBrearty and Brooks 2000). Clear evidence of 'behavioural modernity' also appears (albeit piecemeal) in sub-Saharan Africa many thousands of years earlier than in other parts of the world, with particularly striking examples at least $\sim$ 77 000 years ago at Blombos in southern Africa (D'Errico et al. 2001; Henshilwood et al. 2004). In both morphology and behaviour—from Early Stone Age to Middle Stone Age industries $\sim$ 250 000–300 000 years ago, and from Middle Stone Age to Late Stone Age, from $\sim$ 50 000 years ago onwards—African populations show evidence of gradual transition rather than the abrupt discontinuities often found elsewhere, and the appearance of modern behaviour, including the use of symbols, is associated with the Middle Stone Age, rather than the Late Stone Age (Clark 1970; Deacon 1997; McBrearty and Brooks 2000). There were modern or virtually modern humans in the Levant, at Skhul and Qafzeh, sometime $\sim$ 100 000 years ago (Valladas et al. 1988; Stringer et al. 1989; Grün and Stringer 1991). These were found in a Middle Palaeolithic context indistinguishable from that of Neanderthals, and this population may have been replaced in the region by Neanderthals by $\sim$ 60 000 years ago. They seem unlikely to have emigrated east and south towards Southeast Asia and Australia, as suggested by Kingdon (1993), because the route would have been closed by desert at this time.

By contrast, recent archaeological evidence has indicated that humans with a Middle Stone Age technology were living on the coast at Abdur, on the opposite side of the Red Sea in Eritrea, during the warm Eemian interglacial, $\sim$ 125 000 years ago (Stringer 2000; Walter et al. 2000). This is the earliest archaeological evidence for the exploitation of marine resources, such as shellfish, predating evidence for a similar 'beachcombing' lifestyle at Klasies River Mouth by tens of thousands of years, and substantially strengthens the case for a coastal dispersal from the Horn of Africa (Sauer 1962; Lahr and Foley 1998). Middle Palaeolithic coastal sites are also found on the opposite side of the Arabian Sea, in the Arabian Peninsula, although there is little chronological information (Petraglia and Alsharek 2003), and sites classed as Middle Palaeolithic are also found on the Indian subcontinent (Joshi 1994). In neither Arabia nor India is there much evidence of the use of Levallois technique, and there are suggestions that these traditions may more resemble those of the African Middle Stone Age than those associated with the Neanderthals in Europe and the Near East—which would be at least consistent with dispersals from the Horn of Africa $\sim$ 70 000 years ago (as well as dispersals at earlier low stands, such as $\sim$ 135 000 years ago) (Rohling et al. 1998).

Evidence of the opportunity, at least, for modern humans to begin spreading east across the Gate of Grief can be found in the climatic record. After $\sim$ 80 000 years ago, the world again begun to plunge into a full glacial period and, as the sea level fell, the salinity of the Red Sea rose dramatically (Dansgaard et al. 1993; Rohling et al. 1998, 2003). This would likely have rendered beachcombing on the Eritrean coast less and less effective, whilst at the same time making Arabia more and more accessible as more land was exposed. As people crossed southwards to the Gulf of Aden, the salinity problems of the Red Sea would be behind them and they would have been able to expand further. Coastal hunter–gatherer communities in more recent times have tended towards sedentism and population growth, since the exploitation of marine resources is so productive, and once across, a further expansion eastwards may well have been driven by the continuing need for more resources as populations grew or as the resources in an area declined to stabilize at a lower level.

These movements may have taken place at any time between ~ 80 000 and 60 000 years ago—Red Sea salinity levels were at their peak ~ 65 000 years ago—and would have involved modern humans with a Middle Stone Age/Middle Palaeolithic technology (Oppenheimer 2003). Whether the dispersal began in earnest ~ 80 000 or rather ~ 70 000 years ago is relevant to the much-discussed question of the impact of the Toba eruption on early modern populations (Ambrose 1998; Oppenheimer 2003). This time period is similarly bracketed (approximately) by the genetic ages of the appearance of haplogroup L3 in Africa (~ 85 000 years ago) and haplogroups M and N, almost certainly in Asia, ~ 65 000 years ago. Haplogroup L3 may have been carried to Arabia ~ 80 000 years ago, followed by a period of diversification and drift (involving the appearance of M and N and the loss of the ancestor type); or it may have been moved somewhat later, ~ 70 000 years ago. In either scenario, however, it seems unlikely (given the ages of M and N) that Toba played a major role, since M and N are probably too young—although, given the approximations involved in genetic dating, it cannot be decisively ruled out. Toba might perhaps, however, have been partly responsible for the delay and period of drift before the emigrants began to forge eastwards to the Indian subcontinent; but both its climatic and its possible human impact remain highly controversial (Gathorne-Hardy and Harcourt-Smith 2002; Oppenheimer 2002).

This narrative would explain the generally accepted appearance of modern humans in Australia by at least ~ 40 000 years ago, and possibly up to ~ 65 000 years ago (Roberts et al. 1994; Thorne et al. 1999), which is to some extent corroborated by recent dates of ~ 45 000 years for the 'deep skull' of the Niah cave in Sarawak, Borneo (Barker et al. 2002), and the (still-disputed) redating of the South China Liujiang skull to more than approximately 67 000 years (Shen et al. 2002). The ancestors of modern Papuans and Aboriginal Australians, who must have crossed at least ~ 100 km of sea by boat, lacked the Upper Palaeolithic technology that exploded in Europe, the Near East and northern Eurasia after ~ 50 000 years ago, but were clearly endowed with the same cognitive talents, including symbolic and artistic abilities, as their western cousins. There is increasing evidence that these abilities had emerged in Africa by at least ~ 75 000 years ago (or much earlier), and were therefore most likely carried throughout the world by the emigrant population (Henshilwood et al. 2004).

Modern human populations did not re-enter the Levantine corridor until ∼ 50 000 years ago, dispersing rapidly then onwards into Europe (van Andel et al. 2003). This may have been due to the presence of desert, blocking the route north, once the early emigrant population had spread as far as the eastern Gulf or the western part of the Indian subcontinent between ∼ 65 000 and ∼ 55 000 years ago (Oppenheimer 2003). With the climatic amelioration after ∼ 55 000 years ago, a corridor opened up between the Gulf and the Levant, and populations were able to disperse into the Near East and into Europe. Populations had dispersed further north and east into the east European plain and south-central Siberia by ∼ 40 000 years ago (Dolukhanov et al. 2002).

With the onset of the Last Glacial Maximum after ∼ 30 000 years ago, populations in Europe and northern Asia were again at the mercy of the climate. However, in this case it was the cold-adapted Neanderthals who suffered the most, becoming extinct by ∼ 27 000 years ago, and the tropically adapted modern humans that survived. Despite being close cousins, it was anatomically modern humans, with their much larger communication and exchange networks (Gamble 1999), who spread across the surface of the world and prospered. Gamble's (1993a) explanation for this phenomenon is that modern humans actively planned their exploration of new worlds. If this were even partly the case, it is curious to consider that the patterns of mtDNA and Y-chromosome phylogenies that we reconstruct today might not be simply the result of random forces and environmental exigencies, but might have been shaped to a large extent by human intentions and desires—perhaps for adventure, perhaps Lebensraum, but perhaps most of all because of an acquired taste for shellfish.

Acknowledgement We would like to thank Robert Simpson for a critical reading of the manuscript.

References

Acharyya SK, Basu PK (1993) Toba ash on the Indian subcontinent and its implications for correlation of late Pleistocene alluvium. Quat Res 40:10–19

Achilli A, Rengo C, Magri C, Battaglia V, Olivieri A, Scozzari R, Cruciani F, Zeviani M, Briem E, Carelli V, Moral P, Dugoujon J-M, Roostalu U, Loogväli E-L, Kivisild T, Bandelt H-J, Richards M, Villems R, Santachiara-Benerecetti AS, Semino O, Torroni A (2004) The molecular dissection of mtDNA haplogroup H confirms that the Franco-Cantabrian glacial refuge was a major source for the European gene pool. Am J Hum Genet 75:910–918

Achilli A, Rengo C, Battaglia V, Pala M, Olivieri A, Fornarino S, Magri C, Scozzari R, Babudri N, Santachiara-Benerecetti AS, Bandelt H-J, Semino O, Torroni A (2005) Saami and Berbers—an unexpected mitochondrial DNA link. Am J Hum Genet 76:883–886

Alonso S, Armour JAL (2001) A highly variable segment of human subterminal 16p reveals a history of population growth for modern humans outside Africa. Proc Natl Acad Sci USA 98:864–869

Ambrose SH (1998) Late Pleistocene human population bottlenecks, volcanic winter, and differentiation of modern humans. J Hum Evol 34:623–651

Ambrose SH (2003) Did the super-eruption of Toba cause a human bottleneck? Reply to Gathorne-Hardy and Harcourt-Smith. J Hum Evol 45:231–237

Anderson S, Bankier AT, Barrell BG, de Bruijn MHL, Coulson AR, Drouin J, Eperon IC, Nierlich DP, Roe BA, Sanger F, Schreier PH, Smith AJH, Staden R, Young IG (1981) Sequence and organization of the human mitochondrial genome. Nature 290:457–465

Arredi B, Poloni ES, Paracchini S, Zerjal T, Fathallah DM, Makrelouf M, Pascali VL, Novelletto A, Tyler-Smith C (2004) A predominantly Neolithic origin for Y-chromosomal DNA variation in North Africa. Am J Hum Genet 75:338–345

Bamshad M, Kivisild T, Watkins WS, Dixon ME, Ricker CE, Rao BB, Naidu JM, Prasad BVR, Reddy PG, Rasanayagam A, Papiha SS, Villems R, Redd AJ, Hammer MF, Nguyen SV, Carroll ML, Batzer MA, Jorde LB (2001) Genetic evidence on the origins of Indian caste populations. Genome Res 11:994–1004

Bandelt H-J (2005) Exploring reticulate patterns in DNA sequence data. In: Bakker FT, Chatrou LW, Gravendeel B, Pelser PB (eds) Plant species-level systematics: new perspectives on pattern & process. Regnum vegetabile 142. Koeltz, Königstein, pp 245–270

Barbujani G, Bertorelle G, Chikhi L (1998) Evidence for Paleolithic and Neolithic gene flow in Europe. Am J Hum Genet 62:488–491

Barker G, Barton H, Beavitt P, Bird M, Daly P, Doherty C, Gilbertson D, Hunt C, Krigbaum J, Lewis H, Manser J, McLaren S, Paz V, Piper P, Pyatt B, Rabett R, Reynolds T, Rose J, Rushworth G, Stephens M (2002) Prehistoric foragers and farmers in southeast Asia: renewed investigations at Niah Cave, Sarawak. Proc Prehist Soc 68:147–164

Bar-Yosef O (1998) On the nature of transitions: the Middle to Upper Palaeolithic and the Neolithic Revolution. Camb Arch J 8:141–163

Blanc H, Chen KH, D'Amore MA, Wallace DC (1983) Amino acid change associated with the major polymorphic *Hinc*II site of Oriental and Caucasian mitochondrial DNAs. Am J Hum Genet 35:157–176

Bosch E, Calafell F, Comas D, Oefner PJ, Underhill PA, Bertranpetit J (2001) High-resolution analysis of human Y-chromosome variation shows a sharp discontinuity and limited gene flow between northwestern African and the Iberian peninsula. Am J Hum Genet 68:1019–1029

Brega A, Gardella R, Semino O, Morpurgo G, Astaldi Ricotti GB, Wallace DC, Santachiara-Benerecetti AS (1986) Genetic studies on the Tharu population of Nepal: restriction endonuclease polymorphisms of mitochondrial DNA. Am J Hum Genet 39:502–512

Brinkmann B, Klintschar M, Heuhuber F, Hühne J, Rolf B (1998) Mutation rate in human microsatellites: influence of the structure and length of the tandem repeat. Am J Hum Genet 62:1408–1415

Brown WM (1980) Polymorphism in mitochondrial DNA of human as revealed by restriction endonuclease analysis. Proc Natl Acad Sci USA 77:3605–3609

Cann RL (1984) Mitochondrial DNA variation in Australian aborigines: the spread of modern populations. Acta Anthropogenet 8:125–135

Cann RL, Stoneking M, Wilson AC (1987) Mitochondrial DNA and human evolution. Nature 325:31–36

Capelli C, Wilson JF, Richards M, Stumpf MPH, Gratrix F, Oppenheimer S, Underhill P, Pascali VL, Ko T-M, Goldstein DB (2001) A predominantly indigenous paternal heritage for the Austronesian-speaking peoples of insular Southeast Asia and Oceania. Am J Hum Genet 68:432–443

Cavalli-Sforza LL, Menozzi P, Piazza A (1994) The history and geography of human genes. Princeton University Press, Princeton

Chappell J (2002) Sea level changes forced ice breakouts in the Last Glacial cycle: new results from coral terraces. Quat Sci Rev 21:1229–1240

Clark JD (1970) The prehistory of Africa. Praeger, New York

Comas D, Calafell F, Mateu E, Pérez-Lezaun A, Bosch E, Martínez-Arias R, Clarimon J, Facchini F, Fiori G, Luiselli D, Pettener D, Bertranpetit J (1998) Trading genes along the Silk Road: mtDNA sequences and the origin of central Asian populations. Am J Hum Genet 63:1824–1838

Comas D, Plaza S, Wells RS, Yuldaseva N, Lao O, Calafell F, Bertranpetit J (2004) Admixture, migrations, and dispersals in Central Asia: evidence from maternal DNA lineages. Eur J Hum Genet 12:495–504

Coon CS (1962) The origin of races. Knopf, New York

Cordaux R, Stoneking M (2003) South Asia, the Andamanese, and the genetic evidence for an "early" human dispersal out of Africa. Am J Hum Genet 72:1586–1590

Cruciani F, Santolamazza P, Shen PD, Macaulay V, Moral P, Olckers A, Modiano D, Holmes S, Destro-Bisol G, Coia V, Wallace DC, Oefner PJ, Torroni A, Cavalli-Sforza LL, Scozzari R, Underhill PA (2002) A back migration from Asia to sub-Saharan Africa is supported by high-resolution analysis of human Y-chromosome haplotypes. Am J Hum Genet 70:1197–1214

Cruciani F, La Fratta R, Santolamazza P, Sellitto D, Pascone R, Moral P, Watson E, Guida V, Beraud Colomb E, Zaharova B, Lavinha J, Vona G, Aman R, Calì F, Akar N, Richards M, Torroni A, Novelletto A, Scozzari R (2004) Phylogeographic analysis of haplogroup E3b (E-M215) Y chromosomes reveals multiple migratory events within and out of Africa. Am J Hum Genet 74:1014–1022

Dansgaard W, Johnsen SJ, Clausen HB, Dahl-Jensen D, Gundestrup NS, Hammer CU, Hvidberg CS, Steffensen JO, Sveinbjörnsdottir AE, Jouzel J, Bond G (1993) Evidence for general instability of past climate from a 250-kyr ice-core record. Nature 364:218–220

Deacon TW (1997) The symbolic species. Penguin, Harmondsworth

Denaro M, Blanc H, Johnson MJ, Chen KH, Wilmsen F, Cavalli-Sforza LL, Wallace DC (1981) Ethnic variation in *Hpa*I endonuclease cleavage patterns of human mitochondrial DNA. Proc Natl Acad Sci USA 78:5768–5772

D'Errico F, Henshilwood C, Nilssen P (2001) An engraved bone fragment from c.70,000 year-old Middle Stone Age levels at Blombos Cave, South Africa: implications for the origin of symbolism and language. Antiquity 75:309–318

Dolukhanov PM, Shukurov AM, Tarasov PE, Zaitseva GI (2002) Colonization of northern Eurasia by modern humans: radiocarbon chronology and environment. J Archaeol Sci 29:593–606

Edwin D, Vishwanathan H, Roy S, Rani MVU, Majumder PP (2002) Mitochondrial DNA diversity among five tribal populations of southern India. Current Sci 83:158–162

Endicott P, Gilbert MT, Stringer C, Lalueza-Fox C, Willerslev E, Hansen AJ, Cooper A (2003a) The genetic origins of the Andaman Islanders. Am J Hum Genet 72:178–184

Endicott P, Macaulay V, Kivisild T, Stringer C, Cooper A (2003b) Reply to Cordaux and Stoneking. Am J Hum Genet 72:1590–1593

Eswaran V (2002) A diffusion wave out of Africa: the mechanism of the modern human revolution? Curr Anthropol 43:749–774
Foley R (1998) The context of human genetic evolution. Genome Res 8:339–347
Foley R, Lahr MM (1997) Mode 3 technologies and the evolution of modern humans. Camb Arch J 7:3–36
Forster P (2004) Ice ages and the mitochondrial DNA chronology of human dispersals: a review. Philos Trans R Soc Lond Ser B 359:255–264
Forster P, Torroni A, Renfrew C, Röhl A (2001) Phylogenetic star contraction applied to Asian and Papuan mtDNA evolution. Mol Biol Evol 18:1864–1881
Gamble C (1993a) Timewalkers: the prehistory of global colonization. Sutton, Stroud
Gamble C (1999) The Palaeolithic societies of Europe. Cambridge University Press, Cambridge
Gamble C, Davies W, Pettitt P, Richards M (2004) Climate change and evolving human diversity in Europe during the last glacial. Philos Trans R Soc Lond Ser B 359:243–254
Gamble CS (1993b) Ancestors and agendas. In: Yoffee N, Sherratt A (eds) Archaeological theory—who sets the agenda? Cambridge University Press, Cambridge, pp 39–52
Garrigan D, Mobasher Z, Sverson T, Wilder JA, Hammer MF (2005) Evidence for archaic Asian ancestry on the human X chromosome. Mol Biol Evol 22:189–192
Gathorne-Hardy F, Harcourt-Smith WEH (2002) The super-eruption of Toba, did it cause a human bottleneck? J Hum Evol 45:227–230
Gilead I (1991) The Upper Palaeolithic period in the Levant. J World Prehist 5:105–154
Grün R, Stringer CB (1991) Electron spin resonance dating and the evolution of modern humans. Archaeometry 33:153–199
Hammer MF, Karafet T, Rasanayagam A, Wood ET, Altheide TK, Jenkins T, Griffiths RC, Templeton AR, Zegura SL (1998) Out of Africa and back again: nested cladistic analysis of human Y chromosome variation. Mol Biol Evol 15:427–441
Harding RM, Fullerton SM, Griffiths RC, Bond J, Cox MJ, Schneider JA, Moulin DS, Clegg JB (1997) Archaic African and Asian lineages in the genetic ancestry of modern humans. Am J Hum Genet 60:772–789
Harpending HC, Sherry ST, Rogers AR, Stoneking M (1993) The genetic structure of ancient human populations. Curr Anthropol 34:483–496
Hawks J, Hunley K, Lee S-H, Wolpoff M (2001) Population bottlenecks and Pleistocene human evolution. Mol Biol Evol 17:2–22
Henshilwood C, d'Errico F, Vanhaeren M, van Niekerk K, Jacobs Z (2004) Middle Stone Age shell beads from South Africa. Science 304:404
Herrnstadt C, Elson JL, Fahy E, Preston G, Turnbull DM, Anderson C, Ghosh SS, Olefsky JM, Beal MF, Davis RE, Howell N (2002) Reduced-median-network analysis of complete mitochondrial DNA coding-region sequences from the major African, Asian, and European haplogroups. Am J Hum Genet 70:1152–1171 (erratum 71:448–449)
Heyer E, Puymirat J, Dieltjes P, Bakker E, de Knijff P (1997) Estimating Y chromosome specific microsatellite mutation frequencies using deep rooting pedigrees. Hum Mol Genet 6:799–803
Hill EW, Jobling MA, Bradley DG (2000) Y-chromosome variation and Irish origins. Nature 404:351–352
Horai S, Gojobori T, Matsunaga E (1984) Mitochondrial DNA polymorphism in Japanese. I. Analysis with restriction enzymes of six base pair recognition. Hum Genet 68:324–332
Horai S, Matsunaga E (1986) Mitochondrial DNA polymorphism in Japanese. II. Analysis with restriction enzymes of four or five base pair recognition. Hum Genet 72:105–117

Huoponen K, Schurr TG, Chen YS, Wallace DC (2001) Mitochondrial DNA variation in an Aboriginal Australian population: evidence for genetic isolation and regional differentiation. Hum Immunol 62:954–969

Ingman M, Gyllensten U (2003) Mitochondrial genome variation and evolutionary history of Australian and New Guinean aborigines. Genome Res 13:1600–1606

Ingman M, Kaessmann H, Pääbo S, Gyllensten U (2000) Mitochondrial genome variation and the origin of modern humans. Nature 408:708–713

Jin L, Underhill PA, Doctor V, David RW, Shen P, Cavalli-Sforza LL, Oefner P (1999) Distribution of haplotypes from a chromosome 21 region distinguishes multiple prehistoric human migrations. Proc Natl Acad USA 96:3796–3800

Johnson MJ, Wallace DC, Ferris SD, Ratazzi MC, Cavalli-Sforza LL (1983) Radiation of human mitochondria DNA types analyzed by restriction endonuclease cleavage patterns. J Mol Evol 19:255–271

Joshi RV (1994) South Asia in the period of *Homo sapiens neanderthalensis* and contemporaries (Middle Palaeolithic). In: De Laet SJ (ed) History of humanity. Vol I, UNESCO, London, pp 162–164

Kaessmann H, Heissig F, von Haeseler A, Pääbo S (1999) DNA sequence variation in a non-coding region of low recombination on the human X chromosome. Nat Genet 22:78–81

Kayser M, Sajantila A (2000) Mutations at Y-STR loci: implications for paternity testing and forensic analysis. Forensic Sci Int 118:116–121

Kayser M, Brauer S, Weiss G, Schiefenhövel W, Underhill PA, Stoneking M (2001) Independent histories of human Y chromosomes from Melanesia and Australia. Am J Hum Genet 68:173–190

Kayser M, Brauer S, Weiss G, Schiefenhövel W, Underhill P, Shen PD, Oefner P, Tommaseo-Ponzetta M, Stoneking M (2003) Reduced Y-chromosome, but not mitochondrial DNA, diversity in human populations from West New Guinea. Am J Hum Genet 72:281–302

Ke Y, Su B, Song X, Lu D, Chen L, Li H, Qi C, Marzuki S, Deka R, Underhill P, Xiao C, Shriver M, Lell J, Wallace D, Wells RS, Seielstad M, Oefner P, Zhu D, Jin J, Huang W, Chakraborty R, Chen Z, Jin L (2001) African origin of modern humans in East Asia: a tale of 12,000 Y chromosomes. Science 292:1151–1153

Kingdon J (1993) Self-made man—and his undoing. Simon & Schuster, London

Kivisild T, Bamshad MJ, Kaldma K, Metspalu M, Metspalu E, Reidla M, Laos S, Parik J, Watkins WS, Dixon ME, Papiha SS, Mastana SS, Mir MR, Ferak V, Villems R (1999a) Deep common ancestry of Indian and western Eurasian mtDNA lineages. Curr Biol 9:1331–1334

Kivisild T, Kaldma K, Metspalu E, Parik J, Papiha S, Villems R (1999b) The place of the Indian mitochondrial DNA variants in the global network of maternal lineages and the peopling of the Old World. In: Papiha S, Deka R, Chakraborty R (eds) Genomic diversity: applications in human population genetics. Plenum, New York, pp 135–152

Kivisild T, Tolk H-V, Parik J, Wang Y, Papiha SS, Bandelt H-J, Villems R (2002) The emerging limbs and twigs of the east Asian mtDNA tree. Mol Biol Evol 19:1737–1751 (erratum 20:162)

Kivisild T, Rootsi S, Metspalu M, Mastana S, Kaldma K, Parik J, Metspalu E, Adojaan M, Tolk H-V, Stepanov V, Gölge M, Usanga E, Papiha SS, Cinnioğlu C, Kinf R, Cavalli-Sforza LL, Underhill PA, Villems R (2003) The genetic heritage of the earliest settlers persists both in Indian tribal and caste populations. Am J Hum Genet 72:313–332

Kivisild T, Reidla M, Metspalu E, Rosa A, Brehm A, Pennarun E, Parik J, Geberhiwot T, Usanga E, Villems R (2004) Ethiopian mitochondrial DNA heritage: tracking gene flows across and around the Strait of Tears. Am J Hum Genet 75:752–770

Kong Q-P, Yao Y-G, Sun C, Bandelt H-J, Zhu C-L, Zhang Y-P (2003) Phylogeny of East Asian mitochondrial DNA linerages inferred from complete sequences. Am J Hum Genet 73:671–676 (erratum 75:157)

Krings M, Stone A, Schmitz RW, Krainitzki H, Stoneking M, Pääbo S (1997) Neandertal DNA sequences and the origin of modern humans. Cell 90:19–30

Kuhn SL (2002) Paleolithic archaeology in Turkey. Evol Anthropol 11:198–210

Labuda D, Zietkiewicz E, Yotova V (2000) Archaic lineages in the history of modern humans. Genetics 156:799–808

Lahr (1996) The evolution of modern human diversity. Cambridge University Press, Cambridge

Lahr MM, Foley R (1994) Multiple dispersals and modern human origins. Evol Anthropol 3:48–60

Lahr MM, Foley R (1998) Towards a theory of modern human origins: geography, demography, and diversity in recent human evolution. Yearb Phys Anthropol 41:137–176

Lambeck K, Chappell J (2001) Sea level change through the last glacial cycle. Science 292:679–686

Latham RG (1851) Man and his migrations. van Voorst, London

Lewin R, Foley RA (2004) Principles of human evolution. Blackwell, Oxford

Luis JR, Rowold DJ, Regueiro M, Caeiro B, Cinnioğlu C, Roseman C, Underhill PA, Cavalli-Sforza LL, Herrera RJ (2004) The Levant versus the Horn of Africa: evidence for bidirectional corridors of human migrations. Am J Hum Genet 74:532–544 (erratum 74:788)

Maca-Meyer N, González AM, Larruga JM, Flores C, Cabrera VM (2001) Major genomic mitochondrial lineages delineate early human expansions. BMC Genet 2:13

Maca-Meyer N, González AM, Pestano J, Flores C, Larruga JM, Cabrera VM (2003) Mitochondrial DNA transit between West Asia and North Africa inferred from U6 phylogeography. BMC Genet 4:15

Macaulay V, Richards M, Hickey E, Vega E, Cruciani F, Guida V, Scozzari R, Bonné-Tamir B, Sykes B, Torroni A (1999) The emerging tree of West Eurasian mtDNAs: a synthesis of control-region sequences and RFLPs. Am J Hum Genet 64:232–249

Macaulay V, Hill C, Achilli A, Rengo C, Clarke D, Meehan W, Blackburn J, Semino O, Scozzari R, Cruciani F, Taha A, Shaari NK, Raja JM, Ismail P, Zainuddin Z, Goodwin W, Bulbeck D, Bandelt H-J, Oppenheimer S, Torroni A, Richards M (2005) Single, rapid coastal settlement of Asia revealed by analysis of complete mitochondrial genomes. Science 308:1034–1036

McBrearty S, Brooks AS (2000) The revolution that wasn't: a new interpretation of the origin of modern human behavior. J Hum Evol 39:453–563

Mellars P (1992) Archaeology and the population-dispersal hypothesis of modern human origins in Europe. Philos Trans R Soc Lond Ser B 337:225–234

Metspalu M, Kivisild T, Metspalu E, Parik J, Hudjashov G, Kaldma K, Serk P, Karmin M, Behar DM, Gilbert MTP, Endicott P, Mastana S, Papiha SS, Skorecki K, Torroni A, Villems R (2004) Most of the extant mtDNA boundaries in south and southwest Asia were likely shaped during the initial settlement of Eurasia by anatomically modern humans. BMC Genet 5:26

Mishmar D, Ruiz-Pesini E, Golik P, Macaulay V, Clark AG, Hosseini S, Brandon M, Easley K, Chen E, Brown MD, Sukernik RI, Olckers A, Wallace DC (2003) Natural selection shaped regional mtDNA variation in humans. Proc Natl Acad Sci USA 100:171–176

Nei M (1987) Molecular evolutionary genetics. Columbia University Press, New York
Nei M, Roychoudhury AK (1993) Evolutionary relationships of human populations on a global scale. Mol Biol Evol 10:927–943
Oppenheimer C (2002) Limited global change due to the largest known Quaternary eruption, Toba ∼ 74 kyr BP? Quat Sci Rev 21:1593–1609
Oppenheimer S (2003) Out of Eden. Constable and Robinson, London
Pakendorf B, Stoneking M (2005) Mitochondrial DNA and human evolution. Annu Rev Genomics Hum Genet 6:165–183
Palanichamy Mg, Sun C, Agrawa S, Bandelt H-J, Kong Q-P, Khan F, Wang C-Y, Chaudhuri TP, Palla V, Zhang Y-P (2004) Phylogeny of mtDNA macrohaplogroup N in India based on complete sequencing: implications for the peopling of South Asia. Am J Hum Genet 75:966–978
Passarino G, Semino O, Quintana-Murci L, Excoffier L, Hammer M, Santachiara-Benerecetti AS (1998) Different genetic components in the Ethiopian population, identified by mtDNA and Y-chromosome polymorphisms. Am J Hum Genet 62:420–434
Pearson OM, Stone AC (2003) On the diffusion-wave model for the spread of modern humans. Curr Anthropol 44:559–561
Penny D, Steel M, Waddell PJ, Hendy MD (1995) Improved analyses of human mtDNA sequences support a recent African origin for *Homo sapiens*. Mol Biol Evol 12:863–882
Pereira L, Richards M, Goios A, Alonso A, Albarrán C, Garcia O, Behar D, Gölge M, Hatina J, Al-Ghazali L, Bradley DG, Macaulay V, Amorim A (2005) High-resolution mtDNA evidence for the late-glacial resettlement of Europe from an Iberian refugium. Genome Res 15:19–24
Petraglia MD, Alsharek A (2003) The Middle Palaeolithic of Arabia: implications for modern human origins, behaviour and dispersals. Antiquity 77:671–684
Phillips MS, Lawrence R, Sachidanandam R, Morris AP, Balding DJ, Donaldson MA, Studebaker JF et al (2003) Chromosome-wide distribution of haplotype blocks and the role of recombination hot spots. Nat Genet 33:382–387
Quintana-Murci L, Chaix R, Wells RS, Behar DM, Sayar H, Scozzari R, Rengo C, Al-Zahery N, Semino O, Santachiara-Benerecetti AS, Coppa A, Ayub Q, Mohyuddin A, Tyler-Smith C, Mehdi SQ, Torroni A, McElreavey K (2004) Where west meets east: the complex mtDNA landscape of the Southwest and Central Asian corridor. Am J Hum Genet 74:827–845
Quintana-Murci L, Semino O, Bandelt H-J, Passarino G, McElreavey K, Santachiara-Benerecetti AS (1999) Genetic evidence for an early exit of *Homo sapiens sapiens* from Africa through eastern Africa. Nat Genet 23:437–441
Rando JC, Cabrera VM, Larruga JM, Hernández M, González AM, Pinto F, Bandelt H-J (1999) Phylogeographic patterns of mtDNA reflecting the colonization of the Canary Islands. Ann Hum Genet 63:413–428
Rando JC, Pinto F, González AM, Hernández M, Larruga JM, Cabrera VM, Bandelt H-J (1998) Mitochondrial DNA analysis of Northwest African populations reveals genetic exchanges with European, Near-Eastern, and sub-Saharan populations. Ann Hum Genet 62:531–550
Rayner D, Bulbeck D (2001) Dental morphology of the "Orang Asli" aborigines of the Malay Peninsula. In: Henneberg M (ed) Causes and effects of human variation, Vol 19–41. Australasian Society for Human Biology, University of Adelaide, Adelaide
Redd AJ, Roberts-Thomson J, Karafet T, Bamshad M, Jorde LB, Naidu JM, Walsh B, Hammer MF (2002) Gene flow from the Indian subcontinent to Australia: evidence from the Y chromosome. Curr Biol 12:673–677

Reidla M, Kivisild T, Metspalu E, Kaldma K, Tambets K, Tolk HV, Parik J et al (2003) Origin and diffusion of mtDNA haplogroup X. Am J Hum Genet 73:1178–1190

Relethford JH (2001) Genetic history of the human species. In: Balding DJ, Bishop MJ, Cannings C (eds) Handbook of statistical genetics. Wiley, New York, pp 813–846

Richards M, Macaulay V (2000) Genetic data and the colonization of Europe: genealogies and founders. In: Renfrew C, Boyle K (eds) Archaeogenetics: DNA and the population prehistory of Europe. McDonald Institute for Archaeological Research, Cambridge, pp 139–151

Richards M, Oppenheimer S, Sykes B (1998) mtDNA suggests Polynesian origins in eastern Indonesia. Am J Hum Genet 63:1234–1236

Richards M, Macaulay V, Hickey E, Vega E, Sykes B, Guida V, Rengo C et al (2000) Tracing European founder lineages in the Near Eastern mitochondrial gene pool. Am J Hum Genet 67:1251–1276

Richards MB, Macaulay VA, Bandelt H-J (2002) Analyzing genetic data in a model-based framework: inferences about European prehistory. In: Renfrew C, Bellwood P (eds) Examining the farming/language dispersal hypothesis. McDonald Institute for Archaeological Research, Cambridge, pp 459–466

Richards M, Rengo C, Cruciani F, Gratrix F, Wilson JF, Scozzari R, Macaulay V, Torroni A (2003) Extensive female-mediated gene flow from sub-Saharan Africa into Near Eastern Arab populations. Am J Hum Genet 72:1059–1064

Roberts RG, Jones R, Smith MA (1994) Beyond the radiocarbon barrier in Australian prehistory. Antiquity 68:611–616

Rogers AR, Harpending H (1992) Population growth makes waves in the distribution of pairwise genetic differences. Mol Biol Evol:552–569

Rogers AR, Jorde LB (1995) Genetic evidence on modern human origins. Hum Biol 67:1–36

Rohling EJ, Fenton M, Jorissen FJ, Bertrand P, Ganssen G, Caulet JP (1998) Magnitudes of sea-level lowstands of the past 500,000 years. Nature 394:162–165

Salas A, Richards M, De la Fe T, Lareu M-V, Sobrino B, Sánchez-Diz P, Macaulay V, Carracedo Á (2002) The making of the African mtDNA landscape. Am J Hum Genet 71:1082–1111

Sauer C (1962) Seashore: primitive home of man? Proc Am Philos Soc 106:41–47

Scozzari R, Cruciani F, Santolamazza P, Malaspina P, Torroni A, Sellitto D, Arredi B, Destro-Bisol G, De Stefano G, Rickards O, Martinez-Lebarga C, Modiano D, Biondi G, Moral P, Olckers A, Wallace DC, Novelletto A (1999) Combined use of biallelic and microsatellite Y-chromosome polymorphisms to infer affinities among African populations. Am J Hum Genet 65:829–846

Semino O, Passarino G, Oefner PJ, Lin AA, Arbuzova S, Beckman LE, De Benedictis G, Francalacci P, Kouvatsi A, Limborska S, Marcikiæ M, Mika A, Mika B, Primorac D, Santachiara-Benerecetti AS, Cavalli Sforza LL, Underhill PA (2000) The genetic legacy of Paleolithic *Homo sapiens sapiens* in extant Europeans: a Y chromosome perspective. Science 290:1155–1159

Semino O, Magri C, Benuzzi G, Lin AA, Al-Zahery N, Battaglia V, Maccioni L, Triantaphyllidis C, Shen P, Oefner PJ, Zhivotovsky LA, King R, Torroni A, Cavalli-Sforza LL, Underhill PA, Santachiara-Benerecetti AS (2004) Origin, diffusion, and differentiation of Y-chromosome haplogroups E and J: inferences on the neolithization of Europe and later migratory events in the Mediterranean area. Am J Hum Genet 74:1023–1034

Shen G, Wang W, Wang Q, Zhao JX, Collerson K, Zhou CL, Tobias PV (2002) U-Series dating of Liujiang hominid site in Guangxi, Southern China. J Hum Evol 43:817–829

Siddall M, Rohling EJ, Almogi-Labin A, Hemleben C, Meischner D, Schmelzer I, Smeed DA (2003) Sea-level fluctuations during the last glacial cycle. Nature 423:853–858

Sirocko F (2003) Global change—ups and downs in the Red Sea. Nature 423:813–814

Slatkin M, Hudson RR (1991) Pairwise comparisons of mitochondrial DNA sequences in stable and exponentially growing populations. Genetics 129:555–562

Song S-R, Chen C-H, Lee M-Y, Yang TF, Iizuka Y, Wei K-Y (2000) Newly disovered eastern dispersal of the youngest Toba tuff. Marine Geol 167:303–312

Stoneking M, Fontius JJ, Clifford SL, Soodyall H, Arcot SS, Saha N, Jenkins T, Tahir MA, Deininger PL, Batzer MA (1997) *Alu* insertion polymorphisms and human evolution: evidence for a larger population size in Africa. Genome Res 7:1061–1071

Stringer C (2000) Coasting out of Africa. Nature 405:24–27

Stringer C (2002) Modern human origins: progress and prospects. Philos Trans R Soc Lond Ser B 357:563–579

Stringer C, McKie R (1996) African exodus: the origins of modern humanity. Cape, London

Stringer CB, Andrews P (1988) Genetic and fossil evidence for the origin of modern humans. Science 239:1263–1268

Stringer CB, Grün R, Schwarcz HP, Goldberg P (1989) ESR dates for the hominid burial site of Es Skhul in Israel. Nature 338:756–758

Swofford DL (1993) PAUP: phylogenetic analysis using parsimony release 3.1.1, Champaign

Sykes B, Leiboff A, Low-Beer J, Tetzner S, Richards M (1995) The origins of the Polynesians—an interpretation from mitochondrial lineage analysis. Am J Hum Genet 57:1463–1475

Takahata N, Lee S-H, Satta Y (2001) Testing multiregionality of modern human origins. Mol Biol Evol 18:172–183

Tanaka M, Cabrera VM, González AM, Larruga JM, Takeyasu T, Fuku N, Guo L-J et al (2004) Mitochondrial genome variation in Eastern Asia and the peopling of Japan. Genome Res 14:1832–1850

Tchernov E (1992) Biochronology, paleoecology and dispersal events of hominids in the southern Levant. In: Akazawa T, Aoki K, Kimura T (eds) The evolution and dispersal of modern humans in Asia. Hokusen-sha, Tokyo, pp 149–188

Templeton AR (2002) Out of Africa again and again. Nature 416:45–51

Templeton AR (2003) Against recent replacement (Box 8.9 opinion) In: Jobling M, Hurles M, Tyler-Smith C (eds) Human evolutionary genetics: origins, peoples & disease. Taylor & Francis, London, p 261

Thangaraj K, Singh L, Reddy AG, Rao VR, Sehgal SC, Underhill PA, Pierson M, Frame IG, Hagelberg E (2003) Genetic affinities of the Andaman Islanders, a vanishing human population. Curr Biol 13:86–93

Thangaraj K, Chaubey G, Kivisild T, Reddy AG, Singh VK, Rasalkar AA, Singh L (2005) Reconstructing the origin of Andaman Islanders. Science 308:996

Thorne A, Grün R, Mortimer G, Spooner NA, Simpson JJ, McCulloch M, Taylor L, Curnoe D (1999) Australia's oldest human remains: age of the Lake Mungo 3 skeleton. Hum Evol 36:591–612

Tishkoff SA, Dietzsch E, Speed W, Pakstis AJ, Kidd JR, Cheung K, Bonné-Tamir B, Santachiara-Benerecetti AS, Moral P, Krings M, Pääbo S, Watson E, Risch N, Jenkins T, Kidd KK (1996) Global patterns of linkage disequilibrium at the CD4 locus and modern human origins. Science 271:1380–1387

Torroni A, Schurr TG, Cabell MF, Brown MD, Neel JV, Larsen M, Smith DG, Vullo CM, Wallace DC (1993) Asian affinities and continental radiation of the four founding Native American mtDNAs. Am J Hum Genet 53:563–590

Torroni A, Miller JA, Moore LG, Zamudio S, Zhuang JG, Droma T, Wallace DC (1994a) Mitochondrial DNA analysis in Tibet: implications for the origin of the Tibetan population and its adaptation to high altitude. Am J Phys Anthropol 93:189–199

Torroni A, Neel JV, Barrantes R, Schurr TG, Wallace DC (1994b) Mitochondrial DNA clock for the Amerinds and its implications for timing their entry into North America. Proc Natl Acad Sci USA 91:1158–1162

Torroni A, Bandelt H-J, D'Urbano L, Lahermo P, Moral P, Sellitto D, Rengo C, Forster P, Savantaus M-L, Bonné-Tamir B, Scozzari R (1998) mtDNA analysis reveals a major late Paleolithic population expansion from southwestern to northeastern Europe. Am J Hum Genet 62:1137–1152

Torroni A, Bandelt H-J, Macaulay V, Richards M, Cruciani F, Rengo C, Martinez-Cabrera V et al (2001) A signal, from human mtDNA, of post-glacial recolonization in Europe. Am J Hum Genet 69:844–852

Underhill PA, Passarino G, Lin AA, Shen P, Mirazón Lahr M, Foley RA, Oefner PJ, Cavalli-Sforza LL (2001) The phylogeography of Y chromosome binary haplotypes and the origins of modern human populations. Ann Hum Genet 65:43–62

Underhill PA, Roseman CC (2001) The case for an African rather than an Asian origin of the human Y-chromosome YAP insertion. In: Jin L, Seielstad M, Xiao C (eds) Genetic, linguistic and archaeological perspectives on human diversity in Southeast Asia. World Scientific Publishing, Singapore, pp 43–56

Underhill PA, Shen P, Lin AA, Jin L, Passarino G, Yang WH, Kauffman E, Bonné-Tamir B, Bertranpetit J, Francalacci P, Ibrahim M, Jenkins T, Kidd JR, Mehdi SQ, Seielstad MT, Wells RS, Piazza A, Davis RW, Feldman MW, Cavalli-Sforza LL, Oefner PJ (2000) Y chromosome sequence variation and the history of human populations. Nat Genet 26:358–361

Valladas H, Reyss Jl, Joron Jl, Valladas G, Bar-Yosef O, Vandermeersch B (1988) Thermoluminescence dating of Mousterian "Proto-Cro-Magnon" remains from Israel and the origin of modern man. Nature 331:614–616

van Andel TH, Davies W, Weninger B (2003) The human presence in Europe during the last glacial period I: human migrations and the changing climate. In: van Andel TH, Davies W (eds) Neanderthals and modern humans in the European landscape during the last glaciation. McDonald Institute for Archaeological Research, Cambridge, pp 31–56

Vigilant L, Stoneking M, Harpending H, Hawkes K, Wilson AC (1991) African populations and the evolution of human mitochondrial DNA. Science 253:1503–1507

Wainscoat J, Hill A, Boyce A, Flint J, Hernandez M, Thein SL, Old JM, J.R. L, Falusi Y, Weatherall DJ, Clegg JB (1986) Evolutionary relationships of human populations from an analysis of nuclear DNA polymorphisms. Nature 319:491–493

Walter RC, Buffler RT, Bruggemann JH, Guillaume MMM, Berhe SM, Negassi B, Libsekal Y, Cheng H, Edwards RL, von Cosel R, Neraudeau D, Gagnon M (2000) Early human occupation of the Red Sea coast of Eritrea during the last interglacial. Nature 405:65–69

Watkins WS, Ricker CE, Bamshad MJ, Carroll ML, Nguyen SV, Batzer MA, Harpending HC, Rogers AR, Jorde LB (2001) Patterns of ancestral human diversity: an analysis of *Alu*-insertion and restriction-site polymorphisms. Am J Hum Genet 68:738–752

Watson E, Forster P, Richards M, Bandelt H-J (1997) Mitochondrial footprints of human expansions in Africa. Am J Hum Genet 61:691–704

Weale ME, Shah T, Jones AL, Greenhalgh J, Wilson JF, Nymadawa P, Zeitlin D, Connell BA, Bradman N, Thomas MG (2003) Rare deep-rooting Y chromosome lineages in humans: lessons for phylogeography. Genetics 165:229–234

Wells RS, Yuldasheva N, Ruzibakiev R, Underhill PA, Evseeva I, Blue-Smith J, Jin L et al (2001) The Eurasian heartland: a continental perspective on Y-chromosome diversity. Proc Natl Acad Sci USA 98:10244–10249

White TD, Asfaw B, DeGusta D, Gilbert H, Richards GD, Suwa G, Howell FC (2003) Pleistocene Homo sapiens from Middle Awash, Ethiopia. Nature 423:742–747

Wilson JF, Weiss DA, Richards M, Thomas MG, Bradman N, Goldstein DB (2001) Genetic evidence for different male and female roles during cultural transitions in the British Isles. Proc Natl Acad Sci USA 98:5078–5083

Wolpoff MH, Caspari R (1997) Race and human evolution. Simon and Schuster, New York

YCC (2002) A nomenclature system for the tree of human Y-chromosomal binary haplogroups. Genome Res 12:339–348

Yu N, Fu YX, Li WH (2002) DNA polymorphism in a worldwide sample of human X chromosomes. Mol Biol Evol 19:2131–2141

Zhivotovsky LA, Underhill PA, Cinnioğlu C, Kayser M, Morar B, Kivisild T, Scozzari R, Cruciani F, Destro-Bisol G, Spedini G, Chambers GK, Herrera RJ, Yong KK, Gresham D, Tournev I, Feldman MW, Kalaydjieva L (2004) The effective mutation rate at Y chromosome short tandem repeats, with application to human population-divergence time. Am J Hum Genet 74:50–61

Subject Index

16189 bias, 130
16p13.3 locus, 241
7sDNA, 108
9-bp deletion, 163

admixture, 192, 194
alignment, 70
Alu insertion, 230, 241
ambiguous nucleotide ("N"), 140
amino-acid racemisation, 96, 206, 210
amplification, 118–120, 129
anatomically modern human (AMH), 182–185, 189, 205, 227, 252
ancestral endosymbiont, 31
ancient DNA (aDNA), 91–93, 96, 97, 109–111, 119, 131, 132, 134, 205–211, 214, 219
ancient mtDNA, 42
archaic *Homo sapiens*, 202, 227, 229, 252
artificial mutation processes, 117
artificial recombination, 128
ascertainment bias, 77
Asian mtDNA tree, 186
Aurignacian, 243
authentication, 214
autosomal markers, 229, 231

Bab el Mandab ('Gate of Grief'), 252, 253
back-migration to Africa, 235, 236, 238
Badegoulian, 251
Bantu expansion, 165
base call, 119, 121, 122, 124, 125
base misreporting, 122
base shift, 126
beachcombing, 253
behavioural modernity, 253
β-globin, 241
BLAST search, 35
bottleneck, 160, 171

'Cambridge' reference sequence, 8
CD4 locus, 241
cell death, 92
Central Asian reservoir, 194
Chelex extraction, 118
chimpanzee outgroup, 228
circularity, 232
climatic selection, 171
climatology, 232, 252
clonal inheritance, 171
cloning, 211
Clovis, 168
coalescence time, 48–50, 74, 76, 79–82, 168
coalescent, 166, 185, 188, 189
coalescent theory, 49
coastal route, 168
coding-region rate, 76, 77
complete sequences, 80, 160, 162
contamination, 92, 97, 119, 120, 122, 123, 127–129, 131, 133, 205, 206, 210–214, 219
continuous-time Markov chain, 73
control element MT5, 108
control region, 6, 152, 160, 162, 163
correction for multiple hits, 57, 79, 82, 84
Cro-Magnon, 203
cross-linking, 93, 96, 110
cybrid fusion, 18
cycle sequencing, 120, 121

data handling, 117
database, 56, 59, 60, 66–69, 117, 120, 122, 127, 128, 132, 140
deamination, 92, 99, 100, 103, 110, 214
decay law, 74, 75
dedicated laboratory, 210
deleterious mutations, 73, 76, 170, 171

deletions, 18
depurination, 92, 94, 110, 207, 214
dinosaur, 32
diploid, 147
diplotype, 147
distance analyses, 166
distance method, 67
D-loop, 6, 9, 10
DNA degradation, 118, 207–209, 213
DNA extraction, 118, 119, 210
DNA preservation, 206–208
DNA sequencing, 125, 127, 129
'Dolly', 24
dot table, 127
double-strand breakage, 94
dye blobs, 121
dye terminator sequencing, 121, 124

Early Stone Age, 253
ecology, 232
Eemian interglacial, 182, 233
Egyptian mummies, 34
electrophoresis, 119–121, 124, 125, 131
erroneous transversions, 55
"Etruscan" data set, 133
extension of desert, 182

false group, 162
familial studies, 104
fast train to ..., 194
faunal expansion, 232, 233
Fertile Crescent, 247, 252
forensic case work, 129
fossil record, 183
founder analysis, 79, 169, 186
founder type, 185, 245
founding lineages, 168, 169
fragmentation, 93, 95, 96
free radicals, 94

gamma-distributed rates, 57, 73
'Garden of Eden', 225
gene flow, 189, 228, 231
gene pool, 185, 189, 194
gene tree, 152
genetic code, 6
genetic distances, 152
genetic drift, 150, 160, 189, 191, 218, 219, 221, 236, 242, 248
geographic scope, 186
germ line mutations, 19
ghost group, 162
Gravettian, 243, 251

haplogroup, 150, 153, 155–157, 160, 162–169, 171
haplogroup nomenclature, 153, 155, 157, 163
haploid, 147
haplotype, 147, 153, 157, 160, 168, 170
haplotype blocks, 231
haplotype tree, 160
heterogeneity parameter, 60, 67
heteroplasmic mice, 21, 25
heteroplasmy, 9, 10, 18, 19, 21, 24, 25, 34, 119, 121
histology, 210
historical linguistics, 232
HIV, 152
HKY85 model, 54, 55, 57–59, 73
hominin, 201–203, 212
Homo erectus, 202, 227–229, 233
Homo heidelbergensis, 202, 221, 229, 233
Homo helmei, 203, 229, 252
Homo sapiens, 202, 204
homoplasmy, 9, 18
homoplasy, 19, 20, 102, 162
Horn of Africa, 234, 236, 237, 240, 242, 247, 252, 253
human/Neanderthal admixture, 205
human/Neanderthal split time, 220, 221
humidity, 182, 208
HvrBase, 66–68, 217
HVS-I, HVS-II, HVS-III, 6
hydrolytic damage, 94, 95, 108, 207
hypervariable segments, 152
hypervariable sequence, 6

Ingman et al. data set, 56, 72, 150
interbreeding, 219, 230, 232, 233
intracellular genetic drift, 19

jumping PCR, 213

karyoplast transfer, 25
Kimura model, 54
Klasies River Mouth, 253

Lake Mungo, 183
'Lake Mungo 3', 33

large-scale deletions, 9
Last Glacial Maximum (LGM), 182, 188
Late Pleistocene, 227, 232
Late Stone Age, 253
late-glacial expansion, 250
LDR1, 108, 109
Leber hereditary optic neuropathy (LHON), 9–11, 171
length heteroplasmy, 130, 131
Levallois, 253
Levantine corridor, 234, 240, 249, 255
likelihood ratio test, 73
Liujiang skull, 254
long C-stretches, 129
Luke, the evangelist, 132
lumping vs splitting, 243

majority consensus, 54, 67
male infertility, 11
marine resources, 253, 254
maternal genealogy, 227, 236
maternal inheritance, 19, 20, 147
maximum likelihood (ML), 47, 48, 50, 54, 57–60, 63, 67, 72, 77, 79, 80, 84, 218
maximum parsimony (MP), 47, 48, 53, 60, 79, 217, 227
median-joining (MJ), 62, 248
Middle Palaeolithic, 184, 185, 244, 253
Middle Pleistocene, 227
Middle Stone Age, 253
misalignment, 127
miscoding lesions, 97, 99, 111
misincorporation rates, 97
mismatch distribution, 237, 246
mitochondria, 3
mitochondrial bottleneck, 20–24
mitochondrial Eve, 50
molecular clock, 47–49, 73, 76, 78, 81, 84, 85, 152, 169, 170, 220, 227
molecular fossils, 42
mosaic mtDNA, 206
mosaic structure, 133
most recent common ancestor (MRCA), 157, 167, 168, 185, 188, 191, 227, 245, 249
Mount Toba eruption, 238, 246, 254
mtDNA from hair, 42
Muller's Ratchet, 24, 171
multifurcation, 162
multiregional evolution, 203, 227
multiregionalist, 80
mutation load, 18
mutation rate, 48, 76–78, 80, 82, 83, 152
mutation spectrum, 160
mutational hotspots, 50, 53, 59, 60, 63, 69, 70, 74, 104, 133, 152, 155
mutational rate spectrum, 49

Nasidze/Stoneking data set, 135
natural selection, 48
Neanderthal, 201, 203–205, 207–221, 228
Neanderthal diversity, 205, 221
neighbor-joining (NJ), 48, 60, 69, 82, 160, 162, 163, 217
nested clade analysis, 229
network, 152, 157
N-glycosylase, 214
Niah Cave, 183, 254
nomenclature violation, 122
non-African deep-rooting lineage, 231
non-recombining part of Y chromosome (NRY), 226, 232
non-synonymous substitution, 76, 77
Northern Asian Route, 181, 185, 191
northern route, 181, 234, 235, 246, 249
nuclear factors, 25
nuclear genes, 5
nuclear inserts, 31
nuclear/mtDNA coevolution, 25
nucleotide diversity (π), 81
numts, 31–34, 36, 39–42, 134

Omo-Kibish, 252
oogenesis, 21
optical isomers, 209
'Oriental' origin, 226
Ötzi, the 'Ice Man', 110
Out of Africa, 181, 183, 189, 225, 227, 232–236, 238, 240, 241, 243, 252
out of Africa replacement, 203
outgroup, 32, 54, 57, 76, 85, 157, 163
oxidative damage, 94, 95, 111, 207
oxidative phosphorylation (OXPHOS), 5, 171
oxygen isotope stage, 233, 246

pairwise method, 67, 68, 72
palaeoclimatology, 181
PAML, 73
panmixia, 167

paraphyly, 163, 164, 229
paternal 'leakage', 20, 32
paternal transmission, 20
pathogenic mutations, 9, 10, 17, 18, 20, 22, 24
PAUP, 227
PCR conditions, 41
PCR inhibition, 118
pedigree rate, 74, 75, 77, 78, 170
pedigree studies, 48, 77, 152
Persian Gulf, 251, 252
phantom mutations, 55, 56, 58, 76, 121, 124–126, 130, 131, 134, 140
phenol/chloroform extraction, 118
phylogenetic analysis, 215
phylogenetic method, 47, 53, 70
phylogenetic rate, 74, 75, 78, 170
phylogenetic resolution, 53, 186
phylogenetic tree, 170
phylogeny, 48–50, 53, 54, 56, 59, 60, 66, 69–71, 73, 75, 76, 78, 82
phylogeography, 150, 157, 167, 172, 188, 191, 193
pincer model, 193
point mutations, 18
polar ice caps, 182
polymerase chain reaction (PCR), 118–120, 129, 132, 134, 201, 205, 206, 210–215
polymorphism count, 55, 66, 69–71
polyphyly, 156
poor genealogical resolution, 229
population tree, 165
postmortem damage, 91, 99, 100, 103, 104, 108, 109, 117, 126, 132
primer, 119, 120, 124, 129, 130
primer binding, 41, 42
proof-reading, 122
protein-encoding genes, 6
purification, 118–121, 124
purifying selection, 73, 75, 76, 84, 170, 171
Pygmy groups, 164

quartet puzzling (QP), 54, 67–69, 217

random sampling, 22
rapid dispersal, 237
rate misestimation, 48
rearrangements, 18
recombination, 20, 102, 152, 171
recurrent mutation, 62
reduced-median (RM), 61, 62
reference sequence bias, 122, 126, 127, 129, 135
regionally autochthonous haplogroup, 185
relaxed replication, 8, 18, 19
replacement hypothesis, 229, 233, 242
replication, 6, 8, 9
respiratory chain, 4–6, 9–11, 18, 25
RFLP, 153, 155, 156, 160
RFLP analysis, 127, 132
ρ estimator, 79, 81, 84
ribosomal RNA genes, 6
root, 32, 48, 50, 51, 53–55, 60, 75, 78–83, 153, 155–157, 160, 164, 167–169

Sahul, 182, 185, 187–189, 246
salinity, 253, 254
sample mix-up, 120, 123, 128, 129
sample preservation, 92, 93, 96, 97, 100
sampling, 118
sampling bias, 129
saturation, 55, 58, 74–76, 83, 160
sea level, 182, 183
secondary structure, 108, 111
segregating units, 23
segregation, 17, 19, 24–26
sequencing artefacts, 20, 121, 123, 126, 132, 134
sequencing errors, 53
settlement of Asia, 181, 194
"Seven Daughters of Eve", 169
Silk Road, 181
short tandem repeat (STR), 237, 248–250
sinodonty, 191
sister clades, 167, 217
skewed variation, 119
Skhul/Qafzeh, 233, 253
slippage, 129
soil pH, 208
somatic mutations, 19, 78
Southern Coastal Route (SCR), 181, 183, 189, 190, 194
southern route, 234, 237, 240, 243, 245, 246, 249, 252
sperm dysfunction, 36
Stage 4 Glacial Maximum, 182
star symbol, 153
Steppe Belt, 192, 194
STR, 237, 248–250
substitution vs divergence rates, 80

sundadonty, 191
Sundaland, 246
SWGDAM database, 122, 126
symplesiomorphy, 152
synapomorphy, 150, 162
synonymous substitution, 58, 76

Taq polymerase, 118
termination associated sequence (TAS), 108
tertiary structure, 103
thermal age, 96, 206, 208–210
third codon positions, 109
time to the most recent common ancestor (TMRCA), 48, 49, 74, 75, 79–83
tissue segregation, 24
transcription, 5, 6, 8
transcription error, 122
transfer RNA genes, 6
transition:transversion ratio (TTRR), 54, 55, 57, 58, 73
translation, 5
translocation, 31
transmission, 17, 20, 22, 24–26
Type 1 transitions, 98
Type 2 transitions, 98, 99, 103

Underhill et al. model, 234
Upper Palaeolithic, 184, 185, 189, 233, 240, 243–245, 254

variability of mutation rate, 59
vegetative segregation, 19

'Weak Garden of Eden', 237, 238
weighting procedure, 59

X chromosome, 229

Y chromosome founders, 241, 249
Y chromosome SNP tree, 50, 54
YAP marker, 238
Yemen, 165, 252

Zagros corridor, 182

Printing: Krips bv, Meppel
Binding: Stürtz, Würzburg